Remote and Robotic Investigations of the Solar System

Remote and Robotic Investigations of the Solar System

By
C. R. Kitchin

CRC Press
Taylor & Francis Group
Boca Raton London New York

CRC Press is an imprint of the
Taylor & Francis Group, an **informa** business

CRC Press
Taylor & Francis Group
6000 Broken Sound Parkway NW, Suite 300
Boca Raton, FL 33487-2742

First issued in paperback 2019

© 2018 by Taylor & Francis Group, LLC
CRC Press is an imprint of Taylor & Francis Group, an Informa business

No claim to original U.S. Government works

ISBN-13: 978-1-4987-0493-9 (hbk)
ISBN-13: 978-0-367-87166-6 (pbk)

Front cover image: Curiosity on the surface of Mars – Artist's impression (reproduced courtesy of NASA/JPL).

Library of Congress Cataloging-in-Publication Data

Names: Kitchin, C. R. (Christopher R.), author.
Title: Remote and robotic investigations of the solar system / Chris Kitchin.
Description: Boca Raton, FL : CRC Press, Taylor & Francis Group, [2017] |
Includes bibliographical references and index.
Identifiers: LCCN 2017009368| ISBN 9781498704939 (hardback) | ISBN 149870493X
(hardback) | ISBN 9781498704946 (e-Book) | ISBN 1498704948 (e-Book)
Subjects: LCSH: Solar system--Observations. | Astronomical instruments.
Classification: LCC QB501 .K62 2017 | DDC 629.43/5--dc23
LC record available at https://lccn.loc.gov/2017009368

Visit the Taylor & Francis Web site at
http://www.taylorandfrancis.com

and the CRC Press Web site at
http://www.crcpress.com

For Rowan, Lottie, Arthur and Bob, and in loving memory of Pip, Badger, Jess, Wills, Spruce, Misty, Chalky, Midnight, Sheba, TC, Satchmo, Monty, Snuffles, Merlin and Bassett.

_____

Contents

PART II The Detection and Investigation of Sub-Atomic, Atomic and Molecular Particles

Preface

The purpose of this book is to explain how our present knowledge of solar system objects, other than the Sun, has been obtained. We are thus concerned with investigations of planets, dwarf planets, asteroids, natural satellites, comets, Kuiper belt objects and the interplanetary medium. The book examines the instruments used to obtain observations, samples, measurements, analyses, etc., of the interiors, surfaces, atmospheres and radiation belts of these solar system objects. It describes the underlying physical principles of the instruments and the way in which these principles have been used to design actual instruments. It is not about the results of those investigations (except occasionally in passing). There are plenty of excellent sources describing what our current knowledge of solar system objects is, but detailed and comprehensive descriptions of the basic methods used to obtain that knowledge are far more difficult to find. This book aims to remedy that situation by gathering all that data together in one easily accessible place, linking descriptions of the physical principles to the design of working instruments and illustrating both with examples of real instruments in use.

The bulk of the book concentrates on the period from the launch of Sputnik 1 (4 October 1957) to the present day and looks forward to possible future developments. The majority of instruments discussed are spacecraft borne, but instruments on rockets, balloons and terrestrial telescopes are also included. Appendix B gives a timeline covering more than 4000 years of observations and theoretical ideas about the solar system for readers who would like to set the modern developments into a historical perspective.

Well over 250 individual spacecraft, rockets, balloons, terrestrial instruments and observatories are mentioned in this book when discussing individual examples. Rather than repeat definitions and details (such as launch dates, objectives, orbits, operator, etc.) every time a particular mission is mentioned, these details are listed in Appendix C along with those of their instruments discussed herein. The number of Earth observation spacecraft is large,* so only a small selection of them has been chosen for discussion, otherwise they would overwhelm the missions to other solar system objects.

Abbreviations and acronyms are defined the first time they are used in the text and then listed alphabetically for convenience in Appendix E. Appendix D lists some definitions and values of physical constants and laws used in the book.

---

* At the start of 2017, for example, India launched 103 Earth-orbiting spacecraft from a single rocket.

This book has the same approach to solar system instrumentation as the author's book *Astrophysical Techniques* (Appendix A, reference A.1.1) regarding the instrumentation used to observe the wider universe. This book, however, is intended to be a stand-alone work and so some material is covered in both books. Generally, the level of treatment when the same topics are covered will be different because terrestrial-based studies play a much bigger role in investigating the wider universe and spacecraft-based instruments are generally more important within the solar system.

Readers will need to be familiar with some physics and mathematics to university entrance level for a physical sciences subject to follow some discussions. But generally, the discussions are not that demanding and many of the physical principles involved are described *ab initio* within this book.

Apart from occasional references, specialised instruments for solar observing have been omitted. This is because they are covered in the author's book *Solar Observing Techniques* (Appendix A, reference A.1.6) and because the Sun is so different in its nature from the other solar system objects that instruments developed for its study do not find many applications elsewhere.

I hope that you will find this book interesting and useful and perhaps learn a few new things from it – I have certainly done so while writing it!

C. R. Kitchin
University of Hertfordshire

I

The Detection and Investigation of Solar System Objects via Electromagnetic Radiation

CHAPTER 1

The Extended Optical Region

1.1 INTRODUCTION

In this first part, the aim is to describe, explain and discuss the ways in which our knowledge of the solar system and its contents has been enlarged and enhanced by remote and robotic investigations utilising the photons emitted or reflected from solar system objects. Later parts will deal with more direct and interactive investigations.

Readers of this book are likely to be familiar with the dual wave-particle nature of electromagnetic (e-m) radiation. A reminder of what this implies for the design of instrumentation, etc. however may not come amiss – thus the wave nature of light enables a telescope lens or mirror to focus a beam of (wave) radiation into an image, whilst the detector, in most cases, will utilise the particle nature of light to pick up individual photons.

When behaving as a wave, e-m radiation is described by its frequency (ν or f^*) and wavelength (λ) and, in a vacuum, the two quantities are related to the velocity of light in a vacuum (c) by

$$\lambda\nu = c \tag{1.1}$$

The e-m radiation in its particle manifestation is described by the energy of the individual photons (quanta) that are involved. The wave and particle descriptions are linked by the Planck formula

$$E = h\nu \tag{1.2}$$

where h is Planck's constant (6.62607×10^{-34} m^2 kg s^{-1}) and E is the photon's energy.

Conversion between the frequency, wavelength and energy of e-m radiation is often needed and the formulae

---

* Definitions and, where appropriate, values of symbols and quantities are listed in Appendix D. There, where possible, the full up-to-date values are quoted together with the (generally) more approximate values used for calculations within this book.

$$\text{Frequency}(\text{Hz}) \approx 3.00 \times 10^8 \div \text{Wavelength}(\text{m})$$
$$\approx 1.50 \times 10^{33} \times \text{Energy}(\text{J}) \tag{1.3}$$
$$\approx 2.40 \times 10^{14} \times \text{Energy}(\text{eV})$$

$$\text{Wavelength}(\text{m}) \approx 3.00 \times 10^8 \div \text{Frequency}(\text{Hz})$$
$$\approx 2.00 \times 10^{-25} \div \text{Energy}(\text{J}) \tag{1.4}$$
$$\approx 1.25 \times 10^{-6} \div \text{Energy}(\text{eV})$$

$$\text{Energy}(\text{J}) \approx 1.60 \times 10^{-19} \times \text{Energy}(\text{eV})$$
$$\approx 6.66 \times 10^{-34} \times \text{Frequency}(\text{Hz}) \tag{1.5}$$
$$\approx 2.00 \times 10^{-25} \div \text{Wavelength}(\text{m})$$

$$\text{Energy}(\text{eV}) \approx 6.24 \times 10^{-18} \times \text{Energy}(\text{J})$$
$$\approx 4.16 \times 10^{-15} \times \text{Frequency}(\text{Hz}) \tag{1.6}$$
$$\approx 1.25 \times 10^{-6} \div \text{Wavelength}(\text{m})$$

may be found useful (where the quantities are in the units usually used).

The e-m spectrum thus theoretically ranges from zero frequency or energy (infinite wavelength) to infinite frequency or energy (zero wavelength). It is customarily subdivided, in terms of its frequency, into the regions:[*]

$0 \leftrightarrow$ Extremely low frequency (ELF) $\leftrightarrow$ Radio $\leftrightarrow$ Microwave
$\leftrightarrow$ Far infrared (FIR[†]) $\leftrightarrow$ Medium infrared (MIR) $\leftrightarrow$ Near infrared (NIR) $\leftrightarrow$ Visible[‡]
$\leftrightarrow$ Ultraviolet $\leftrightarrow$ Soft x-rays $\leftrightarrow$ Hard x-rays $\leftrightarrow$ γ rays $\leftrightarrow \infty$

The approximate extents of these regions are shown in Table 1.1 together with conversions between wavelength, frequency and energy for quick reference.

E-m radiation is generally, but by no means always, discussed in terms of its frequency at low energies (ELF, radio, microwave regions), by its wavelength in the intermediate region (infrared [IR], visible and ultraviolet [UV] regions) and by the photon energy in the x-ray and γ ray regions. This practice is mainly historical in origin, but remains convenient in that the numbers thus being used have relatively small powers of ten in their expressions.

A reminder may also be useful that e-m radiation interacts with matter in different ways depending upon its energy (particle character)/frequency (wave character). Thus, the main interactions are:

---

[*] There are many other ways of dividing up the e-m spectrum. The microwave region, for example, is sometimes subdivided into eight bands such as the S band (2–4 GHz) and the Ku band (12.5–18 GHz). Visible light is, of course subdivided into the colours: red, orange, yellow, . . . , violet, etc.
[†] Sometimes called Terahertz or sub-millimetre radiation.
[‡] The NIR, visible and long-wave ultraviolet regions are often combined and referred to as the optical region.

TABLE 1.1 The Regions of the e-m Spectrum and the Interrelationships between Frequency, Wavelength and Photon Energy

	Wavelength (m)	Frequency (Hz)	Energy (eV)	Energy (J)
ELF	∞	0	0	0
	↑	↑	↑	↑
	1.00×10^5	3.00×10^3	1.24×10^{-11}	1.99×10^{-30}
Radio	↕	↕	↕	↕
	1.00	3.00×10^8	1.24×10^{-6}	1.99×10^{-25}
Microwave	↕	↕	↕	↕
	1.00×10^{-3}	3.00×10^{11}	1.24×10^{-3}	1.99×10^{-22}
FIR	↕	↕	↕	↕
	1.00×10^{-4}	3.00×10^{12}	1.24×10^{-2}	1.99×10^{-21}
MIR	↕	↕	↕	↕
	1.00×10^{-5}	3.00×10^{13}	0.124	1.99×10^{-20}
NIR	↕	↕	↕	↕
	3.80×10^{-7}	4.00×10^{14}	1.65	2.65×10^{-19}
Visible	↕	↕	↕	↕
	3.80×10^{-7}	7.89×10^{14}	3.26	5.24×10^{-19}
Ultraviolet	↕	↕	↕	↕
	1.00×10^{-8}	3.00×10^{16}	1.24×10^2	1.99×10^{-17}
Soft x-rays	↕	↕	↕	↕
	1.00×10^{-10}	3.00×10^{18}	1.24×10^4	1.99×10^{-15}
Hard x-rays	↕	↕	↕	↕
	1.00×10^{-11}	3.00×10^{19}	1.24×10^5	1.99×10^{-14}
γ rays	↓	↓	↓	↓
	0	∞	∞	∞

ELF and radio radiation – direct induction of electric currents, synchrotron and free-free emissions and absorptions

Microwave and Infrared – molecular rotational and vibrational emissions and absorptions

Optical – interactions with the outer electrons of atoms, ions and molecules

UV and x-ray – interactions with the inner electrons of atoms and ions, ionisation and recombination

X-ray and γ rays – interactions directly with nuclei

Thermal radiation can be emitted by materials at any temperature when their constituent atoms or molecules undergo changes of velocity during their interactions with other nearby atoms or molecules. The change in velocity results in bremsstrahlung* radiation. In liquids and gases, the particles are free to move around. In solids, the particles are more or less fixed in place, but vibrate about their mean positions within the material so that

---

* The German word for 'braking radiation'. Both synchrotron and free–free radiation are also forms of bremsstrahlung radiation.

the individual radiation patterns have a dipole nature. Thermal radiation's spectrum* is described by the Planck equation and its emissions peak at a wavelength or frequency for a particular temperature given by Wien's displacement law.

Planck equation:

$$I(\nu,T)d\nu = \frac{2h\nu^3\mu^2}{c^2\left(e^{h\nu/kT}-1\right)} \tag{1.7}$$

$$I(\lambda,T)d\lambda = \frac{2hc^2\mu^2}{\lambda^5\left(e^{h\nu/\lambda kT}-1\right)} \tag{1.8}$$

and

$$I(\nu,T)d\nu = I(\lambda,T)d\lambda \tag{1.9}$$

where $I(\nu,T)$ and $I(\lambda,T)$ are the thermal radiation intensities at temperature T and in terms of frequency and wavelength, respectively, and μ is the refractive index of the material forming the transmission medium (equal to 1 for a vacuum).

Wien's displacement law:

$$\lambda_{\text{Max}} = \frac{2.898\times10^{-3}}{T} \quad \text{m} \tag{1.10}$$

$$\nu_{\text{Max}} = 5.879\times10^{10}T \quad \text{Hz} \tag{1.11}$$

where λ_{Max} and ν_{Max} are the wavelength and frequency of the maximum thermal emission per unit wavelength and unit frequency intervals.[†]

Measurements of e-m radiation can, theoretically, determine its frequency (or wavelength or photon energy), its intensity at a particular frequency (and also the variation of its intensity as the frequency changes – i.e. the spectrum), its variation in intensity in one, two or three dimensions (i.e. the direction of its sources/imaging), its state of polarization and its phase. Generally, only at low frequencies can all these properties of radiation be determined and at the highest frequencies, measurements are limited largely to determining the radiation's intensity and direction.

Particularly for the spacecraft-borne instrumentation used for studying the objects within the solar system, the main observing region of the spectrum ranges from microwaves to soft x-rays. Over much of this region instruments' operating principles are clearly related to each other, although the details vary. We shall therefore start by looking at the instrumentation used over the MIR to UV parts of the spectrum before going on to consider instruments operating at lower or higher frequencies. Since, as already mentioned, the term

---

* Also often called the blackbody spectrum.

† Since the wavelength unit interval is 1 m and the frequency unit interval is 1 Hz the maxima occur at different parts of the spectrum (e.g. at 1000 K, $\lambda_{\text{Max}} = 2.898 \times 10^{-6}$ m ($= 1.05 \times 10^{14}$ Hz), $\nu_{\text{Max}} = 5.879 \times 10^{13}$ Hz ($= 5.10 \times 10^{-6}$ m)).

'optical region' is generally used for the NIR, visible and near UV regions, it is convenient to use the term 'extended optical region' (EOR), to include the MIR and far UV regions as well.

1.2 TELESCOPES AND CAMERAS

Most people find that the direct images of objects, whether black and white, colour or false colour, are the most vivid and impressive results from spacecraft solar system missions. The implications of such images may normally be understood immediately, without (or without much) specialist training. From a scientific point of view, images contain a wealth of information and form an invaluable reference archive even after they have been used for their primary purpose(s). Obtaining images over the EOR customarily involves using telescopes – or since the instruments do not use eyepieces, what should more properly be called cameras.

Apart from the lack of an eyepiece, the cameras on spacecraft are often identical in their designs to terrestrial telescopes, or are minor variants of such designs. Both refracting (lens-based, dioptric) and reflecting (mirror-based, catoptric) instruments are in use, with the latter being the more common. There are also some custom-built optical systems that have been designed (usually) because of constraints on mass or size required within a particular spacecraft.

Almost all of these camera designs may include additional optics to fold the light beam in order to shorten the instrument's length or to direct the image to a convenient place. Most cameras have interchangeable filters to enable images to be obtained within different parts of the spectrum and/or be multipurpose instruments obtaining direct images or spectra (Section 1.3) or both simultaneously.

The reader is assumed to be familiar with the basics of the optics of light – laws such as those of reflection and refraction and terms such as focal plane, field of view, focal length, focal ratio, etc. The basic principles of usual designs of cameras (telescopes) are likely to be familiar to most readers of this book, but if not, Appendix A lists various sources suitable for background reading. Here just a brief summary of the main designs is given with examples of their use in actual missions.

1.2.1 Reflecting Cameras

Some cameras are based just upon a single concave parabolic mirror with the detector or detector array placed at its focus – a system known as a prime focus instrument when used for terrestrial telescopes (Figure 1.1a). Subsidiary lenses or mirrors will generally be needed to correct the aberrations* inherent in a simple parabolic mirror, providing a wider, sharply focussed, field of view. The Cassegrain design (Figure 1.1b) uses a concave parabolic primary mirror and a convex hyperbolic secondary mirror. The closely related Ritchey–Chrétien instruments have a wider field of sharp focus than a similar Cassegrain and use a concave hyperbolic primary mirror and a convex hyperbolic secondary mirror. Some Gregorian-based designs (Figure 1.1c) are also to be encountered with a concave elliptical secondary mirror placed after the primary focus.

---

* See reference A.1.1 for a detailed discussion of the Seidel aberrations of optical systems, or other sources listed in Appendix A.

(a) Prime focus

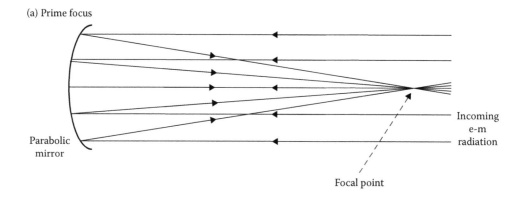

(b) Cassegrain and Ritchey–Chrétien

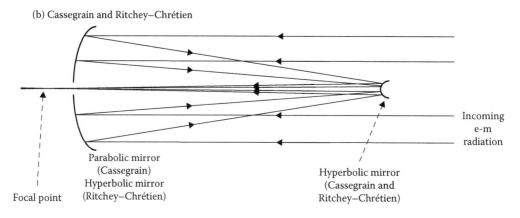

(c) Gregorian

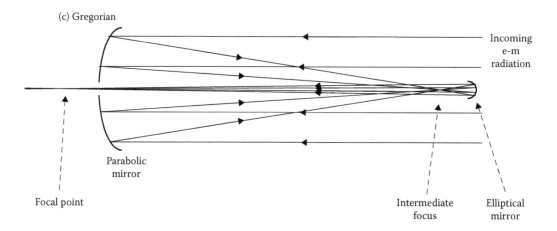

FIGURE 1.1 Light paths in the main camera designs: (a) prime focus, (b) Casegrain and Richtey–Chrétien, and (c) Gregorian.

(*Continued*)

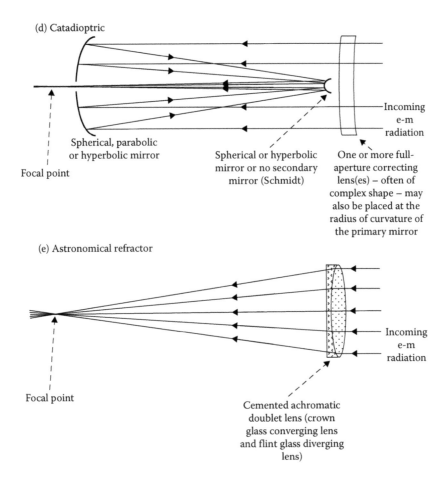

FIGURE 1.1 (*Continued*) Light paths in the main camera designs: (d) Catadioptric and (e) Astronomical refractor.

Catadioptric instruments, such as the Schmidt camera and the very popular commercially produced compact Schmidt–Cassegrains aimed at the amateur astronomy market, use both mirrors and lenses as their primary radiation-gathering elements (Figure 1.1d). The Schmidt design has a spherical primary mirror and an aspherical correcting lens placed at its radius of curvature and provides a very large field of view. The original Schmidt–Cassegrain design uses a spherical primary, an aspherical correcting lens and a spherical secondary mirror. In the compact Schmidt–Cassegrain, the correcting lens is placed at or just inside the primary mirror's focal point, thus halving the tube length. The primary and secondary mirrors may be spherical, parabolic or hyperbolic – there are many variations. The Makzutov design is very similar to the compact Schmidt–Cassegrain but uses a spherical primary mirror and a meniscus correction lens with both surfaces spherical, and has the secondary mirror aluminized at the centre of the correcting lens on its inner surface – the secondary mirror is therefore also spherical. Examples of some of these designs, mostly small, may occasionally be used within instruments on board spacecraft.

A number of individual cameras that have been used or are being used for solar system investigations are discussed below to illustrate the range of the possibilities. Detector types which operate within the EOR and their theory and practice are covered in Section 1.4. They are mentioned however here, as well, in order to complete the instrument' descriptions.

1.2.1.1 Prime Focus Systems

Mars Atmosphere and Volatile Evolution Mission (MAVEN)* is a Mars orbiter launched in November 2013 by the National Aeronautics and Space Administration (NASA) and still operating at the time of writing. Its Imaging Ultraviolet Spectrometer (IUVS), although principally functioning as a spectrograph (Section 1.3), has a direct imaging mode as well. The spherical primary mirror has a 13.3 mm × 20 mm aperture and a 100 mm focal length. The spectrograph entrance slit is at the prime focus of the primary mirror. The primary mirror is fed by a plane scanning mirror and can point towards the Martian limb with a field of view of 12.5° × 24° or towards the Martian surface below the spacecraft (nadir) with a field of view of 12.5° × 60°. The spectrograph's entrance slit measures 0.06° × 212.5°. IUVS observes over the 110–340 nm spectral region using two detectors. A beam splitter separates the radiation into two beams, one covering the 110–180 nm region, the other the 180–320 nm region. The detectors are 1024 × 1024 CMOS† arrays that are fed by image intensifiers. The image intensifiers use UV optimised photocathodes – caesium telluride for the longer wavelengths and caesium iodide for the shorter wavelengths.

Direct imaging is accomplished through the nadir field of view when the spacecraft is near apareon (the highest point in its orbit – about 6000 km above the Martian surface). The scanning mirror sweeps the field of view seen through the spectrograph slit over a swath of the surface to build up an image at a particular UV wavelength with a surface resolution of ~160 km (cf. pushbroom scanning, Section 1.2.1.3).

1.2.1.2 Cassegrain Systems

The two Japan Aerospace Exploration Agency (JAXA) asteroid sample-return spacecraft, Hayabusa 1 and 2, use Cassegrain-design cameras for their LIDAR‡ systems. The LIDAR is used to measure the distance of the spacecraft from its target over the range ~30 m to ~25 km and for surface and topographical measurements of the asteroid. The silicon-carbide primary mirrors have 127 mm diameters and focal ratios of about 1.5. The laser uses a 17 W neodymium-doped yttrium aluminium garnet (Nd:YAG) crystal with a pulse length of 14 ns operating at 1064 nm. Detection of the returned pulse is by a silicon avalanche photodiode (APD – Section 1.4.1.4). The Cassegrain optics are used for the range 1–25 km. A separate small receiver is used for the 30 m to 1 km range.

---

* Acronyms and abbreviations are listed within the text when first introduced and also listed alphabetically in Appendix E. Details of spacecraft and their instrumentation described in this book are listed in Appendix C.

† Complementary metal-oxide semiconductor – see Section 1.4.1.3.

‡ Light detection and ranging (cf. RADAR – Section 2.4).

The lunar mapper, Clementine, carried several imagers (see also Section 1.2.2.2). Its f/2.67 Long Wave Infrared Camera (LWIC) was of the Cassegrain design with a 131 mm diameter primary mirror and relay lenses. With a 1° field of view and a 128 × 128 pixel HgCdTe detector (Sections 1.4.1.1 and 1.4.1.4) its surface resolution on the Moon was ~50 m at best. It operated over the 8.0–9.5 μm spectral region.

The Earth imaging Deep Space Climate Observatory (DSCOVR) observes Earth from the L1* point using a 300 mm Cassegrain telescope (EPIC – Earth Polychromatic Imaging Camera). It has 10 narrow band (1–3 nm) filters distributed over the 317–780 nm spectral range and uses a 2048 × 2048 pixel charge-coupled device (CCD) (Section 1.2.1.2) array as its detector.

A spacecraft originally built for surveying infrared galaxies was later repurposed into discovering near-Earth asteroids. The remarkable transformation happened to the Wide field Infrared Survey Explorer (WISE) spacecraft which, for its new mission, was named Near Earth Object Wide field Infrared Survey Explorer (NEOWISE). Its telescope is a 400 mm Cassegrain with a gold coating. Its field of view is 47′ across and its angular resolution 5″. A further 13 mirrors follow the main telescope. All the mirrors are diamond turned aluminium with gold coatings. Until its coolant ran out, the optics were cooled to 17 K. A beam splitter separates the incoming light into four channels centred on 3.4, 4.6, 12 and 22 μm and two cameras each accept two of the channels for imaging. The first two channels use a 1024 × 1024 pixel HgCdTe array (Sections 1.4.1.1 and 1.4.1.4) and the second two channels use a SiAs blocked impurity band (BIB) array (Section 1.4.1.4) of the same size. A scan mirror counteracts the spacecraft's rotation during each 8.8-second exposure. WISE imaged some 750 million objects (including many asteroids), while NEOWISE has since discovered ~250 near-Earth asteroids and two possible comets.

The aircraft-borne f/1.3, 2.5 m SOFIA (Stratospheric Observatory for Infrared Astronomy) is also a Cassegrain-type instrument with a third 45° flat mirror to reflect its light beam to a Nasmyth† focus. It can operate from 300 nm to 655 μm wavelengths using a variety of instruments. Two weeks before New Horizon's fly-by of Pluto it was able to observe an occultation by Pluto with its high speed photometer revealing the structure of Pluto's atmosphere. It has also been used to detect atomic oxygen in the atmosphere of Mars and to map Jupiter in the FIR.

1.2.1.3 Ritchey–Chrétien Systems

The comet probe, Giotto, sent by the European Space Agency (ESA) to fly by Halley's comet in March 1986 carried the Halley Multicolor Camera (HMC). This instrument had Ritchey–Chrétien optics with a 160 mm aperture and 1 m effective focal length. The camera was mounted within the spacecraft body for protection and viewed the comet via a plane

---

* The Lagrange points (L1, L2, L3, L4 and L5) are points in space where a small object can maintain a stable position with respect to much larger bodies that are orbiting each other. The L1, L2 and L3 point lie along the line joining the two main bodies, with L1 between the two main bodies and L2 and L3 outside them. L4 and L5 share the same orbit as the smaller of the two main bodies and lie 60° ahead and 60° behind that body. The Trojan and Greek asteroids occupy the L5 and L4 points, respectively, for several planets, especially Jupiter.

† A third, plane, mirror set at 45° sends the light beam of the telescope through a hollow altitude axis of the telescope's Alt-Az mounting to a focus whose position is fixed whatever altitude the telescope may point to (cf. Coudé focus for equatorially mounted telescopes).

45° mirror which was the only part of the instrument directly exposed to impacts by dust particles.* The instrument had a field of view of 1.5° and could be moved through 180° in a plane that was parallel to the spacecraft's spin axis. The spin, at 15 rpm, then allowed the HMC to image any part of the sky – an observing method known as pushbroom scanning.

Two 390 × 584 pixel CCDs (see Section 1.4.1.2) were used as the detectors and divided into two sections each. One of the sections was clear, two had fixed filters – one red and one blue. The final section imaged through a filter wheel containing 11 filters and polarisers. The image was split using two mirrors set at 90° to each other, placed just before the focal plane and which reflected the light beam out to the sides and onto the two CCDs. The images from each of the CCD sections were thus of slightly different parts of the comet. A 2 × 936 pixel Reticon™ silicon photodiode detector (Section 1.4.1.4) placed to the side of the mirrors received the image directly and was used for pointing the camera towards the comet.

Light entered the CCD detector sections through narrow slits so that only a line of the CCD section, some four to eight pixels wide, was illuminated at any given instant. The spacecraft's rotation moved the illuminated line across the CCD and the accumulating electric charges within the detector array were moved to keep pace with it (time-delayed integration – TDI – see Section 1.4.1.2). The resolution at the comet's nucleus was 10 m from a distance of 500 km.

The Pluto fly-by mission, New Horizons, has a Long-Range Reconnaissance Imager (LORRI) which is also of a Ritchey–Chrétien optical design, with a 208 mm diameter, f/1.8 primary mirror and an effective focal length of 2.63 m (f/12.6). The instrument's mirrors are formed from silicon carbide and it uses three fused silica correcting lenses to produce a flat image plane. Its field of view is 0.29° and its resolution is 1″/pixel over a 350–850 nm range (equal to 60 m on Pluto at its closest approach). The detector is a 1024 × 1024 pixel CCD array rear illuminated and thinned and operated in a frame-transfer mode (Section 1.4.1.2). The planned Lucy mission to Jupiter's Trojan asteroids will carry an updated version of LORRI, called L'LORRI.

A 150 mm, f/3.75 Ritchey–Chrétien is currently operating as the first robotic lunar telescope. Whilst its targets are deep space objects, it is included here as a possible forerunner of lunar-based instruments that are likely to be investigating solar system objects at some time in the future. The lunar-based ultraviolet telescope (LUT) is on board the Chinese National Space Administration's (CNSA) Chang'e 3 lunar lander. The LUT is fixed in position on the lunar lander and is fed by a flat mirror† that tilts in two dimensions to acquire the desired targets. A small flat mirror at 45° to the optical axis and placed just in front of the primary mirror reflects the radiation out to a focus at the side of the instrument (an arrangement known as the Nasmyth focus for terrestrial telescopes). Its field of view is 1.36°. The LUT observes over the 245–320 nm waveband using a 1024 × 1024 pixel CCD detector, illuminated from its back and with enhanced UV sensitivity. The advantage of using a lunar-based telescope lies principally in the long periods of continuous

---

* The relative velocity of Giotto to Halley's comet during the fly-by was 68 km s^{-1}, so a 1 mm^3 particle would hit with an energy of ~2.5 kJ – about the same energy as a rifle bullet.
† Termed a siderostat or coelostat for similar Earth-based systems.

observations of a target that are permitted by the Moon's slow rotation – up to 250–300 hours are possible. The Moon also provides a stable base, allowing high pointing accuracy without the need to consume much power or thruster fuel, as would be needed for a spacecraft-based instrument.

The Cassini Saturn orbiter spacecraft carried two cameras – a narrow angle (i.e. high resolution), 2 m focal length, f/10.5, Ritchey–Chrétien reflector (narrow angle camera [NAC]) and a wide angle, 200 mm focal length, f/3.5 refractor (wide angle camera [WAC]). Details of the latter are discussed below. The NAC had a field of view of 0.35° square and a resolution of about 1.3″ per pixel (Figure 1.2). Its mirrors were composed of fused silica while the (refractive) field correctors and the detector entrance window were made from fused silica or calcium fluoride. The instrument had two filter wheels with a design derived from the Hubble Space Telescope's (HST) Wide Field Planetary Camera (WFPC)*. There were a total of 24 filters covering the range 200 nm to 1.05 μm with 12 filters in each wheel. Exposures were made using a focal-plane shutter similar to those used for both the Voyager and Galileo missions (the same design of shutter is also used for the WAC). The detector was a 1024 × 1024 CCD array.

The Mercury orbiter, MESSENGER, carried a 24 mm, f/22 off-axis Ritchey–Chrétien NAC (see also Section 1.2.2.1) with a field of view of 1.5°. The instrument was constructed from aluminium with gold coatings for the mirrors and used a 1024 × 1024 pixel frame transfer CCD detector. It observed over the 700–800 nm wavelength range and its filters rejected thermal radiation at longer wavelengths in order to minimise the heat entering the instrument. At the spacecraft's lowest perihermion,[†] its ground resolution could be as good as 2 m.

While most detailed observations of solar system objects come from spacecraft physically close to those objects, some terrestrial telescopes can still make contributions. Thus the Keck and Gemini telescopes – all of Ritchey–Chrétien design – have recently been able to track Io's active volcanoes over periods of several years. Operating in the NIR, they use adaptive optics (reference A.1.1) to sharpen the blurring caused by Earth's atmosphere until they can resolve down to a tenth of a second of arc – a linear distance of ~300 to ~400 km on the surface of Io.

1.2.1.4 Catadioptric Systems

The Super Resolution Camera (SRC – Figure 1.3) on the Mars Express Orbiter uses the Makzutov variation on the Cassegrain design. This has a full-aperture correcting lens which has the secondary mirror aluminised on its centre. The camera has a 90 mm aperture and a 970 mm effective focal length. The instrument uses a 1024 × 1032 CCD (Section 1.2.1.2) array with 9 μm pixels. From a height of 250 km its resolution[‡] of the Martian surface is 2.3 m.

Also using an all-spherical catadioptric design, although not actually Makzutovs, were the closely similar imagers on the Mariner 10 (Television Photography Experiment [TPE]),

---

* One of the HST's launch instruments, replaced in 1993.
† The lowest point in its orbit.
‡ Sometimes called the Instantaneous Geometrical Field of View (IGFoV).

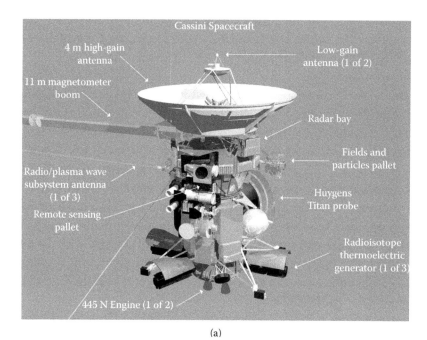

(a)

(b)

FIGURE 1.2 (a) The Cassini spacecraft showing the high-resolution Ritchey–Chrétien camera which is the light-coloured tube just left of centre; the WAC (see Section 1.2.2) is just below it. (Courtesy of NASA/JPL-Caltech.) (b) A raw, NAC, Cassini image of part of Iapetus. The large crater at the bottom of the image is about 1.4 km in diameter. (Courtesy of NASA/JPL.)

Voyager 1 and 2 (their NACs) and Galileo spacecraft. In fact, the instrument on Galileo was a modified flight spare from Voyager. The Solid State Imager (SSI) on NASA's Galileo Jupiter orbiter had a 250 mm diameter f/3.5 primary mirror and an effective focal length of 1500 mm (f/8.5). There were two full-aperture correcting lenses placed inside the primary

FIGURE 1.3 The Super Resolution Camera of the Mars Express Orbiter spacecraft. The correcting lens with the secondary mirror aluminised on its inner surface can clearly be seen as the lower half of the instrument. The High Resolution Stereo Camera (HRSC) forms the upper half of the instrument and its aperture is above the Makzutov barrel – see Section 1.2.2.1. (Courtesy of Deutsches Zentrum für Luft und Raumfahrt (DLR)/ESA.)

focal point with the secondary mirror attached to the centre of the innermost lens surface. Two further small correcting lenses were placed just before the final focus. The instrument had a 0.46° field of view and eight filters which covered the spectral band from 375 nm to 1.1 μm. Galileo was one of the first spacecraft to use a CCD array as its detector. The CCD had 800 × 800 pixels and was front illuminated. It was of the buried channel design and used a single set of electrodes (virtual phase CCD – Section 1.4.1.2) with a resolution of 2″/pixel.

The Kepler mission's primary aim is the detection of exoplanets (Appendix A, reference A.1.3). However, it was also used to observe Rosetta's comet (67P/Churyumov–Gerasimenko) during a period when it was invisible to terrestrial telescopes. Kepler uses an f/0.95 Schmidt camera system with a 1.4 m primary mirror. A mosaic of forty-two 1024 × 2000 pixel CCDs acts as its detector. Its field of view is thus some 12° across.

Although neither remotely nor robotically operated, the Far Ultraviolet Camera/ Spectrograph (FUVCS) (Figure 1.4) used by the Apollo 16 astronauts on the lunar surface is sufficiently similar to the other cameras discussed here to be worthy of inclusion. The Apollo 16 mission to the Moon included a 75 mm, f/1.0 Schmidt camera that could obtain images over the 105–155 nm spectral region (and also spectra over the 50–135 nm region by rotating the camera through 90° to point to a reflective diffraction grating and collimator – see Section 1.3). The camera was mounted on a tripod and deployed onto the lunar surface in the landing module's shadow in order to help keep it cool and to reduce solar glare. It was used to observe UV radiation from Earth's upper atmosphere and aurorae as well as from stars and galaxies.

FIGURE 1.4   The Apollo 16 far ultraviolet camera/spectrograph. (Courtesy of NASA.)

The instrument had two correcting lenses – one composed from lithium fluoride which transmitted down to a wavelength of 105 nm thus allowing the Lyman-α line* at 121.6 nm through. The second lens was made from calcium fluoride with a cut-off at 123 nm and so excluded Ly-α emissions. For detecting the shortest wavelengths (in the spectroscopic mode) the corrector lenses were removed. The spherical primary mirror had a reflective coating of rhenium (which has a much higher short wave UV reflectivity than aluminium) and the field of view was 20°.

From a modern perspective, the far UV camera's detector was a rather Heath Robinson system called an electronographic tube (see also Section 1.4.3.5.3). This had a potassium bromide photocathode coated onto a supporting plate curved to match the curved image surface of the mirror and placed at the focal surface. Potassium bromide's electron emission cut-off is at 155 nm, so the system did not respond to the much more plentiful longer wavelength photons. Electrons emitted by the photocathode were accelerated by a –25 kV electrostatic field and magnetically focussed to pass through a central hole in the primary mirror. Most of the electrons then passed through a thin (7.7 nm) plastic membrane at the electron image plane and were absorbed by nuclear grade photographic emulsion (Section 1.4.3.5.3). The emulsion was coated (like conventional photographic film) onto a plastic tape which was pressed against the membrane during an exposure and then moved on for the next exposure. After alignment of the instrument onto a desired target by an astronaut the exposures were obtained automatically. The film cassette was brought back to Earth by the astronauts for processing.

The backup camera to the one used during the Apollo 16 mission was used in a slightly adapted form during the Skylab-4 mission to obtain UV images of comet Kohoutek.

---

* A spectrum line arising from electron transitions between the first and second energy levels in the hydrogen atom – see Section 1.3.3.

1.2.1.5 Gregorian Systems

Several solar spacecraft, such as Hinode, which observed the 2012 transit of Venus, have carried Gregorian-type cameras. The Solar Optical Telescope (SOT) on board Hinode has a 0.5 m aperture and 0.2″ resolution over a 0.11° square field of view. Stratoscope 2,* a balloon-borne telescope that flew missions between 1963 and 1971 to study, amongst other topics, planetary atmospheres at infrared wavelengths, was a 0.91 m variant on a Gregorian telescope. The primary, fused silica mirror operated at f/4 and the secondary converted this to f/33. A flat mirror set at 45°, just before the primary mirror, reflected the light beam sideways into the optical train of the detector and the guidance system. The camera operated over the visible and NIR range using photography initially and later TV cameras.

1.2.1.6 Custom-Designed Systems

The Mars Odyssey Orbiter carries a Thermal Emission Imaging System (THEMIS) that covers the range 425 nm to 14.88 μm and provides mainly mineralogical information about Mars. The anastigmatic† instrument is based upon three aspherical off-axis mirrors feeding separate infrared and visual detector arrays via a beam splitter. The mirrors are diamond-turned from aluminium blocks and subsequently polished and nickel plated. THEMIS has an effective focal length of 200 mm at f/1.6 (aperture 120 mm). The field of view is 3.5° × 4.6° and its resolution at the Martian surface is 18 m/pixel (visual) and 100 m/pixel (infrared). Ground truth data (Section 7.2.7) to calibrate its measurements have been provided by Opportunity's Miniature Thermal Emission Spectrometer (Mini-TES – Sections 7.2.1 and 7.2.2.1).

Nine filters cover the spectral region from 6.78 to 14.88 μm. The thermal radiation is detected by an uncooled 240 × 320 pixel microbolometer array (Section 1.4). The filters are in the form of strips, each covering a line 16 pixels wide, and the images are built up by pushbroom scanning as the spacecraft's orbital motion sweeps the field of view over the Martian surface. The signals from the detectors are read out by TDI (see Section 1.4.1.2) to fit in with the scanning method. Five filters in the forms of 192-pixel wide strips are used to cover the visible and NIR and their images are obtained by a 1024 × 1024 pixel frame scan CCD giving a 2.6° field of view (Figure 1.5).

Rosetta's Optical, Spectroscopic and Infrared Remote Imaging System (OSIRIS) NAC was also based upon a three-mirror anastigmatic design. The mirrors were formed from silicon carbide with the primary and secondary mirrors having aspherical surfaces whilst the tertiary was spherical. The instrument had an effective focal length of 717 mm, a 90 mm aperture, a 2.2° field of view and an angular resolution of 3.8″ – corresponding to a surface resolution of about 20 mm at best. It used an UV-enhanced 2048 × 2048 pixel CCD array and covered the 250 nm to 1 μm spectral range with 12 wide to medium band filters.

The f/5.6 WAC within the OSIRIS package had a 12° field of view provided by two aspherical off-axis mirrors. Its focal length was 130 mm and its angular resolution 20″. Its 14 narrow band filters covered from 250 to 630 nm and its detector was a 2048 × 2048 pixel CCD.

---

* Operated by NASA, the Office of Naval Research (ONL) and the Naval Research Laboratory (NRL).
† An optical system in which both astigmatism and field curvature are corrected. See reference A.1.1 for a detailed discussion of the Seidel aberrations of optical systems, or other sources listed in Appendix A.

FIGURE 1.5 Mars Odyssey's THEMIS camera. The entrance aperture is at the top of the smooth rectangular tube that slants slightly to the right. (Courtesy of NASA/JPL-Caltech/Arizona State University.)

Another custom-designed reflective camera is Ralph on board the New Horizons spacecraft. Ralph operates over the 400–975 nm region (visible and NIR) and can obtain both direct images and spectra (Section 1.3). Its design is anastigmatic with three aspherical, off-axis mirrors. The primary mirror has a concave elliptical shape and a 75 mm aperture. The secondary mirror is a convex hyperbola and the tertiary mirror is a concave ellipse. The final focal length is 658 mm and the field of view 5.7° by 1.0°. A dichroic* beam splitter sends radiation over the range 400 nm to 1.1 μm to the imager and the longer wave radiation to the spectrograph.

---

* A mirror with a coating that reflects at some wavelengths and is transparent at others. For example, a thin gold coating reflects at 90% or better for wavelengths longer than 600 nm and is transparent for wavelengths shorter than 450 nm.

Images are obtained as the spin of the spacecraft scans the field of view over the target. The detectors are six 32 × 5024 CCD arrays operated in TDI mode (Section 1.4.1.2). Two of the arrays produce panchromatic images over the 400–975 nm range. The other four obtain images through blue, red and infrared filters and a narrow band methane filter centred on 885 nm (see also JunoCam below). There are also two wide band filters, one of which acts as a backup to the main navigation instrument (LORRI – see above). A seventh 128 × 5024 pixel CCD operating in frame transfer mode (Section 1.4.1.2) aids the spacecraft's navigation.

The mirrors for Ralph were formed by computer-controlled diamond turning of aluminium blocks with special processing to reduce the surface roughness and grooving normally left by the process – giving their surfaces a roughness of about 6 nm. To achieve the Rayleigh limit of resolution, α,

$$\alpha = \frac{1.220\lambda}{d} \text{radians} \tag{1.12}$$

(where d is the aperture diameter) of an optical system requires a surface accuracy for the mirrors in the region of $\lambda/8$ – which at 400 nm would be about 50 nm. So, neither any additional polishing of the mirror surfaces nor separate reflective coatings were needed.

The ExoMars Trace Gas Orbiter[*] has Colour and Stereo Surface Imaging System (CaSSIS) on board to look for sources of trace gases and to investigate surface processes – such as erosion, sublimation and volcanism – as possible contributors of gases to the atmosphere. Its telescope is a three-mirror anastigmat with a 135 mm aperture and an 880 mm focal length. Its field of view is 0.88° by 1.34° and its angular resolution is 1′ per pixel. The detector is a 2024 × 2024 CMOS array (Section 1.4.1.3). It has filters enabling images to be obtained in the wavelength ranges 400–550 nm, 525–775 nm, 790–910 nm and 875 nm–1.025 μm.

The Earth resources spacecraft, Landsat-7, carries the Enhanced Thematic Mapper plus (ETM+) in which a scanning mirror feeds a collimator based upon a Ritchey–Chrétien design followed by two correcting mirrors focussing at the primary focus (for the visible and NIR bands) with some of the radiation redirected through a relay system to a secondary, cold, focus (for the MIR bands).

Three identical three-mirror off-axis aspherical anastigmatic telescopes are carried by the Proba-V spacecraft. They each have fields of view of 34° and are angled to each other to give a total coverage of 102° – providing an observed swath 2285 km across at Earth's surface. Their effective focal lengths are 110 mm and best ground resolutions 100 m. They have four filters each and operate over the 462 nm to 1.6 μm spectral range in order to survey vegetation on Earth. The long wave filter uses a separate light path through the telescope and is diverted by a 45° mirror to a separate detector. Three linear 6000 pixel CCD (Section 1.2.1.2) detectors are used for the shorter wavelengths and three staggered linear 1024 pixel InGaAs photodiode detectors (Section 1.4.1.4) for the long wavelength band.

---

[*] ESA/RKA. RKA – Federal'noye kosmicheskoye agentstvo Rossii (Russian Federal Space Agency).

1.2.2 Refracting Cameras

Today's space cameras based upon lenses are either mostly similar to the basic astronomical refracting telescope with a single objective lens (usually achromatised for its operating wavelengths – Figure 1.1e) projecting its images directly onto the detector instead of into an eyepiece or more like a high-quality digital commercial single lens reflex (SLR) camera as sold to the general public. Their detector arrays are usually CCDs or other solid state detector arrays.

1.2.2.1 Telescope-Type Systems

In the early days of space exploration, photographic cine cameras and vidicon TV cameras were used (Section 1.4.3.5). Indeed, the very first images obtained from space – of parts of Earth – used an off-the-shelf 35 mm cine camera. The camera was launched to a height of 105 km in October 1946 by an ex–World War II German V2 rocket and its main modifications were firstly a reinforced film cassette so that it could survive the 150 m·s^{-1} impact when the rocket fell back to Earth and secondly slowing the frame rate to one frame every second and a half.

Film photography was used by several of the Moon-fly-by/Earth-return Zond spacecraft since their returns to Earth meant that the film could be retrieved. Zond-5 was the first such successful mission although, due to a fault, only images of Earth were obtained. Zond-7 carried a 300 mm focal-length camera and obtained 56 mm × 56 mm colour and panchromatic images. Zonds 6 and 8 both used 400-mm-focal-length cameras and their isopanchromatic images were 130 mm × 180 mm in size. Zond-6 however crash landed and only one image was retrieved.

The lunar impactor, Ranger 7, used six slow-scan vidicon cameras – two WACs and four NACs–to obtain the first close-up images of the Moon's surface. The WACs used 25 mm focal length lenses and had fields of view of 25°, while the NACs used 76 mm focal length lenses and had 8.5° fields of view. The vidicon detector windows were coated with an antimony-sulphide/oxy-sulphide sensitive layer and scanned 1150 lines in 2.5 seconds or 300 lines in 0.2 seconds. Over 4000 images were obtained in the last 14 minutes before the spacecraft hit the Moon's surface at 2.6 km·s^{-1}.

The Dawn spacecraft, which is currently orbiting Ceres* having previously visited Vesta, carries two refracting telescopes for imaging and to aid navigation when close to its targets. The telescopes, called the framing cameras (FCs), are identical and each is a complete instrument in its own right. They are physically separate and having two of them provides for backup in case one of them fails. The objective lens has a 20 mm aperture and a 150 mm focal length (f/7.5) and is essentially an achromatic triplet. The elements of the triplet though are widely separated and the third element is split into two (i.e. there are a total of four lenses in the system). The first element is a crown glass converging lens with an aspherical front surface to compensate for spherical aberration. It is followed by two flint glass diverging lenses and the final element, also of flint glass, is again a converging

---

* A possible visit by Dawn to asteroid 145 Adeona has been abandoned in favour of continuing observations of Ceres during its perihelion passage.

lens. The focal position is maintained to within ±15 μm over a wide temperature range by mounting the two central lenses separately in a tube with a different thermal expansion coefficient from the rest of the camera. The cameras have 5.5° square fields of view and operate from 400 nm to 1.05 μm using one of the eight available filters (one broad band, the others all narrow band). The detectors are 1024 × 1024 CCDs with a best resolution of 17 m/pixel at Vesta and 66 m/pixel at Ceres.

The Mercury fly-by and orbiter spacecraft, MESSENGER, carried a reflective NAC and a refractive WAC. The WAC used two biconvex converging lenses and two biconcave diverging lenses in two groups separated by about 20 mm. Its focal length was 78 mm and its aperture 7 mm. It used 12 filters and operated over the 395 nm to 1.04 μm spectral range. The detector was a 1024 × 1024 CCD array giving a resolution of 36″ – corresponding to about 20 m on Mercury's surface at perihermion. In 2014, it showed the presence of water ice in permanently shadowed craters at Mercury's North Pole.

As mentioned above, the Cassini spacecraft carried a WAC with refracting optics (Figure 1.2). The lenses were spares originally manufactured for the Voyager missions. The camera was f/3.5 with a focal length of 200 mm and a 3.5° square field of view, giving a resolution of 13″/pixel. The lenses were fabricated from radiation-hardened glass or lithium fluoride. It had two filter wheels, each carrying nine filters enabling images over the 380 nm to 1.05 μm region to be obtained. The detector was a 1024 × 1024 CCD array.

The Eros orbiter, NEAR Shoemaker, carried a 50 mm, f/3.4 refractor as its imaging instrument. The camera, called the Multi-Spectral Imaging system (MSI), used a five-element lens and had a 2.3° × 2.9° field of view. It used one wide band and seven narrow band filters, covering the spectral range from 400 nm to 1.1 μm, chosen to facilitate the identification of silicate minerals containing iron. The detector was a 244 × 537 pixel CCD array that was operated at the ambient temperature (typically −30°C to −40°C) with a best resolution of 20″/pixel. From NEAR Shoemaker minimum height before landing of 120 m, this resolution corresponds to a distance of about 10 mm on the surface of Eros.

1.2.2.2 Camera-Type Systems

Examples of the 'commercial digital SLR' type of camera are to be found widely. Thus, JunoCam on the Juno spacecraft uses a 14-element lens. Its focal length is 11 mm and it has a field of view of 58°. The lens' T/number* is 3.2. A 1200 × 1600 pixel CCD is used as the detector and this has a field of view of 3.4° by 18°. The point of view is scanned across the planet by the spacecraft's rotation to build up the full image with TDI (Section 1.4.1.2) read-out. It is designed to obtain colour images of Jupiter with a best resolution of 3 km per pixel (a nearly 40 times improvement of the best Jovian images obtained by the HST[†]). Colour filters in the form of strips and covering the visible region and into the NIR are bonded to the detector

---

* A similar measure to a lens' f/number or focal ratio, but including the effects of absorption within the elements of the lens.

† The HST has a yearly programme (Outer Planet Atmospheres Legacy [OPAL]) of obtaining global maps of all the outer planets.

MARDI

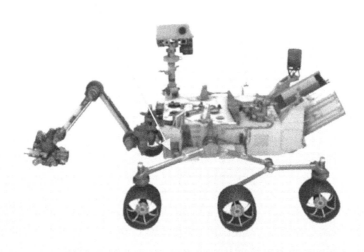

FIGURE 1.6   MARDI mounted on the Curiosity rover. (Courtesy of NASA/JPL-Caltech.)

and are pushbroom scanned across the planet by the spacecraft's rotation. Additionally, there is a narrow band filter centred on the methane absorption line at 889 nm.

JunoCam is a development of the Mars Descent Imager camera (MARDI) mounted on the Mars Science Laboratory's (MSL) Martian rover, Curiosity (Figure 1.6). The main changes from that instrument are radiation hardening (including making the first five elements of the lens from radiation-proof glass) and a 6 mm thick titanium housing. MARDI also used a Bayer* arrangement of filters similar to those used in most publically available digital cameras to generate its colour images.

The Lunar Orbiter/Asteroid fly-by mission, Clementine (see also Section 1.2.1.2), had several cameras. The UV and Visible camera covered from 250 nm to 1 μm and had five narrow band filters and one broadband filter. It used fused f/1.96 silica lenses with an aperture of 46 mm. The detector was a 288 × 384 pixel CCD array. Its field of view was 4.2° × 5.6° and its lunar surface resolution was between 100 and 325 m. The NIR camera operated between 1.1 and 2.8 μm with six filters and a 256 × 256 pixel InSb photovoltaic detector array (Section 1.4.1.4). Its f/3.3 lens had a focal length of 96 mm giving it surface resolution between 150 and 500 m.

In addition to Mars Express' SRC, Figure 1.3 shows the HRSC. This is a seven-element apochromatic† lens with a flat image plane and very little distortion. This is a lens design that is often used for high-quality consumer telephoto lenses. It has a 31 mm aperture and

---

* For each set of four pixels, two have green filters, one a red filter and one a blue filter.
† An optical system in which chromatic aberration (different focal lengths for different colours) has been corrected at three different wavelengths. See reference A.1.1 for a detailed discussion of the aberrations of optical systems, or other sources listed in Appendix A.

a 175 mm focal length (f/5.6). Five filters (blue, green, red, NIR and semi-panchromatic) enable it to image over the 400 nm to 1 μm region. The detector is a CCD (Section 1.2.1.2) of somewhat unusual configuration – it has nine lines, one for each channel and each 1 × 5184 pixels in size (cf. Ralph, above). Stereo images are produced by observing the same region three times over as the spacecraft moves around its orbit using forward-facing, nadir-facing and backward-facing sensors. Its resolution of the Martian surface is about 10 m. The HRSC was originally built as a duplicate instrument for the RKA's Mars 96 mission (destroyed during its launch in 1996).

The Indian Space Research Organisation's (ISRO) Mars Orbiter Mission (MOM) uses a commercial, off-the-shelf, 105 mm focal length, f/4.0 telephoto lens (Mars Colour Camera – MCC). It has a field of view of 8.8° and a surface resolution varying from 15 m (when at periareon – 420 km above the Martian surface) to 4 km (when at apareon – 70,000 km above the Martian surface). The detector is a 2048 × 2048 CCD array using Bayer-pattern filters operating over the 400–700 nm spectral range. The failed Mars lander, Schiaparelli's descent camera (DECA), would have obtained monochromatic images of its landing area. The camera had a focal length of 6.65 mm, a field of view of 60° and used a CMOS array (Section 1.4.1.3) as its detector.

The SMART-1* lunar orbiter carried a small camera constructed from off-the-shelf components with a 1024 × 1024 CCD detector. Asteroid-Moon Micro-Imager Experiment (AMIE) imaged in white light plus through filters at 750, 915 and 960 nm with the filters deposited directly onto the CCD pixels. Its best surface resolution was 27 m and its objective was to look at lunar surface minerology.

JAXA's Akatsuki's (also known as Venus Climate Orbiter) 2 μm NIR camera is designed to study the middle to lower part of Venus' atmosphere using windows in the carbon dioxide absorptions at 1.74 and 2.3 μm. It is based upon a triplet lens comprising a zinc sulphide and two quartz elements and has five narrow band filters. Its focal length is 84 mm with a 12° field of view and its angular resolution is 1°. A 1040 × 1040 pixel platinum silicide array is used as the detector.

The ExoMars Rover mission is planned for a launch in 2020 with a landing on Mars in 2021. The rover will carry three cameras mounted as a single rotatable and tiltable unit, called PanCam, at the top of a mast. The two identical WACs are planned to have fields of view of 32.3° supplied by 22 mm focal length f/10 lenses. Each camera will have 11 filters covering the range 400–1050 nm. The detectors are to be 1024 × 1024 CMOS arrays (Section 1.4.1.3) with an angular resolution of 2′ per pixel (corresponding to 1.3 mm on the surface at a distance of 2 m). The two WACs will be separated by 500 mm in order to enable stereoscopic images to be obtained and are sighted on the same spot at a distance of about 5 m from the rover.

The Hayabusa-2 asteroid sample and return mission's lander, Mobile Asteroid Surface Scout (MASCOT) carries a lens-based camera designed on the Scheimpflug† principle so that its images are in focus from 150 mm to infinity. The Scheimpflug design may be used

---

* Small Missions for Advanced Research in Technology.
† Theodor Scheimpflug, a captain in the Austrian army, used the principle for aerial photography in the early twentieth century.

when the plane containing the object and the plane containing the image are angled to each other. The two planes thus meet somewhere in a line. Scheimpflug showed that the object would be in focus at almost all distances when the plane containing the camera lens also intersected the same line. MASCOT's camera has a 1024 × 1024 pixel CMOS detector array (Section 1.4.1.3) that is sensitive to the 400 nm to 1 μm spectral region and has red, green and blue light emitting diodes (LEDs) to illuminate the asteroid's surface when needed.

Small cameras, little different from webcams, are often mounted on spacecraft to enable Earth-based controllers to see that manoeuvres such as the deployment of solar panels or the release of a separate module such as a lander have been accomplished as expected. Often called Visual Monitoring Cameras (VMCs), the one on board ESA's Mars Express Orbiter is fairly typical. This camera was installed in order to check the separation of the Beagle-2 Mars lander from the mother craft. It uses a 2.5 mm diameter, f/5, three-element lens plus an entrance window and an infrared filter. It has a field of view of 31° × 40° and covers the spectral range, 400–650 nm. Its detector is a 480 × 650 pixel Complementary Metal-Oxide Semiconductor Active Pixel Sensor (CMOS APS) array with a Bayer-pattern filter integrated with the array and used for colour imaging. Although its only purpose initially was to check the lander's deployment, Mars Express' VMC has now a second life: obtaining images of Mars itself. From 2007 onwards it has been imaging Mars at resolutions between 3.5 and 11.5 km/pixel and the images are made available on the Internet, sometimes almost in real time, for anyone to use.

1.2.2.3 Panoramic Systems

Cameras designed to obtain panoramic images have been used on many missions, especially during the first few decades of planetary space exploration. The term 'cycloramic camera' is often used for such instruments and there are two basic approaches to their design. The first is to scan the camera or lens or detector around, usually in a horizontal direction.* This design is the one that has been used for most space-based applications. When using a single detector scanned in two dimensions, the process is similar to that of early facsimile (fax) machines and facsimile camera is an alternative name for these instruments. If using a linear array detector, scanned in one dimension, then the process is the same as pushbroom scanning. A second approach is to use a more-or-less conventional camera and to point it towards a reflecting hemisphere. In the latter case, the panorama automatically covers 360° and includes the camera obtaining the image at its centre.

The Luna-19 Moon orbiter carried a facsimile-type camera that scanned in one direction only (hence it is also called a linear camera) and used pushbroom scanning arising from its orbital motion to obtain the panorama. A pair of lenses focussed the light from about a 100 m × 100 m area of the lunar surface onto the entrance window of a photomultiplier. The point of view was scanned 180° over the lunar surface by a rotating rectangular

---

* Many modern digital commercial cameras have the ability to obtain panoramas in this fashion with the photographer scanning the camera around. Recently, mobile phones and related devices have been designed to rotate themselves automatically to obtain 360° panoramas.

FIGURE 1.7   Luna-19. (Courtesy of NASA.)

prism. The scan rate was four lines per second. The surface resolution along the line of the orbital motion was about 400 m (Figure 1.7).

Scanning panoramic cameras have been mounted on numerous surface landers and rovers, including on many of the USSR's (Veneras, Lunas and the two Lunokhod* rovers) and on NASA's Viking Mars landers.

The two cycloramic cameras carried by the Venera-9 lander sent back the first images of Venus' surface. The cameras each had a six-element, 28 mm focal-length lens and used a single photomultiplier tube (PMT) as the detector. The cameras were mounted inside the main pressure hull of the lander, for protection against the 460°C surface temperature and 9 MPa pressure of Venus' atmosphere. The surface was viewed via periscopes whose entrance windows projected outside the body of the lander. The entrance windows to the periscopes were made from 10 mm thick quartz and compensating lenses were used to correct the distortion produced by the windows. The cameras were thermally insulated, as far as possible, from the periscopes which had to function at Venus' surface temperature. The instruments had a 37° field of view and pointed down at an angle of 35° to the vertical. Scanning was accomplished by using wires to rotate the top mirror of the periscope through an angle of 180°. The images thus extended from horizon to horizon,

---

* Russian for 'Moon Walker'.

whilst at their mid-points pointed down at the surface close to the lander. A single horizontal scan took 3.5 seconds and was divided into 512 individual exposures (i.e. what would now be called pixels) and the images were 125 such scans deep. Four lamps illuminated the surface close to the lander and the lenses were focussed to provide acceptably sharp images from 0.8 m to infinity. Although only one camera worked (the lens cover of the second camera failed to release), it obtained almost two complete 180° panoramas before contact with the lander was lost. The identical Venera-10 lander also sent back two panoramas.

Four cycloramic cameras were carried by both Lunokhods 1 and 2, with two on either side of the rovers. They obtained 180° horizontal panoramas stretching from the horizon to 1.4 m from the spacecraft and 360° images in a vertical plane. They scanned at four strokes per second and used a single photomultiplier in each camera as the detector.

1.3 SPECTROSCOPY AND SPECTROGRAPHS

1.3.1 Introduction

vPassing mention of some spectrographs has been made in Section 1.2 because many imaging instruments also have spectrographic modes. Thus, MAVEN's IUVS (Section 1.2.1.1) principally works as an UV spectrograph. It operates over the 110–320 nm spectral region with a spectral resolution of 220–250 when using a plane reflective grating and a spectral resolution of 14,500 over the spectral range 116–131 nm when using an echelle grating with a prism cross-disperser.

While it is likely that some readers of this book will be familiar with the terms used in the previous paragraph, others may not have encountered them before. A brief summary of the basic physics of spectroscopy is therefore given below. Further details may be found in references A.1.1, A.1.9 and other sources listed in Appendix A.

1.3.2 Basic Spectroscopy

Most people will have seen a rainbow or a sun dog where water droplets or ice crystals respectively in Earth's atmosphere split sunlight into the seven colours – violet, indigo, blue, green yellow, orange and red. Many will also be familiar with the action of a triangular prism in producing a spectrum. Whilst prisms are still used in spectrographs, diffraction gratings of various types are much more commonly encountered. We therefore limit this introduction to the latter.

Diffraction gratings function through the interference of e-m radiation behaving in its wave manifestation. The fundamental process is exemplified by Thomas Young's double slit experiment. A beam of radiation incident onto a narrow slit partially passes straight through the slit and parts of it have their paths bent sideways at the edges of the slit (diffracted). If two such slits are used that are parallel to each other and separated by a short distance (Figure 1.8), then the emerging fans of radiation will cross each other. At some points the waves from the slits will both be crests or both be troughs. At these points the waves will add together (interfere constructively) and a screen placed at one of those points would show a bright spot. At other points the wave from one slit will be a crest and that

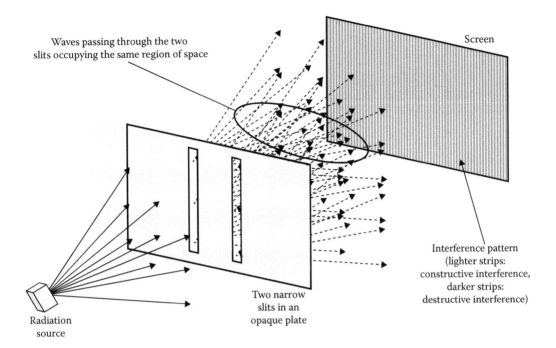

Waves passing through the two
slits occupying the same region of space

Screen

Radiation
source

Two narrow
slits in an
opaque plate

Interference pattern
(lighter strips:
constructive interference,
darker strips:
destructive interference)

FIGURE 1.8   Young's double slit experiment.

from the other be a trough. At these points the waves will cancel each other out (interfere destructively*) and a screen placed at one of those points would be dark.

The significance of Young's double slit experiment for spectroscopy lies in the observation that the nature of the interference pattern depends upon the wavelength of the radiation involved. In Figure 1.8, the interference pattern is shown as a series of light and dark lines when projected onto a screen (because the two apertures involved are slits – other shaped apertures would produce other patterns). If, say, the radiation involved in Figure 1.8 were blue light and the light source was then replaced with red light then the interference pattern would remain as a series of light and dark lines but the separation of the lines would increase. If the source had both red and blue light, then both patterns would be superimposed. At some points, therefore, we would see a bright red line next to a bright blue line – in effect a very simple spectrum.

The spectrum produced by a double slit is of little practical use, but if a third slit were to be added then the bright lines (or fringes as they are usually called) would become brighter and narrower; weak fringes would also appear where the dark lines had been previously (Figure 1.9b). Adding fourth and fifth slits would brighten and narrow the fringes even further (Figure 1.9c). Continuing this process until we have the thousands of apertures used in practical diffraction gratings would result in the bright fringes becoming very narrow. A radiation source producing monochromatic red and blue light would then be seen to produce very narrow blue lines separated by wide dark regions from very

---

* The waves are not actually destroyed – they continue onwards through each other and will interfere both constructively and destructively again at other points with other waves.

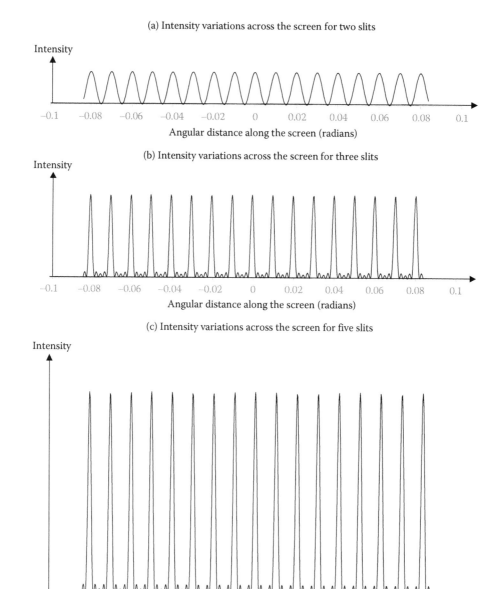

FIGURE 1.9 A plot of the intensity of the fringes on the screen in Figure 1.8. The change in fringe shape and intensity for interference patterns from two, three and five slits is shown. The peak intensities are scaled to show the actual differences between the fringes in the three cases.

narrow red lines on the screen in Figure 1.8. If we add intermediate colours to the light source, then these would appear as bright lines between the blue and red lines and a source emitting over the whole visible spectrum would fill in the gaps between the blue and red lines with all the colours, thus producing a continuous spectrum, like a rainbow.

The device just described is called a transmission diffraction grating. Nothing about the interference process, however, is changed if the apertures shown in Figure 1.8 are replaced by narrow mirrors, producing a reflection diffraction grating. Both types of gratings are to be found in actual spectrographs.

One problem with either type of grating that will be apparent from Figures 1.8 and 1.9 is that many spectra are produced so that the available radiation is spread very thinly (i.e. it produces spectra very inefficiently). At the centre of the interference pattern on the screen in Figure 1.8 will be a bright fringe where the radiation passing through each of the two slits has travelled the same distance to the screen. The two waves therefore interfere constructively. Since the path difference between these two waves is zero, the fringe is referred to as the zero-order fringe.

Moving to the left or right from the zero-order fringe will produce a small difference in the length of the light paths of the waves coming through each aperture. When that path difference reaches half the wavelength of the radiation involved, then the waves will interfere destructively, producing a dark line. However by continuing to move left or right we reach a point where the path difference becomes one wavelength. The waves will thus again be in step and we have another bright fringe. Since the path difference is one wavelength, this is called the first-order fringe. Continuing to move left or right we reach another bright fringe produced when the path difference between the waves is two wavelengths – and this is the second-order fringe. Third-order fringes, fourth-order fringes, fifth order fringes, etc. result from constructive interference with path differences of three, four, five, etc. wavelengths.

However, what is not shown in Figure 1.8 is that there is a second interference process going on. If the opaque plate has only one slit, then the screen will show a rectangular illuminated area. However, if that area is examined closely it will be seen to have faint light and dark fringes along its sides and top and bottom (Figure 1.10a). These fringes arise from interference between waves passing through different parts of the single aperture.

Astronomers reading this book will also be familiar with this effect from the circular fringes surrounding the images of stars produced by even high-quality telescopes and which arise from the circular aperture of the instrument through the same process. The bright central part of the stellar image, called the Airy disk, is the zero-order fringe in this situation. It is the angular size of the Airy disk that determines the Rayleigh resolution (Equation 1.12) of the telescope and so this is often also called the diffraction limit of the telescope.

Returning to the diffraction grating, the interference pattern shown on the screen in Figure 1.8 will be modulated by the interference pattern of the single aperture (Figures 1.10a and 1.11a). The peak intensities of the five-slit interference pattern shown in Figure 1.9c will thus be limited by the envelope of the single slit diffraction pattern (Figures 1.10b and 1.11b). As may be seen from those figures, much, but not all, of the energy is now going into the zero-order fringe.

Detail of the interference pattern produced by a single slit

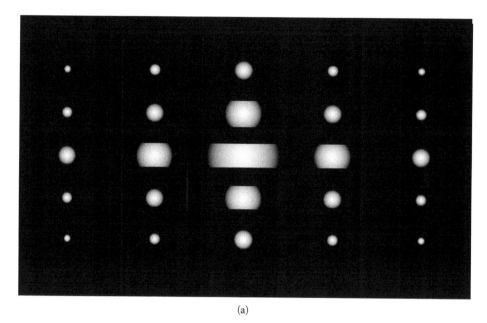

(a)

Detail of the interference pattern produced by two slits

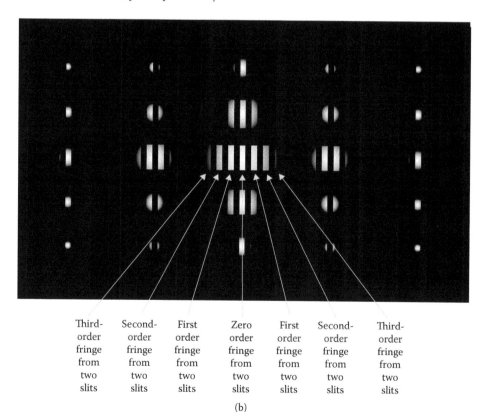

| Third-order fringe from two slits | Second-order fringe from two slits | First order fringe from two slits | Zero order fringe from two slits | First order fringe from two slits | Second-order fringe from two slits | Third-order fringe from two slits |

(b)

FIGURE 1.10   The fringe patterns for one and two slits.

(a) Intensity variations across the screen for a single slit with a width of 15 μm, for radiation with a 500 nm wavelength

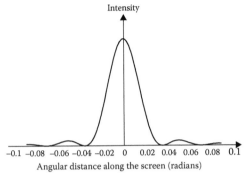

Angular distance along the screen (radians)

(b) Intensity variations across the screen for <u>five slits</u>, each with a width of 15 μm and with a separation of 50 μm, for radiation with a 500 nm wavelength

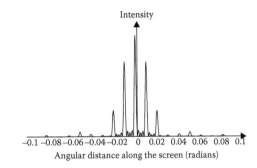

Angular distance along the screen (radians)

(c) Intensity variations across the screen for five slits, each with a width of <u>50 μm</u> and with a separation of 50 μm, for radiation with a 500 nm wavelength

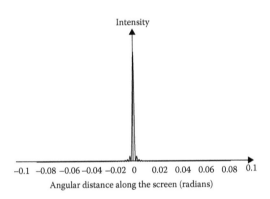

Angular distance along the screen (radians)

(d) Intensity variations across the screen for five slits, each with a width of 50 μm and with a separation of 50 μm, for radiation with a 500 nm wavelength and <u>blazed for the third order</u>

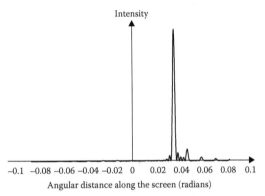

Angular distance along the screen (radians)

FIGURE 1.11 Fringe intensities. (The changes from one graph to the next are underlined.) (a) Intensity variations across the central horizontal line of fringes in Figure 1.10a, for a single slit with a width of 15 μm, for radiation with a 500 nm wavelength (the vertical [intensity] scale is expanded by a factor of 25 compared with the scale of Figure 1.9). (b) Intensity variations across the central horizontal line of fringes in Figure 1.10b, for five slits, each with a width of 15 μm and with a separation of 50 μm, for radiation with a 500 nm wavelength (the vertical [intensity] scale is the same as that in Figure 1.9). (c) Intensity variations across the central horizontal line of fringes in Figure 1.10b, for five slits, each with a width of 50 μm and with a separation of 50 μm, for radiation with a 500 nm wavelength (the vertical [intensity] scale is the same as that in Figure 1.9). (d) Intensity variations across the central horizontal line of fringes in Figure 1.10b for five slits, each with a width of 50 μm and with a separation of 50 μm, for radiation with a 500 nm wavelength and blazed for the third order (the vertical [intensity] scale is the same as that in Figure 1.9).

(Continued)

(e) Intensity variations across the screen for five slits, each with a width of <u>35 μm</u> and with a separation of 50 μm, for <u>radiation with 480 nm (dashed line) and 520 nm (dotted line) wavelengths</u> and blazed for the third order

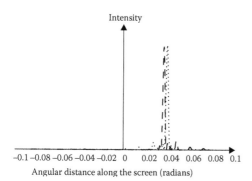

(f) Intensity variations across the screen for <u>25 slits,</u> each with a width of 35 μm and with a separation of 50 μm, for radiation with 480 nm (dashed line) and 520 nm (dotted line) wavelengths and blazed for the third order. NB −The maximum intensities should be 25 times higher than those shown in Figure 1.11e, but they are scaled down in order to get the graph onto the page.

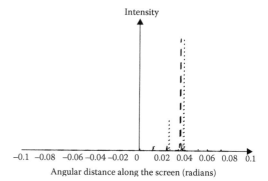

FIGURE 1.11 (*Continued*)   Fringe intensities. (The changes from one graph to the next are underlined.) (e) Intensity variations across the central horizontal line of fringes in Figure 1.10b for five slits, each with a width of 35 μm and with a separation of 50 μm, for radiation with 480 nm (dashed line) and 520 nm (dotted line) wavelengths and blazed for the third order (the vertical [intensity] scale is the same as that in Figure 1.9). (f) Intensity variations across the central horizontal line of fringes in Figure 1.10b for 25 slits, each with a width of 35 μm and with a separation of 50 μm, for radiation with 480 nm (dashed line) and 520 nm (dotted line) wavelengths and blazed for the third order (the vertical [intensity] scale is reduced by a factor of 25 compared with Figure 1.9).

Now the angular distance between the two first *minima* (i.e. the angular width of the central bright area) for the single slit case is given by*

$$2 \times \sin^{-1}\left(\frac{m\lambda}{D}\right) \text{radians} \tag{1.13}$$

where m (= 1) is the order of the minimum, λ the wavelength of the radiation and D the slit width.

Thus, the angular width of the zero-order maximum for the single slit depends inversely upon the slit width and by increasing the slit width we may decrease that width.

However (rather confusingly), the angular separation of the first-order *maxima* ($m = 1$) from the centre of the zero-order maximum for slits separated by a distance, d, is

$$2 \times \sin^{-1}\left(\frac{m\lambda}{d}\right) \text{radians} \tag{1.14}$$

---

* This equation is based upon the assumption that the screen onto which the interference pattern projects is a long way from the slit when compared with the radiation's wavelength, so that the wave fronts are essentially linear (or that a lens is used to give the same practical effect). This is known as the Fraunhofer or far-field regime. If the curvature of the wave fronts is significant on a scale comparable with the wavelength, then we have the Fresnel or near-field regime and different equations will apply.

Thus when $D = d$, the first order maxima for the basic interference pattern from two or more slits (Figure 1.9) are superimposed upon the first minima of the diffraction pattern for a single slit (Figures 1.10a and 1.11a). Only the zero-order fringe from the slits' interference pattern therefore contains any significant amount of energy (Figure 1.11c) and, except for one further problem we have an efficient grating.

The further problem is that the zero-order fringe from two or more slits is *not* a spectrum! Since the situation is symmetrical at the centre of the fringe pattern, the paths for radiation of any wavelength from slits equidistant from the centre is the same and all wavelengths will have bright fringes at that point.

Fortunately, the diffraction pattern from the single slit can be moved with respect to the interference pattern from the multiple slits. It is easiest to comprehend how this can be accomplished for reflective diffraction gratings – the individual strip mirrors are tilted slightly from being perpendicular to the incoming beam of radiation until the reflected beams coincide with the angular position of one of the higher order spectra. This process is called blazing the grating and gives the surface of the grating a saw-tooth shape in cross section (Figure 1.12). Transmission gratings can be blazed by making the apertures from narrow angle prisms. The effect of blazing the grating to one of the third-order fringes is shown in Figure 1.11d.

So far we have just looked at what happens with a monochromatic beam of radiation. Equation 1.13, however, shows that the position of the fringes is wavelength dependent. If two wavelengths are present in the beam of radiation then two sets of fringes will be produced with a slight angular separation (Figure 1.11e – the slit width has been decreased slightly to expand the envelope of the single slit diffraction pattern so that it encompasses both wavelengths).

With five slits, the two wavelengths (480 and 520 nm) shown in Figure 1.11e are just about resolved. The fringe widths, though, are reduced (spectral resolution improved) by increasing the number of slits and Figure 1.11f shows the result for 25 slits where the two wavelengths are now clearly separated. In practice, diffraction gratings have from 100 to 10,000 slits (Figure 1.13) so that the fringes become very narrow indeed (i.e. the spectral resolution becomes very high).

With the above discussion of the basic physics of how diffraction gratings function, we may now go on to look at the equations describing the process. The intensity within the modulated interference pattern (e.g. Figure 1.11b) is given by the equation

$$I(\theta) = I(0) \left[\frac{\sin^2\left(\frac{\pi D \sin\theta}{\lambda} \right)}{\left(\frac{\pi D \sin\theta}{\lambda} \right)^2} \right] \left[\frac{\sin^2\left(\frac{N\pi d \sin\theta}{\lambda} \right)}{\sin^2\left(\frac{\pi d \sin\theta}{\lambda} \right)} \right] \tag{1.15}$$

where θ is the angle perpendicular to the plane of the grating, $I(\theta)$ is the intensity of the interference pattern at that angle, $I(0)$ is the intensity at the centre of the zero-order fringe and N is the number of slits forming the grating. The term in the first set of square brackets calculates the diffraction pattern of the slit; the second term in square brackets calculates the interference pattern of the multiple slits.

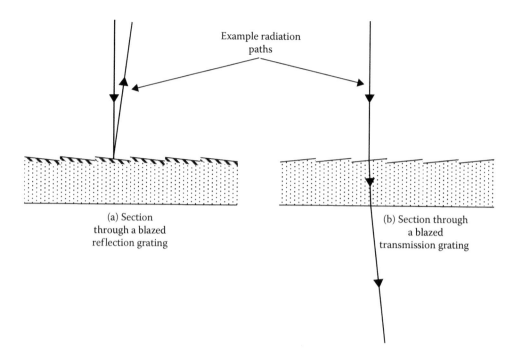

FIGURE 1.12   Cross sections through parts of blazed gratings.

FIGURE 1.13   An example of a large reflection diffraction grating for the European Southern Observatory (ESO)'s Echelle Spectrograph for Rocky Exoplanet and Stable Spectroscopic Observations (ESPRESSO) spectrograph. (Courtesy of ESO/M. Zamani.)

The angular positions of the main maxima within the slits' interference pattern (Figure 1.9) when the beam of radiation is incident perpendicularly to the plane of the grating occurs for

$$\theta = \sin^{-1}\left(\frac{m\lambda}{d}\right) \tag{1.16}$$

However within spectrographs, gratings are often placed at an angle to the beam. If the beam is at an angle φ to the perpendicular to the plane of the grating, then the main maxima are shifted in position to

$$\theta = \sin^{-1}\left(\frac{m\lambda}{d} - \sin\varphi\right) \tag{1.17}$$

In this form, the relationship is often called the grating equation.

The spectral resolution provided by a grating is its ability to separate two slightly different wavelengths clearly. There are two ways of measuring it – one is the actual minimum such wavelength difference, W_λ, given by

$$W_\lambda = \frac{\lambda}{Nm} \tag{1.18}$$

The second measure is the ratio of the operating wavelength to W_λ, denoted by R:

$$R = \frac{\lambda}{W_\lambda} = Nm \tag{1.19}$$

Thus, in the example of MAVEN given in the first paragraph of this section, the plane grating has $R \approx 235$ and $W_\lambda \approx 0.9$ nm, while the echelle grating (see below) has $R \approx 14,500$ and $W_\lambda \approx 0.0009$ nm.

The rate of change of wavelength with θ is called the angular dispersion of the spectrum and is given by

$$\frac{d\lambda}{d\theta} = \frac{d}{m}\cos\theta \tag{1.20}$$

so that any two given wavelengths will have a larger angular separation for higher order spectra than for lower order spectra. This is usually (somewhat confusingly since the numbers involved for, say nm radian^{-1} will be smaller at the higher orders) called having a higher dispersion.

If the spectral resolution is to be realised, then two wavelengths separated by W_λ must be detected by separate pixels within the detector array. The physical separation of the two wavelengths may be increased by using a higher order spectrum, so increasing the dispersion, or by using a longer focal length imaging element (Section 1.3.3). Limitations on space and mass sometimes mean that for spacecraft-borne instrumentation the detector pixel size limits the spectral resolution, not the grating properties.

The intensity at the centre of a bright fringe is proportional to N^2, since clearly for identical slits, the more such slits that there are, the more radiation will be gathered. However, from Equation 1.17, the widths of the fringes decrease with N, so this increase in the gathered energy is concentrated into smaller regions.

Thus, both the spectral resolution and the image brightness are improved by having more lines* to the grating. There are, however, practical limits to that number. Many reflective gratings are manufactured by a machine that scores a diamond point, shaped to produce the blaze angle, across a metallic (usually aluminium) reflective coating on a substrate. Wear on the diamond point limits the number and length of the grating lines in this instance. Transmission gratings can be made by a similar method using a transparent material or by photolithography, but limitations still apply. Large gratings are often now produced holographically although they differ somewhat in the way they operate from the 'simple' slit or strip mirror devices just discussed. In holographic gratings, the lines are produced by a varying refractive index within a layer of gelatine.

Spectral resolution however depends upon both the number of lines and the spectral order (Equations 1.18 and 1.19) and so it can be improved by blazing the grating to operate at a higher spectral order. Echelle gratings are often used in this fashion and can operate at spectral orders of several hundred. They are reflective gratings illuminated at a low angle to the plane of the grating and in cross section look rather like a flight of stairs (Figure 1.14). The 'riser' part of the grating can easily be made (say) 200 wavelengths deep – and the spectrum then produced will be of the 200th order. That spectrum, though, will only cover a few wavelengths because, from Equation 1.13, the envelope of the single slit diffraction pattern will be very narrow for a high value of m. In fact, slightly different wavelengths from the . . . 197th, 198th, 199th, 201st, 202nd, 203rd . . . spectral orders will all be superimposed upon the 200th order spectrum.

While the superimposition of different spectral orders might seem to be a severe disadvantage of the echelle grating, it actually enables such gratings to produce a spectrum covering a wide range of frequencies. The various superimposed spectral orders are, of course, of slightly different wavelengths. A second grating (or often a prism) is set with its dispersion perpendicular to that of the echelle grating and it separates out the individual short spectra (Figure 1.15). By designing the echelle grating carefully it is possible to make the short wavelength end of the spectrum from one order overlap the long wavelength end of the next order. A complete spectrum can thus be obtained by joining together the short individual spectrum segments. The second grating or prism in this design is called the cross-disperser. The example of the two spectroscopic modes in the MAVEN IUVS spectrograph has already been mentioned and the echelle mode has about 60 times better spectral resolution than the simple grating mode.

The Atmosphere Chemistry Suite (ACS) package on board the ExoMars Trace Gas Orbiter contains three separate spectrometers – two of which are based upon echelle gratings (the third is a Fourier transform spectrometer – see Section 2.5). The mid-IR spectrometer operates over the 2.3–4.2 µm spectral region with a spectral resolution of 50,000. The instrument has a periscopic feed incorporating a lens-based telescope with a reflective collimator and final imaging system. The main reflection grating has three grooves per millimetre over a 107 mm × 240 mm area and has two gratings as its cross-dispersers. Up to 17 spectral orders covering a total of 300 nm can be imaged at one time onto the HgCdTe detector (Sections 1.4.1.1 and 1.4.1.4).

---

* We use the term 'lines' instead of 'slits' generally from this point onwards since it is the term in common use and includes both the actual slits forming a transmission grating and the thin mirror strips of a reflective grating.

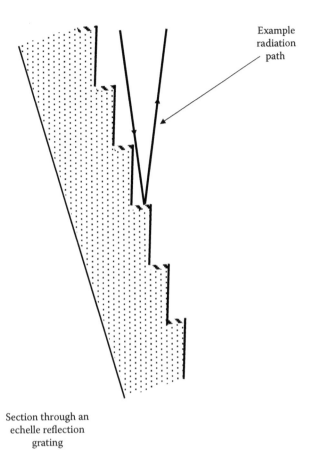

Example radiation path

Section through an echelle reflection grating

FIGURE 1.14   Cross section through a part of an echelle grating.

FIGURE 1.15   An example of an echelle spectrum after the individual short spectrum segments have been separated by a cross-disperser. The spectrum segments run horizontally and slightly upward towards the right. Absorption and emission spectrum lines (Section 1.3.4) run vertically across the segments. (Courtesy of ESO.)

The NIR spectrometer covers the 700 nm to 1.6 μm range at a spectral resolution of 20,000. The echelle grating has 24.35 grooves per millimetre, covering a 42 mm × 102 mm area. An acousto-optical tunable filter (AOTF) (Section 1.3.8) is used to select the different spectral orders. It also has a periscopic feed and an off-axis mirror acts as both collimator and final imager. The detector is a 512 × 640 InGaAs array (Section 1.4.1.4).

Gratings need not be necessarily be flat – reflective gratings can be produced on curved surfaces. In some designs of spectrographs (Section 1.3.3), the use of a curved grating (say with a concave parabolic or spherical shape) can replace some of the other optical components generally needed within the instruments. The design may thus be made simpler and more efficient.

1.3.3   Spectrographs

While the diffraction grating (or prism) separates out the differing wavelengths within a beam of radiation, it does not directly produce a useable spectrum – additional optical components are needed to do that. The components and layout of a basic reflection grating-based spectrograph are shown in Figure 1.16. A slit is placed at the focal plane of the telescope feeding the spectrograph in order to define the field of view, to reduce scattered radiation and to ensure that the spectral features are seen as thin lines within the spectrum (Section 1.3.4) (NB – this entrance slit is not to be confused with slits (or strip mirrors) previously discussed which form the diffraction grating). A converging lens or mirror, known as the collimator, converts the expanding cone of radiation that has passed through the entrance slit into a parallel beam. That parallel beam is then directed onto the grating. The grating reflects the differing wavelengths at slightly differing angles, but still as parallel beams. An imaging lens or mirror then focuses the reflected beams onto the detector.

There are many other designs for spectrographs other than the basic layout shown in Figure 1.16. However, most of these have the five components shown there (entrance slit, collimator, grating (or prism), imager and detector) in some form or other. The use of a concave reflection grating, for example, can mean that the grating also acts as the collimator or imager or even both. Multi-object and integral field spectrographs are able to obtain spectra of numerous objects and/or of extended objects in a single exposure. Spectra can also be obtained by quite different approaches such as scanning narrow band filters, Fabry–Perot etalons and Michelson interferometers. Details of these instruments may be found from sources in Appendix A and brief summaries are included with the instruments' descriptions (Sections 1.3.5 and 2.5) when needed.

1.3.4 Spectral Features

The aim of this book is to describe the physical principles behind the instruments used to investigate solar system objects. It is not therefore really a part of its brief to look at the reasons why the data from the instruments are obtained. However, so much information is potentially contained in a spectrum that a brief discussion of this topic is included here for completeness and (it is hoped) just as a reminder. Sources in Appendix A may be used for further study by interested readers.

Information on the source of radiation is contained in the way that the intensity of radiation of a particular wavelength varies as that wavelength changes. Even with an otherwise featureless continuous spectrum, the slow changes in intensity over broad spectral

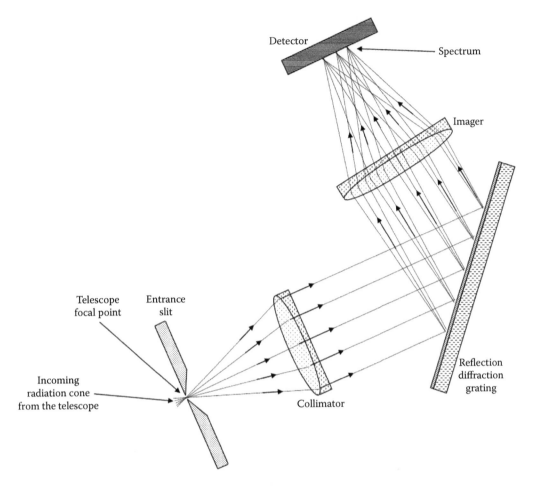

FIGURE 1.16 A cross section through a simple reflection grating-based spectrograph showing the optical layout. Lenses are shown as the collimator and imager in this illustration (for simplicity of drawing), but mirrors are often used in their place.

regions will enable the temperature of a thermal source to be estimated or the magnetic field strength and electron energies for a synchrotron radiation source to be determined. Mostly, however, the spectral features of interest are rapid and sometimes large changes in intensity over a very narrow range of wavelengths. Such features are usually called spectrum lines.* The linear shape, however, is simply due to the use of a linear slit as the entrance aperture to the spectrograph; other shapes of entrance aperture would produce other shapes for the spectrum 'lines'. When the intensity within the spectral feature decreases compared with the intensity of its neighbouring wavelengths, a dark (absorption) line appears in the spectrum; when the energy increases, a bright (emission) line is seen (Figure 1.15).

The spectrum lines arise because energy is absorbed whenever electrons in atoms, ions or molecules jump from a lower energy level to a higher one (see Appendix A for further reading on atomic structure, etc.). The energy needed for a jump between two particular levels within a particular atom, ion or molecule always takes the same, or very nearly the same, value. The energy for the jump is provided by the absorption of a photon with that specific

---

* Not to be confused with the lines of a diffraction grating itself.

energy from the beam of radiation passing by the particle. From Equation 1.2, however, we see that such an energy corresponds to a particular frequency or wavelength for the radiation. If many identical particles have electrons making that same jump at the same time, then sufficient photons, all with the same wavelength, will be lost from the beam of radiation to reduce the intensity noticeably at that wavelength. Rarely will every single photon at the wavelength involved be lost from the radiation beam, but the intensity will be sufficiently reduced to show up as a dark line in comparison to the brighter wavelengths nearby.

Emission lines arise through the inverse process to the one just described – because electrons jumping down from the higher energy level to the lower release (emit) their stored energy in the form of photons which have the same wavelength as those lost during the absorption process. The added photons increase the intensity at the wavelength involved and are seen as a brighter region (line) within the spectrum.

Both absorption and emission lines can also arise, especially in the MIR and FIR regions, through a slightly different process. In the same way that the energies which electrons can take up within atoms, ions and molecules are only allowed to take certain values, so the energies of vibrating or rotating molecules are also only allowed take certain values. If a molecule changes its rate of vibration or rotation it will therefore absorb or emit a photon of a specific wavelength.

Electron transitions generally produce spectral features over the UV, visible and NIR spectral regions. Vibrational transition features tend to occur in the MIR and rotational transition features in the FIR and microwave spectral regions – because the energies of the vibrations or rotations are usually much smaller than the energies of the electrons. The spectral features have very narrow widths; however, some molecules produce hundreds of such overlapping lines, so that they appear quite broad and are then termed bands.

Now although the energy (wavelength/frequency) of a photon absorbed or emitted as described above is initially determined by the properties of the atom, ion or molecule, a very large number of other processes can affect that energy both during the transition and afterwards. Conversely by observing the photons lost or produced during transitions, information on those processes can be obtained; this is why spectroscopy is such a powerful tool for astrophysics (and for many other branches of science as well).

Absorption lines in the solar spectrum were noticed by William Wollaston in 1802 and studied more extensively by Joseph von Fraunhofer between 1814 and 1826. Fraunhofer also observed absorption lines in stellar spectra. The origin of the absorption lines however remained a mystery until Gustav Kirchhoff and Robert Bunsen realised in 1860 that the emission lines that they observed when looking at the spectra of chemicals burning in hot flames coincided exactly in wavelength with some of the solar absorption lines.

Within two years of publishing this result Kirchhoff had shown that the Sun contained the elements barium, calcium, chromium, copper, iron, magnesium, nickel, sodium and zinc and Anders Ångstrom had detected hydrogen. By 1864 William Huggins had found calcium, hydrogen, iron and sodium in Sirius and other stars. Huggins also showed that comets and some nebulae had emission line spectra and so must be composed of thin hot gases. In 1868 he found that in Sirius' spectrum, the line due to hydrogen at 486.1* nm

---

* The Fraunhofer F line or the hydrogen β line in the Balmer spectrum.

was displaced towards the longer wavelengths by 0.109 nm. He interpreted this to mean that Sirius was moving away from us at 41.4 miles per second. After allowing for Earth's orbital motion, this was an actual recessional velocity of 29.4 miles per second (47.3 km·s^{-1}). Huggins' measurements were adrift (Sirius' true velocity is 5.5 km·s^{-1} towards us) but the principle of using line shifts to determine velocities was correct.

Thus from its very beginnings, spectroscopy has enabled the compositions and line-of-sight velocities of astronomical objects to be determined and it is still used for these purposes today. However much more information is available, especially from the detailed study of the shapes (profiles) of spectrum lines. For example, the pressure, temperature and turbulent or other motions of the material producing the spectrum lines broadens them, but in different ways so that these quantities may be separately determined. Since we see stars as point sources their spectra come from the whole disk – when the star is rotating, one limb will be approaching us and the other moving away; this also broadens the spectrum lines, producing yet another shape. For angularly resolved objects such as most of the bodies within the solar system, their rotation can be measured by determining the separate velocities of the approaching and receding limbs.

Ions have different spectra from their parent atoms and so levels of ionisation may be observed directly. Different isotopes of the same element also have slightly different spectra. The intensities of lines are related, though not in a simple way, to the relative proportions of the atoms, ions and molecules present in the line-producing material. Magnetic fields cause spectrum lines to split or broaden and can induce polarization, so providing information on the fields' strengths and directions. The spectra of high density gases, liquids and solids do not, in general, show the narrow lines of the spectra just considered, but they have broader spectral features (green grass, red rubies, yellow bananas, etc.) that provide some information about their natures (see also Section 7.2.2.2) These few examples will suffice to show just why spectrographs form one of the main means of solar system investigation.

1.3.5 Examples of Spectrographic Instruments

Three imagers that also incorporated spectrographs have been mentioned in Section 1.2. Of these, MAVEN's spectrographs have been considered in previous sections. The other two instruments are Apollo 16's FUVCS (Figure 1.4) and New Horizon's Ralph.

The spectrographic part of the FUVCS had a very simple design. No entrance slit was used; instead a grid collimator restricted the field of view to a strip of the sky 0.25° × 20°. A grid collimator is simply a stack of grids arranged so that only radiation from the desired point of view has a clear path through the device – radiation from other parts of the sky is blocked by one or more of the opaque parts of the grids. The beam of radiation (still a parallel beam since the collimator was not optically active) was directed onto a reflection grating set at 45° and so the resulting spectra are turned through 90°. To obtain spectra, a motor turned the whole Schmidt camera and detector assembly through 90° so that it pointed at the grating. With the lithium fluoride corrector plate in place, spectra were obtained over the 105–160 nm spectral range with a resolution of 3 nm ($R \approx 40$). Without a corrector plate spectra from 50 to 160 nm could be obtained with a resolution of 4 nm ($R \approx 30$).

Ralph uses a dichroic beam splitter that transmits radiation with wavelengths longer than 1.1 µm and reflects radiation at shorter wavelengths. The transmitted radiation is directed into the spectrograph called the Linear Etalon Imaging Spectral Array (LEISA). LEISA uses a tapered etalon filter* to obtain spectra rather than a conventional diffraction grating.

The basic etalon forms spectra by the interference of many beams of radiation. Those beams, however, are produced by multiple reflections between two parallel partially reflecting mirrors rather than by passing through slits or reflecting off strip mirrors (Figure 1.17). The beams emerging on the right in Figure 1.17 will have consecutive path differences (when the substance between the mirrors has a refractive index of 1) of

$$2t\cos\theta \tag{1.21}$$

When this path difference equals a whole number of wavelengths there will be constructive interference and that radiation will emerge as a bright fringe. Nearby wavelengths will undergo destructive interference and not be transmitted. However, those wavelengths will emerge for slightly different values of θ when the path differences (Equation 1.21) are equal to integer multiples of their wavelengths, thus producing a spectrum. Etalons are used mainly in laboratory-based spectrographs, however by making the separation of the mirrors small (1 or 200 nm), the fringes produced become a few nanometres wide and the device acts as a narrow band filter – usually called an interference filter. The interested reader is referred to references A.1.1 and A.1.9 and other sources listed in Appendix A for further information on etalons and interference filters.

The interference filter used in LEISA has a further refinement – the partially reflecting mirrors are *not* parallel to each other, but inclined at a small angle. The separation of the mirrors ('t' in Figure 1.17) thus changes along the length of the filter and so the transmitted wavelength also changes along the length of the filter. Two filters are used that are bonded together. The

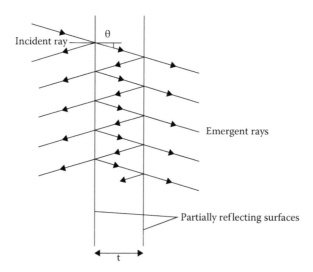

FIGURE 1.17   Light paths in a basic etalon.

---

* Also called a tapered Fabry–Perot filter or a linear variable optical filter.

first covers the spectral range from 1.25 to 2.5 μm with a spectral resolution of $R = 240$. The second covers the range from 2.1 to 2.25 μm at a resolution of $R = 560$.

A 256 × 256 pixel HgCdTe array (Sections 1.4.1.1 and 1.4.1.4) acts as LEISA's detector and is sensitive from 1.2 to 2.5 μm. The array is illuminated from the back and the silicon substrate has been thinned (Section 1.4) to just 200 μm. The filters are positioned about 100 μm above the detector. Each pixel has an angular resolution of 12.5″ and the instrument's field of view is 0.9° across. LEISA uses New Horizon's rotation in a pushbroom fashion to sweep the point of view across all the spectral bands and so build up spectral maps of the surface.*

The Bepi-Colombo mission to Mercury is planned for an October 2018 launch. It will comprise two separate spacecraft – the Mercury Planetary Orbiter (MPO) and the Mercury Magnetospheric Orbiter (MMO). One of the instruments on board the MMO will be the Mercury Sodium Atmosphere Spectral Imager (MSASI), designed to map Mercury's atmospheric sodium. MSASI will isolate the sodium D2 line at 589 nm using a very narrow band interference filter and a high resolution etalon (Figure 1.17). Its spectral resolution will be 0.007 nm ($R = 84,000$). Maps of the planet will be built up using the spacecraft's spin to scan in one direction and a scanning mirror covering 30° for the other dimension. A complete map of the disc of the planet and the sodium tail should be obtainable in about 85 seconds of observation with a spatial resolution ranging from about 5 km at perihermion to about 40 km at aphermion. The detector will be a radiation-hardened CMOS device (Section 1.4.1.3) fed by an image intensifier (cf. MAVEN's IUVS – Section 1.2.11).

The Orbiting Carbon Observatory-2 (OCO-2 – the first version of this mission had a launch failure) uses an f/1.8, 110 mm Cassegrain telescope to feed three spectrometers via beam splitters. Each spectrometer has a reflective holographic grating fed by a two-lens collimator and uses a two-lens imager to focus the image onto a 1024 × 1024 pixel array detector cooled to near 150 or 180 K. The spectrometers are designed to monitor the oxygen band at 765 nm and the carbon dioxide bands at 1.61 and 2.06 μm in order to measure the carbon dioxide levels in Earth's atmosphere from the reflected/scattered solar radiation that has passed twice through Earth's atmosphere. The array detectors for the NIR spectrometers are HgCdTe photodiodes (Sections 1.4.1.1 and 1.4.1.4) whilst the oxygen line is detected by a silicon CMOS array (Section 1.4.1.3). The spectrometers have spectral resolutions of 20,000 and individual pixels cover a 1.29 km × 2.5 km rectangle on the ground. A 10.6 km wide swath is normally imaged by using the spacecraft's orbital motion for pushbroom scanning, but when higher accuracy measurements are needed, the spacecraft can rotate so that a single area of the ground is observed for up to 7 minutes.

The Lunar Reconnaissance Orbiter carries the Lyman-α Mapping Project (LAMP) instrument. Rather remarkably, LAMP monitors the lunar dark side in the UV using just the illumination from stars and sky glow. Its primary mission is to confirm the presence of ice in permanently shadowed craters near the Moon's poles. Although hydrogen can be detected

---

* A system that produces an image of the field of view for every resolution element of the spectrum is often called an image cube. It can be visualized as a stack of two-dimensional (2D) direct images with the wavelength changing from one image to the next.

using neutrons (Section 7.2.3.5.1) and then is likely to be in the form of water, this cannot be guaranteed. Water ice, however, has a reflectivity of less than 2% at wavelengths shorter than ~170 nm and a reflectivity of ~35% at longer wavelengths. Observation of this absorption edge provides very strong confirmation that the hydrogen is present in the form of water ice.

LRO's FUV spectrometer has been developed from those on New Horizons and Rosetta. It covers the 57–196 nm spectral range with a spectral resolution of 0.36 nm and an angular resolution of 35′. The spectroscope slit is at the prime focus of the f/3, 40 mm, off-axis paraboloidal primary mirror and is 0.3° × 6° in shape. All the instrument's mirrors are aluminium with a nickel coating. It operates by pushbroom scanning of the lunar surface. Its optical design is a conventional one based upon a toroidal holographic diffraction grating and the detector is a 1024 × 32 pixel Z-stack CsI microchannel plate (MCP; Sections 1.4.3.2, 3.2.4 and 4.2.5).

Mars Express' Observatoire pour la Minéralogie, l'Eau, les Glaces et l'Activité (OMEGA) observes over the spectral range from 350 nm to 5.1 μm using three spectrographs. The Visible and Near Infrared (VNIR) instrument covers from 350 to 1000 nm while the Short Wavelength Infrared (SWIR) instrument has two channels covering 930 nm to 2.73 μm and 2.55–5.1 μm.

The VNIR instrument uses a 15.6 mm, f/3.7 refracting telescope with a variant of the double Gauss lens* design to project the image onto the spectroscope's slit. The expanding beam from the slit falls directly onto a tilted concave holographic reflecting grating with 65 lines per millimetre which fulfils the roles of collimator, disperser and imager. The tilt of the grating sends the spectrum beam to the side of the instrument where a 45° flat mirror reflects it to focus on the detector located on the other side of the instrument. The detector is a 288 × 384 CCD (Section 1.2.1.2) array with the spectrum along the shorter axis. The longer axis of the detector obtains one dimension of the direct image and the second is obtained by pushbroom scanning using the spacecraft's orbital motion. Angular resolution is about 4′ and spectral resolution either 4 or 7 nm depending on the sampling method.

A 50 mm f/4 Cassegrain telescope focuses an image onto the single slit used by both SWIR channels. A fixed mirror and a moving mirror placed before the telescope enable the point of view to be scanned through 8.8°. After the slit, a 91 mm focal-length off-axis parabolic mirror acts as the collimator. A dichroic beam splitter then separates the shorter wavelengths from the longer ones forming the two channels of the instrument. The longer wavelengths are transmitted through the dichroic beam splitter and fall directly onto a plane reflective grating blazed for 3.8 μm and ruled with 120 lines per millimetre. The shorter wavelengths reflect from the beam splitter and so must be reflected a second time using a flat folding mirror to be directed towards their grating. This is also a plane reflective grating, blazed for 1.7 μm and with 180 lines per millimetre. The two channels are then focussed onto the detectors by two spherical mirrors and four zinc-selenide lenses each.

The detectors are each 1 × 128 indium antimonide photodiode arrays (Section 1.4.1.4) that image the spectrum of a single spot on Mars' surface in a single exposure. The other two spatial dimensions of the image cube are built up by scanning the moving mirror in

---

* The single Gauss lens is an achromatic doublet with a converging meniscus lens closely followed by a diverging meniscus lens. The double Gauss lens uses two single Gauss units separated by a short distance with the second unit reversed. This design and its variants were widely used as camera lenses for much of the twentieth century.

front of the telescope and by pushbroom scanning using the spacecraft's orbital motion. The angular resolution is 4′ and the spectral resolution between 13 and 20 nm.

Mars Express also carries the Ultraviolet and Infrared Atmospheric Spectrometer (SPICAM). The UV channel covers the spectral range from 118 to 320 nm, while the infrared channel covers from 1.0 to 1.7 μm. The UV spectrograph is fed by a 40 mm, f/3 off-axis parabolic mirror. A holographic concave toroidal grating with 290 lines per millimetre and blazed at 170 nm then focuses the spectrum onto an image intensifier with a solar blind* caesium telluride photocathode and a 290 × 408 pixel CCD detector. Its angular resolution is 0.7′ over a field of view, when the spectroscope slit is in place, of 0.24° × 0.95° and it has a spectral resolution of 0.5 nm per pixel.

The infrared channel has a 30 mm, f/1.3 refractor as its telescope. A 1 mm diameter diaphragm at the telescope's focal plane feeds light to a refractive collimator. The dispersing element is an AOTF.

In an AOTF, the equivalents to the slits of a diffraction grating are produced within a crystal by ultrasound waves. The sound waves are generated by a piezo-electric crystal bonded to one face of the AOTF crystal and driven by a radio frequency signal. The refractive index within the AOTF crystal changes its value through the photoelastic effect as the high- and low-pressure zones of the sound wave pass through it (this is similar to the way in which holographic gratings operate). The device is usually operated in the Bragg regime where radiation is mostly concentrated into the first order. Bragg reflection is likely to be more familiar to most readers as a means of studying crystal structure and in that application a crystal illuminated by a beam of x-rays at an angle ϕ to the crystal lattice structure will reflect only those x-rays for which

$$\lambda = \frac{2d \sin \phi}{m} \tag{1.22}$$

where d is the separation of the layers within the crystal and m the spectral order.

Since m is usually equal to 1, the reflected beam will be monochromatic, but the reflected wavelength may be changed by altering ϕ. The equivalent equation for an AOTF is

$$\lambda = \frac{2\delta \sin \phi}{m} \tag{1.23}$$

where δ is the wavelength of the ultrasound.

The transmitted beam will again be monochromatic but can be scanned over wavelength by changing either ϕ or δ. Usually δ is changed by altering the frequency of the signal driving the piezoelectric crystal.

For SPICAM-IR, the AOTF crystal is tellurium dioxide[†] and the driving frequency scans from 85 to 150 MHz. This produces a spectrum over the 1.0–1.7 μm spectral range with a

---

* The term 'solar blind' means that the photocathode does not respond to visible and longer wavelength radiation. The detector is thus not swamped by the far greater energy from the Sun at visible and NIR wavelengths when being used to detect the less numerous ultraviolet photons. Caesium telluride only detects photons with wavelengths shorter than 320 nm.
† Also known as paratellurite.

spectral resolution ranging from 0.5 nm ($R = 2000$) to 2 nm ($R = 850$). Tellurium dioxide has a high level of birefringence* so the ordinary and extraordinary rays exit from the AOTF at an angle of 5° to each other and are detected separately. The two polarised beams are focussed by lenses onto two separate detectors that are both single indium–gallium–arsenide photodiodes (Section 1.4.1.4). The ExoMars Trace Gas Orbiter carries a similar instrument to SPICAM, though with better spectral resolution.

SPICAM-IR's main scientific purpose is monitoring atmospheric water vapour via its 1.38 μm absorption band. It has a ground resolution of ~4 km at periareon and uses the spacecraft's orbital motion to scan its point of view over the planet.

Jupiter currently has the Juno spacecraft orbiting it. The planet's auroral emissions (Sections 2.3.1 and 3.1.2) are being studied in the infrared by the Jovian Infrared Auroral Mapper (JIRAM) package. JIRAM has been developed from similar instruments on Cassini (Visible and Infrared Mapping Spectrometer [VIMS-V] – see below), Dawn and Rosetta. It has a refracting telescope using two doublet lenses that feeds, via a beam splitter, both an IR imager and an IR spectrometer. The spectrometer operates over the 2–5 μm region to observe the water absorption bands at 2.7 and 2.9 μm and the methane bands at 2.3 and 3 μm. Since Juno is rotating, a mirror spinning in the opposite direction precedes the telescope so that it receives a stationary image. The spectrograph is of conventional design (Figure 1.16) and uses a reflection grating 30 mm × 60 mm in size with 30 lines per mm providing a spectral resolution of ~400 and a spatial resolution at the 10 Pa pressure level in Jupiter's atmosphere of between 1 and 200 km.

The HST's Space Telescope Imaging Spectrograph (STIS) has, like Cassini's Ultraviolet Imaging Spectrograph (UVIS) (see below), detected water vapour plumes, although these were being emitted by Jupiter's moon Europa. The STIS is fed by the main telescope (a 2.4 m Ritchey–Chrétien – Section 1.2.1.3) and contains numerous interchangeable components. The basic layout, though, is still slit – collimator – grating – imager – detector. In addition, filters and the optics to correct HST's aberration are included.

The slit wheel contains 15 slits and apertures for spectroscopic and imaging purposes. The grating wheel contains 10 primary gratings, cross-disperser gratings for the four echelle gratings (installed near the detectors), a prism and mirrors for imaging. The three detectors are all 1024 × 1024 pixel arrays – a CCD for the UV to NIR (164 nm to 1.03 μm), a CsTe Multi Anode Microchannel Array (MAMA†) for 160–310 nm and a CsI MAMA for 115–170 nm. Their fields of view are 52″, 25″ and 25″, respectively and their angular resolutions 0.05″, 0.025″ and 0.025″. Spectral resolutions vary between 10 and 114,000. Finally, there are 5 filters for the CCD, 13 filters for the CsTe MAMA and 10 filters for the CsI MAMA.

Two Jupiter missions, Jupiter Icy Moons Explorer (JUICE) and Europa Multiple Flyby Mission‡ (Figure 1.18), are in their early stages of planning for launches around 2022–2025 and arrivals at Jupiter around 2030–2033. They are expected to carry almost identical UV

---

* See Appendix A for sources of information on birefringence and optical activity.

† A variety of MCP (Sections 1.4.3.2, 3.2.4 and 4.2.3) with the electrons being detected by an array of anodes.

‡ Previously known as the Europa Clipper. Current planning is for some 45 fly-bys at altitudes ranging from 25 to 2700 km so that most of the surface is mapped to a resolution of 50 m, but problems with the thrusters may alter this. A small lander may be included if the possibility of the contamination of Europa by terrestrial organisms can be eliminated.

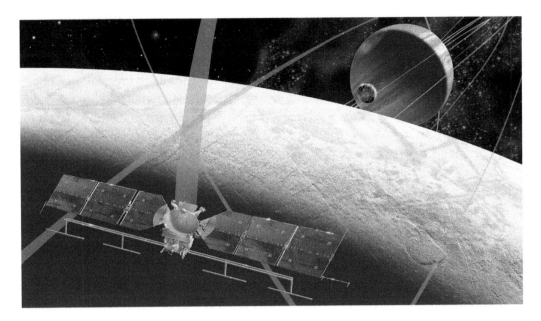

FIGURE 1.18   An artist's impression of the NASA Europa Multiple Flyby Mission spacecraft near Europa showing some of the possible complex orbital paths. (Courtesy of NASA/JPL-Caltech.)

spectrographs developed from the design of the Juno UV spectrograph and earlier spectrographs on the New Horizons and Rosetta spacecraft.

The spectrographs are planned to use single off-axis 120 mm paraboloid primary mirrors feeding concave 50 mm × 50 mm reflection gratings. Their spectral range will be from 55 to 210 nm with best spectral resolutions of 0.6 nm and best spatial resolutions of 120″. The detectors will be solar blind MCPs (Sections 1.4.3.2, 3.2.4 and 4.2.3) using caesium iodide as the photoemitter. Their read-outs are by cross delay lines (Section 1.4). The spectrographs will be able to build up one-dimensional images by pushbroom scanning (i.e. single-line tracks across the target's surface).

The Cassini Saturn Orbiter spacecraft carried three spectrographs covering parts of the EOR – the UVIS, the Visible and Infrared Mapping Spectrometer (VIMS) and the Composite Infrared Spectrometer (CIRS), each with several channels. The last of these will be considered in Chapter 2; the designs for the other two instruments are reviewed here.

UVIS (Figure 1.2 below the NAC) comprised two almost identical telescope and spectrograph combinations. One operated from 56 to 118 nm (the extreme UV – EUV) and the other from 110 to 190 nm (the far ultraviolet – FUV). Both instruments used off-axis parabolic primary mirrors with the spectrograph entrance slits at their prime foci. Each primary mirror had a 20 mm aperture and a 100 mm focal length. The reflective coatings were aluminium plus magnesium fluoride for the EUV instrument and boron carbide for the FUV instrument. Toroidal* concave gratings combined the roles of collimator, grating and imager and had focal lengths of ~150 mm with 1066 lines per millimetre (EUV instrument) and 1371 lines per millimetre (FUV instrument). Three slits were available

---

* A concave grating with differing radii of curvature in the 'horizontal' and 'vertical' directions.

with widths from 75 to 800 μm giving best spectral resolutions of 0.275 nm in both cases. The detectors were MCPs (Section 1.4.3.2) with caesium iodide and potassium bromide photoemitters for the EUV and FUV instruments, respectively. Each detector was a 64 × 1024 array with the spectrum aligned along the long axis. The angular resolution along the shorter axis was about 200″ – corresponding to a spatial resolution of 300 km from a distance of 300,000 km above Saturn's surface. UVIS has been used to study the water plumes (geysers) emitted from Saturn's moon, Enceladus, in particular monitoring the occultation of stars and the Sun by plumes to determine their compositions.

VIMS (Figure 1.2 above the NAC) also had two separate telescope-spectrograph systems. The spectral region from 300 nm to 1.05 μm was covered by VIMS-V using a 45 mm Shafer telescope. The Shafer telescope is a variant of the Makzutov design that uses a small correcting lens placed close to the secondary mirror. The primary mirror can be tilted to point the telescope towards objects within a 3.6° field of view without needing to move the whole spacecraft. The angular resolution is 35″ per pixel.

The spectroscope had a Cassegrain-like configuration in which the incoming beam hit one side of the primary mirror and was reflected towards the secondary mirror which was also the grating. After reflection from the secondary mirror/grating the light beam hit the other side of the main mirror and was reflected to a focus at the detector. This design is sometimes called an Offner relay system. The holographic reflection grating had a spherically convex shape so that it could also act as the secondary mirror. The best spectral resolution was 1.46 nm. The detector was a 512 × 512 CCD that operated in a frame-transfer mode (Section 1.4.1.2) so that the actual sensitive area was 256 × 480 pixels. The spectrum was dispersed along the long axis of the array and the telescope's primary mirror was moved to scan the field of view down the shorter axis in a variant of pushbroom scanning.

VIMS-IR's design was derived from the Near Infrared Mapping Spectrometer (NIMS) on the Galileo spacecraft and used some of the spares from that instrument. It obtained spectra over the 850 nm to 5.1 μm region. A 230 mm, f/3.5 Ritchey–Chrétien telescope with a scanning secondary mirror fed the spectroscope. The spectroscope's collimator was a Cassegrain variant called the Dall-Kirkham design that uses an elliptical primary mirror and a convex spherical secondary mirror. A conventional reflection grating produced the spectrum. This had 28 lines mm^{-1} and had three zones blazed for 1.3, 3.25 and 4.25 μm, respectively. The spectrum was imaged onto the detector by a reflective imager. The detector was a 1 × 256 linear array of indium antimonide photodiodes giving a spectral resolution of 16.6 nm per pixel. The secondary mirror scanned the 3.6° field of view over the entrance slit of the spectroscope obtaining the spectrum of a single 100″ × 100″ area of the target with each exposure. This type of scanning has previously been called facsimile scanning (Section 1.2.2.3). In this context, it is often called whiskbroom scanning.

The James Webb Space Telescope (JWST), like the HST, is certain to be used to probe solar system objects as well as its main targets in the rest of the universe. The telescope is a three mirror anastigmat (Section 1.2.1.6) with a diameter of 6.5 m made up from 18 hexagonal segments. Its effective focal length is 131 m. JWST's Near Infrared Spectrograph (NIRSpec) will be passively cooled to 38 K and will cover the spectral range from 600 nm to 5 μm. Its spectral resolution will range from 100 to 2700. The spectrograph will use 14 mirrors and will have six reflection gratings and a calcium fluoride prism. Another mirror will provide an imaging mode for target acquisition. There will be eight interchangeable filters. The detectors will be two 2048 × 2048 pixel HgCdTe arrays (Sections 1.4.1.1 and 1.4.1.4). The most innovative part of NIRSpec will be its ability to obtain spectra of up to 100 objects simultaneously and to slice 3″ × 3″ areas of the sky into thirty 0.1″ strips so that the complete spectrum of an extended object can be obtained. Both these operational modes of the spectrograph will use an array of small shutters in the image plane that can be opened or closed individually to allow the images of the targets into the instrument. There will be four such arrays in a 2 × 2 mosaic, each with 62,400 shutters. The individual shutters are ~100 μm × ~200 μm in size (equivalent to ~0.2″ × ~0.4″) and are assembled into a 171 × 365 array.

1.4 DETECTORS

A number of different detectors operating within the EOR have been mentioned in the previous sections of this chapter. Some may be familiar to the reader, others less so. A brief summary of their operating principles is given here, concentrating on those in current use and possible future devices. Further details may be found from sources in Appendix A.

The operating principles of the majority of EOR detectors involve the transfer of the energy of a photon to an electron within an atom, ion or molecule and the subsequent detection of the changed properties of that electron. Many detectors operating on the same or similar principles are used to detect charged particles. Further discussions on many of the topics in this section may therefore be found in Chapter 4. In particular, the outline in that chapter of the band theory of solids may be found to be useful reading for this section as well.

1.4.1 Electron–Hole Pair Production-Based Detectors

1.4.1.1 Introduction

In some semiconductors (Section 4.2.6), the absorption of an EOR photon may release an electron and/or a hole* that is/are free to move within the solid's structure. Such an absorption always produces an electron and a hole and so the interaction is often known as electron–hole pair production, but in some cases either the electron or hole may not be free to move. Detectors based upon electron–hole pair mostly utilise the free electron. The hole, whether free or trapped, is just eliminated in some way.

Silicon, often doped (Section 4.2.6), is the most widely used semiconductor material for electron–hole pair-based detectors. Photons with wavelengths as long as 1.1 μm potentially may produce electron–hole pairs in silicon. The peak detection interaction efficiency is around 750 nm and it tails off towards shorter wavelengths as the opacity of silicon increases, with a practical limit at around 350 nm. Shorter wavelengths, to about 190 nm, may be detectable by using very thin layers of silicon and/or by coating the silicon with a layer of a phosphorescent material to convert the short wave radiation to a longer wavelength. Silicon may also form the basis for x-ray detectors (Section 3.2).

---

* That is, the absence of an electron which behaves as though it were a positively charged particle within the solid – Section 4.2.6.

Other materials that may form the basis for electron–hole pair production detectors include the following:

Germanium (spectral response ~400 nm to ~1.6 μm)

Indium gallium arsenide (spectral response ~800 nm to ~2.6 μm)

Lead sulphide (spectral response ~1.0 to ~3.5 μm)

Indium arsenide (spectral response ~1.0 to ~3.8 μm)

Indium antimonide (spectral response ~1.0 to ~7 μm)

Mercury–cadmium–telluride (spectral response ~1.0 to ~12 μm)

Lead tin telluride (spectral response ~2.0 to ~18 μm)

1.4.1.2 CCDs

CCD stands for Charge-Coupled Device (or Detector), however the charge coupling has nothing to do with the detection of radiation – it is just the way that the detected signal is read out to other electronics.

The basic structure of a CCD is a layer of n-type silicon superimposed upon a layer of p-type silicon together with a thin insulating layer, anodes, a cathode and (possibly) an anti-reflection layer (Figure 1.19a). The anodes are grouped in threes, with the central anode maintained at about +10 V and the ones on each side at about +2 V during an exposure. Photons with wavelengths shorter than about 1.1 μm penetrate into the silicon layers and upon absorption produce electron–hole pairs (Figure 1.19b). The electrons are attracted towards their nearest +10 V anode and accumulate beneath it (Figure 1.19c). The holes are repelled and disperse into the p-type silicon.

When the exposure has been completed, the groups of electrons are moved physically through the device to an output electrode by charge coupling. If the motion is to be towards the right as seen in Figure 1.19, then the voltage on the right-hand anode in each triplet is raised to +10 V. The electrons then spread out below the two +10 V anodes (Figure 1.19d). The voltage on the central anode of each triplet is then lowered to +2 V and the electrons move over until entirely gathered below the right-hand anode which is still at +10 V (Figure 1.19e). The electron groups have thus all been moved to the right by a distance equal to the size of one anode.

Now raising the voltages on anodes to the right of those anodes now storing the electrons to +10 V and then lowering the voltages on the left—hand anodes to +2 V will move the electron group another step to the right. Continuing this process gradually moves each electron group to the discharge electrode where it is picked up and amplified, etc. to form the measured signal. The original position of the discharged electron group within the device may be inferred from the number of voltage cycles necessary to bring it to the discharge electrode.

As just described, the CCD would be a linear array and, though such arrays are still used (e.g. in shops' bar code readers), most CCDs are now two-dimensional (2D) arrays. 2D arrays are obtained just by stacking several linear arrays side by side. They may then

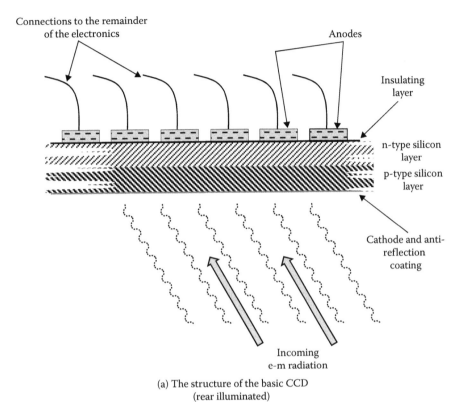

Connections to the remainder of the electronics

Anodes

Insulating layer

n-type silicon layer

p-type silicon layer

Cathode and anti-reflection coating

Incoming e-m radiation

(a) The structure of the basic CCD (rear illuminated)

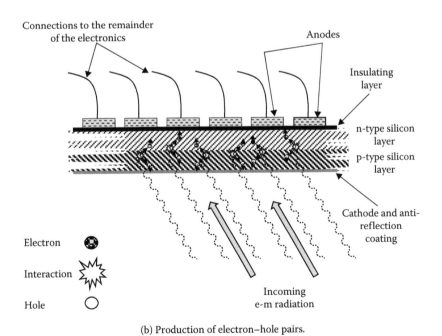

Connections to the remainder of the electronics

Anodes

Insulating layer

n-type silicon layer

p-type silicon layer

Cathode and anti-reflection coating

Electron

Interaction

Hole

Incoming e-m radiation

(b) Production of electron–hole pairs.

FIGURE 1.19 The basic structure and operating principles of a rear-illuminated buried channel CCD.

(*Continued*)

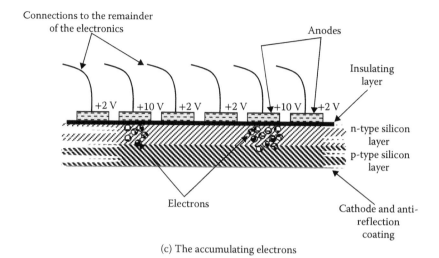

(c) The accumulating electrons

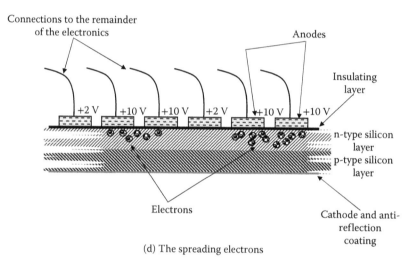

(d) The spreading electrons

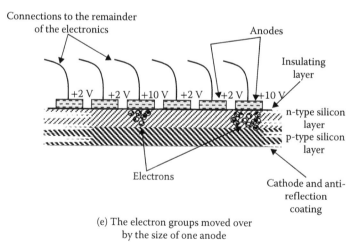

(e) The electron groups moved over
by the size of one anode

FIGURE 1.19 *(Continued)* The basic structure and operating principles of a rear-illuminated buried channel CCD.

be read out sequentially, or if a faster response is needed then other read-out systems may be used as discussed below. The CCD as described would also be a three-phase, rear-illuminated device (since the anodes are in triplets); other designs are described below.

CCD arrays are essentially very large scale integrated circuits (VLSI) and are manufactured in the same way. Their physical sizes are limited by the sizes of the silicon boules produced for making VLSI wafers and this is typically 200–300 mm across, although 450 mm wafers may become available in a few years. The largest available individual CCD arrays are thus about 4096 × 4096 pixels (16 Megapixels) in size. Although 9,000–10,000 pixel square arrays are now technologically feasible, if detectors larger than 4096 × 4096 pixels are needed then they are usually made by mosaicking several smaller arrays. Thus the Large Synoptic Survey Telescope will employ a 32 Gigapixel camera formed from one hundred and eighty-nine 4096 × 4096 pixel arrays. For most spacecraft-borne instruments, though, such large arrays are not required.

Originally CCDs had transparent anodes and the radiation passed through the anodes (i.e. the front of the device). Nowadays, to improve their optical efficiency, many are illuminated from the rear (back) of the device. Since the original CCD wafer is typically 200–300 μm thick and the pixel size generally between 5 and 30 μm, electrons originating near the rear surface are likely to be gathered up by the 'wrong' +10 V anode. Rear-illuminated CCDs are therefore thinned after production to be just 10–20 μm thick.

CCDs in which the pixels have just one or two anodes each can be produced by additional n and p doping of regions of the silicon layers (virtual phase CCDs) or by having one anode on top of the insulating layer and the second buried inside the insulating layer (two phase CCDs).

The basic read-out of the image obtained with a 2D CCD array as outlined above is a relatively slow process – up to tens of seconds for the larger arrays. The use of the n-type layer of silicon as shown in Figure 1.19 rather than a layer of intrinsic silicon produces an electric field that forces the electrons slightly away from the interface with the insulator. This is known as a buried channel CCD and it has an improved read-out speed over simpler designs.

When rapid read-outs are required, such as to enable many short exposures to be obtained in quick succession, then there are various ways of speeding up the process. The simplest is to have a separate read-out channel for each line of the CCD (a column parallel CCD) – and a further factor of two improvement may be had by reading out from both ends of the line. A second approach results in the frame-transfer CCD (see LORRI, Section 1.2.1.3, Ralph, Section 1.2.1.6 and Cassini's UVIS, Section 1.3.5). Half the CCD array is covered over during an exposure. At the end of the exposure, the electrons in the exposed pixels are transferred to the pixels in the unexposed half of the array. A new exposure can then be started in the active half of the array, while the previous image is read out from the storage half. Even faster exposures can be obtained with the interline CCD. The principle is the same as for the frame-transfer CCD, but each active line has an adjacent covered-over storage line so that the electrons have to be moved by only the size of a single pixel before the next exposure can begin. This does mean though that every alternate column of the image is not observed.

A method of operating CCDs that is closely related to the interline CCD is TDI (see Giotto, Section 1.2.1.3, Mars Odyssey Orbiter, Section 1.2.1.6, Ralph, Section 1.2.1.6 and Juno, Section 1.2.2.1). This is a method of improving the signal-to-noise ratio in an image when the field of view is moving over the detector. The electrons in each pixel in a line of the CCD are moved over to the next line of the CCD at the same rate as the motion of the image over the CCD. The exposure time is thus the time taken for a point in the image to cross the CCD array, not the time that it is on a single pixel.

1.4.1.3 Complementary Metal-Oxide Semiconductor Detectors

CMOS detectors such as those found on MAVEN's IUVS (Section 1.2.1.1) and Mars Express Orbiter's VMC (Section 1.2.1.2) and planned for use on the ExoMars 2020 Lander and Rover (Sections 1.2.1.6 and 1.2.1.2) and Bepi-Colombo's MSASI (Section 1.3.5) are essentially CCDs without the charge coupling read-out system; instead, the pixels are read out individually.

1.4.1.4 Photodiodes*

Photodiodes detect radiation though the production of electron–hole pairs. They are just the junction between a p-type and an n-type semiconductor. Sometimes, to increase the physical size of the detecting region, a layer of intrinsic semiconductor is sandwiched between the two doped layers.

The junction acts as a radiation detector because of its band structure (Section 4.2.6). When the junction is formed there are electrons in the conduction band of the n-type material and holes in the valence band of the p-type material. The electrons are at higher energies than the holes and so they diffuse across the junction to fill the latter. The movement of the electrons produces a voltage across the junction with the p-type material becoming negative and the n-type positive. The induced voltage thus opposes the movement of the electrons and equilibrium is quickly established with no further electron movement. The junction itself becomes a region containing neither free electrons nor free holes and is called the depletion zone.

The absorption of a photon with a high enough energy boosts electrons from the valence band to the conduction band and leaves holes in the valence band. The voltage across the junction pulls the electrons into the n-type material and the holes into the p-type material – that is an electric current flows across the junction. That electric current can be measured and used to infer the intensity of the radiation falling on the junction.

The induced current can be measured directly or passed through a high resistance and the voltage over the resistance measured (photovoltaic mode). The junction may also be back biased – that is an external voltage is applied that enhances the induced voltage over the junction – so reducing the dark current.[†] Back-biased photodiodes are often called photoconductors.

Silicon is often used for the manufacture of photodiodes, but HgCdTe is becoming widely used, especially in spacecraft instrumentation. The reason for this is that its detection region is tunable from wavelengths of 800 nm or so to hundreds of microns (cf. CZT detectors – Section 3.2.5.2.2.) The tuning is possible because the material is an alloy of CdTe (which has a band gap of 1.5 eV – section 4.2.6.1) and HgTe (which has a band

---

* Also known as pn diodes, pin diodes, photovoltaic cells, photoconductors and barrier junction detectors. See also Section 4.2.6.2.

† The current that flows through the device when it is unilluminated. It is one of many noise sources in detectors – reference A.1.1.

gap of zero). Changing the proportion of the two alloys in the mix changes the composite material's band gap from zero to 1.5 eV and the centre of the detection region from an infinite wavelength (in theory) to around 800 nm. The devices though, especially at longer wavelengths, need liquid nitrogen cooling (77 K).

Photodiode arrays can be manufactured but until recently they have been significantly smaller than CCD arrays due to the more complex connections required. Thus Mars Express SPICAM (Section 1.3.5) uses just two single indium gallium arsenide photodiodes as its detectors, while OMEGA on the same spacecraft has a 1 × 128 indium antimonide photodiode array. Cassini's VIMS (Section 1.3.5) had a 1 × 256 indium antimonide array and Giotto used a 2 × 936 silicon photodiode array (Section 1.2.1.3). Landsat-7's ETM+ uses an HgCdTe detector with 8 pixels for its 10.4–12.5 μm waveband. However, OCO-2 (Section 1.3.5) has recently been using 1048 × 1048 pixel HgCdTe array detectors.

BIBs are used by the WISE/NEOWISE spacecraft (Section 1.2.1.2). These respond to radiation in a similar manner to reverse-biased photodiodes. They absorb the radiation in a thin, heavily doped, semiconductor layer, whose normal high dark current is supressed (blocked) by a layer of undoped semiconductor. Their spectral response extends out to 200 μm or longer.

By increasing the back bias voltage to around half the breakdown voltage* an APD (see also Section 1.2.1.2) is obtained. The electrons from the electron–hole pairs produced by incident radiation are accelerated by the high voltage sufficiently to produce further electron–hole pairs as they travel through the material. The new electrons may also gain sufficient energy to produce yet more electron–hole pairs and so on. The single original electron from the photon interaction is thus multiplied up to many thousands of electrons and the device acts as an amplifier (cf. Geiger counters and proportional counters – Section 4.2). Silicon APDs are used, for example, on the Hayabusa spacecraft (Section 1.4).

1.4.2 Thermal Detectors

Detectors based upon the heating effect of the absorption of radiation were once used throughout much of the infrared region. Known as bolometers, they are now largely replaced by photoconductors but still find some applications in the FIR. They operate by monitoring the change in the electrical resistance of the material as its temperature changes. The Mars Odyssey Orbiter, for example, carries a 240 × 320 microbolometer array as the detector for its THEMIS instrument (Section 1.2.1.6). The Europa Multiple Flyby mission is planned to incorporate the Europa Thermal Imaging System (E-THEMIS), an uncooled microbolometer array, to obtain high resolution thermal images and to analyse surface structures such as the dust particle sizes. It will operate over three wavebands within the 7–70 μm region with thermal resolutions between <0.1 and 0.2 K (at temperatures over 220 K and less than 90 K, respectively) using TDI (Section 1.4.1.2.) for the lowest temperature measurements. Its angular resolution will be about 30″ and its surface spatial resolution will be between 5 and 35 km.

---

* The breakdown voltage for a non-conductor is the voltage at which the substance is forced to conduct a current. For silicon it is about 3×10^7 V·m^{-1} and for germanium about 10^7 V·m^{-1}.

The early Earth resources spacecraft, European Remote Sensing satellite-1* (ERS-1), observed Earth's surface at 1.6, 3.7, 10.8 and 12.0 μm using its Along Track Scanning Radiometer (ATSR). The instruments measured the temperatures of the sea's surface and the cloud tops. The detectors were placed at the prime focus (Section 1.2.1.1) of an off-axis parabolic mirror which viewed two 500 km wide swathes of Earth's surface via a plane scanning mirror. Each portion of the surface was thus observed twice – once directly below the spacecraft (nadir) and once at a slant angle of 47°. The full image was then built up by pushbroom scanning. The first two channels used single InSb photodiode thermal detectors, the second two channels, single HgCdTe photodiode detectors (see previous section). The detectors were cooled to less than 95 K by a Stirling cycle cooler. The surface temperatures could be measured to better than 0.5 K accuracy with a ground resolution of 1 km.

The current successor to ERS-1, called Sentinel-3 (see also Section 2.4.4.2), carries an instrument similar to ATSR-1: Sea and Land Surface Temperature Radiometer (SLSTR). This has a swathe width for the nadir view of 1400 km and of 740 km for the slant view. Its ground resolution is 0.5 km and it has two f/7.2, 110 mm aperture telescopes and nine channels. The thermal channels include those for ERS-1 plus ones centred on 1.38 and 2.25 μm. Its radiometric accuracy is 0.2 K. The detectors for the infrared bands are all small (1 × 2 and 2 × 4 pixel) HgCdTe arrays of photodiodes – working in the voltaic mode at the shorter wavelengths and the conductive mode at the longer wavelengths. The detectors are all cooled by a Stirling engine to 80 K.

1.4.3 Photoelectric Effect-Based Detectors

1.4.3.1 Introduction

Although essentially based upon electron–hole pair production, the detectors utilising the photoelectric effect are sufficiently different in their operating principles to be treated separately from the detectors covered in Section 1.4.1. It has been known since Heinrich Hertz's experiments in 1887 that some materials, when exposed to e-m radiation, emit electrons from their surfaces. The explanation for the phenomenon was provided by Albert Einstein in 1905.† It arises when the electron from an electron–hole pair reaches the surface of a photoelectric material with sufficient energy to escape from that material.

1.4.3.2 Microchannel Plates

The main discussion of MCPs may be found in Section 4.2.5. Here, we just note that photoemitters generally used for EOR MCPs include the following:

Potassium bromide (cut-off wavelength – 155 nm)

Caesium iodide (cut-off wavelength – 200 nm)

---

* Not to be confused with Landsat-1, which at one time was called ERTS-1, though this stood for Earth Resources Technology Satellite-1.

† It was for his explanation of the photoelectric effect that Einstein was awarded the Nobel physics prize, not for his relativity theories.

Rubidium telluride (cut-off wavelength – 300 nm)

Caesium telluride (cut-off wavelength – 350 nm)

Cassini's UVIS instrument (Section 1.3.5) used 64 × 1024 pixel MCPs and MCPs are planned to be used for the JUICE and Europa Multiple Flyby missions (Section 1.3.5).

1.4.3.3 Photomultipliers

PMTs were used for primary imaging purposes early in the exploration of the solar system by spacecraft – see Luna-19 and Venera-9 (Section 1.2.2.3) for example – however they have now mostly been replaced by other, usually array-type, detectors. They may still be found though in some star sensors that enable a spacecraft's orientation to be determined, and as a part of scintillator-based x- and γ ray and high-energy charged particle detectors (Sections 3.2.3 and 4.2.4). For example, Interstellar Boundary Explorer (IBEX) uses a PMT as the detector in one of its two star trackers. PMTs are also still used in some terrestrially based neutrino and cosmic-ray detectors because they can easily and relatively cheaply be made physically large so that large areas or volumes can be monitored.

Here then just a very brief outline of the operating principles of a PMT is given for completeness; more details may be found from sources listed in Appendix A. The initial detection mechanism for a PMT, like that of an MCP, is the absorption of a photon by a photoemitting substance and the emission of an electron. Commonly used photoemitters include the following:

Sodium–potassium–antimony–caesium (multi-alkali) (spectral range from UV to NIR)

Antimony–potassium–caesium or antimony rubidium caesium (bi-alkali) (wavelength cut-off at around 650–700 nm)

Antimony–caesium (spectral range from UV to visible)

Silver–oxygen–caesium (spectral range 300–850 nm)

Caesium telluride (solar blind) (spectral range from UV to 320 nm)

Caesium iodide (solar blind) (spectral range from UV to 200 nm)

Like the MCP, the initial photoelectron is multiplied up by the production of further electrons after acceleration through a high voltage (the PMT should more accurately be called an electron-multiplier tube). The structure of the electron-multiplying section of the PMT however differs from that of the MCP. The photoemitter is enclosed in a vacuum tube and an electrode (called a dynode) that is positively charged with respect to the photoemitter by some 100 V is placed physically close to the photoemitter. The photoelectron is thus attracted towards that dynode. The dynode is coated with a secondary electron-emitting substance (which is often the same as one of the photoemitters). After acceleration through the 100 V potential differences between the photoemitter and the dynode, the photoelectron collides with the secondary electron emitter with sufficient energy to enable up to 10 or 20 secondary electrons to be produced. A second dynode that is around 100 V more positive than the first, collects these secondary electrons and multiplies them up by another factor of 10–20. The process continues with further dynodes until up to a million electrons emerge from the PMT for every incoming photon and are detected as an electric current.

1.4.3.4 Image Intensifiers

Image intensifiers are essentially amplifiers rather than detectors and are usually coupled with a true detector. Thus Bepi-Colombo's MSASI (Section 1.3.5) will use an image intensifier to feed a CMOS detector, SPICAM on Mars Express (Section 1.3.5) uses a CCD detector fed by an image intensifier and MAVEN's IUVS (Section 1.2.1.1) has two image intensifiers feeding separate CMOS arrays. In the past, image intensifiers have been used in conjunction with TV cameras as detectors. Image intensifiers may also be familiar to some readers from their terrestrial usages in civilian and military applications.

An image intensifier has a photoemitter (see above) coated on the inner side of a flat transparent plate which forms the front end of a vacuum tube. The photoelectrons are accelerated down the tube by a high voltage. Their paths through the tube are kept coherent by magnetic and/or electric fields so that the electrons coming from one point on the photoemitter are focussed onto a specific point on the target. The original photon image that was focussed onto the photoemitter plate is thus converted to an equivalent image comprising high-energy electrons. Those electrons will then normally fall directly onto the detector and be picked up normally. Alternatively, the target may be a phosphor that converts the energy of the electrons back into radiation for a photon detector (including the eye) – with many photons being emitted for each incident photon.

1.4.3.5 Detectors of Historical Interest

Brief descriptions of some detectors used in the early stages of the spacecraft exploration of the solar system are included here for completeness and for their historical interest.

1.4.3.5.1 Photography   Imaging using photographic emulsions has a long and honourable history but will now rarely be encountered. Its detection mechanism is electron–hole pair production (Section 1.4.1), usually in silver bromide. The silver bromide is in the form of small (~1 μm) crystals and the hole is consumed in separate chemical reactions. The electron causes a dislocation in the crystal structure so that if a second photon is absorbed within the crystal, its electron is attracted towards the first. The second electron reinforces the crystal dislocation and the third, fourth, fifth, etc., electrons from further interactions will soon join the first two. When around 5–20 electrons have accumulated, the dislocation becomes stable against dispersion by other reactions and is known as the latent image. A chemical reaction (developing) then converts those silver bromide crystals with latent images in the emulsion to silver. A separate chemical reaction (fixing) dissolves away the remaining silver bromide. The final image appears as a negative because the silver crystals are densest where the original photon image was brightest and so absorb radiation most strongly. A positive image can be obtained by re-photographing the negative, but most scientific work was carried out directly on the negatives.

Apart from having a low efficiency (only about 1% of the incident photons would be detected), for space applications, photography generally required the exposed emulsion to be returned to Earth for processing. Photography was thus employed on sub-orbital missions such as the V2 sounding rocket mentioned in Section 1.2.2.1 and during manned missions (e.g. Apollo 16's FUVC – Section 1.2.1.4). Orbiting spy satellites used photographic film in their early days and this was returned to Earth in small re-entry capsules for processing, but this system was never used for scientific purposes. Luna 3, however, did obtain the first images of the far side of the Moon by developing its photographs on board, scanning them and then transmitting the images back to Earth.

1.4.3.5.2 Television Tubes   The operating principle of a Vidicon TV tube is based upon a material that has a high electrical resistance in the dark but which develops a lower resistance when illuminated by light (a photoconductor). Typical photoconductors include antimony tri-sulphide, lead oxide and zinc selenide. The photoconductor is deposited on the inside of the flat end window of a vacuum tube with the end window being made from a transparent conductor. A heated wire at the other end of the tube provides a beam of electrons whose direction is controlled by orthogonal pairs of electrically charged plates. The beam is scanned over the inside of the end window. When it hits a part of the photoconductor that has been exposed to light a current flows through the (now) conducting segment of the window and the magnitude of that current is proportional to the light exposure received. The changing current emerging from the conducting window as the electron beam scans over the whole imaging area forms the video signal.

Vidicon and related TV cameras were used on many early spacecraft missions including the Ranger lunar impactors (Section 1.2.2.1) and Voyagers 1 and 2.

1.4.3.5.3 Electronographic Cameras   The electronographic camera, as used for the Apollo 16 FUVC (Section 1.2.1.4), was a hybrid of an image intensifier and a photographic emulsion camera. The high-energy electron image produced by the image intensifier was focussed onto a very thin plastic film. Many of the electrons were able to pass through the film where they were then absorbed into a nuclear emulsion film.

NuclearNuclear emulsion is similar to ordinary photographic emulsion except that the density of the sensitive silver bromide crystals is much higher. It was mainly used for detecting primary and secondary cosmic-ray particles. An energetic charged particle, such as an electron, entering the emulsion produces electron–hole pairs within the crystals in a similar manner to photons of light. The final image is then developed (relatively) conventionally and is made up of the short tracks of the electrons within the emulsion.

CHAPTER 2

Microwave and Radio Regions

2.1 INTRODUCTION

Low-frequency/long wavelength e-m radiation can provide much information about solar system objects in several different ways.

First, there is direct emission from the object itself – thermal emission (Section 1.1) from its surface, emissions due to lightning and other atmospheric phenomena and synchrotron, and other radiations from the outer atmosphere and magnetosphere (Section 2.3).

Second, the radio emissions from spacecraft, whether from the communication system or from purpose-built transmitters, when the spacecraft is on the far side of the object from Earth, will be affected by the object's upper atmosphere, radiation belts, coma (in the case of comets), etc. as they pass by the object. The ways in which the spacecraft's long wave emissions have thus been affected can be analysed and used to infer the properties of the regions through which the radiation has passed (Section 2.4.4.6).

Third, the objects may be studied actively using radar – sometimes terrestrially based and sometimes carried on spacecraft (Section 2.4).

Finally, spectrographs may be used to identify molecules etc. from their vibrational and rotational transitions (Section 1.3.4), to study the energy spectra of charged particles and the minerals forming the objects (Section 2.5).

A brief reminder here of the basics of non-thermal radiation (thermal radiation is described in Section 1.1 and discrete emission and absorption – i.e. spectrum lines – in Section 1.3.4) at this point may be useful to some readers. Non-thermal radiation* mostly originates when a charged particle changes its velocity (accelerates/decelerates/changes direction). The (classical) equation for the rate of change of particle energy (dE/dt) is

---

* Maser/laser emissions could come under this heading. Pair production (the mutual annihilation of a particle and its anti-particle) can also produce pairs of photons though these would normally be γ rays. Čerenkov radiation is produced when a charged particle passes through a substance faster than the speed of light in that substance (Section 4.2.9). Scattering of photons off dust and charged particles, including the Compton and inverse Compton effects, changes the energies and/or directions of existing photons, but does not produce them in the first place. However, none of these processes have much importance in a solar system context when the Sun itself has been excluded.

$$\frac{dE}{dt} = -\frac{e^2 a^2}{6\pi\varepsilon_0 c^2} \tag{2.1}$$

where a is the charged particle's acceleration.

The radiation produced via this process is given different names depending upon the cause of the charged particle's acceleration, but all the mechanisms are essentially the same.

Thus, Bremsstrahlung or free-free radiation results when a charged particle (usually an electron) travels close to an ion. The ion's electric field causes the electron to move in a curved path emitting photons. An alternative way of regarding the process is that the passing electron occupies one of the ion's energy levels in the region above the ionisation limit. Such an energy level is known as a free level as opposed to the bound levels discussed in Section 1.3.4. Like the electron transitions between bound levels, however, the electron can undergo a transition to another, lower energy, free level. Its transition is thus from one free level to another – i.e. a free-free transition.

The second main type of non-thermal radiation is called synchrotron (sometimes gyro-synchrotron) radiation. The name arises from the synchrotron charged particle accelerators of subatomic physicists wherein the particles lose energy via the process. The charged particles, again usually electrons, interact with a magnetic field and spiral around the magnetic field lines (cf. Figure 5.10) emitting photons as they do so.

When there are many charged particles contributing to Bremsstrahlung or synchrotron radiation, the overall spectrum is a continuous one whose energy distribution depends upon the kinetic energy distribution of the charged particles and the strengths and distributions of the electric or magnetic fields. A commonly observed form for the synchrotron radiation spectrum is

$$F(\nu) \propto \nu^{-\alpha} \tag{2.2}$$

where α is called the spectral index.

When the electron kinetic energy spectrum takes the form

$$N_e(v) \propto E^{-n} \tag{2.3}$$

where $N_e(v)$ is the number of electrons with velocities in the range v to $v + dv$ and E the electron's kinetic energy, then α and n are related by

$$\alpha = \frac{n-1}{2} \tag{2.4}$$

The instruments (see below) used to detect and measure microwave and radio radiation can be generally used over quite wide spectral ranges. They can thus normally be used to study all types of radiation, whatsoever its source may have been.

2.2 TELESCOPES, RECEIVERS AND DETECTORS

Many terrestrial radio telescopes are huge ponderous physical structures towering a hundred or more metres into the sky with massive banks of electronics and computers backing them up and costing more than a jumbo jet. However, in most cases, their basic operating principles differ little from those of the cheapest of home radios. The super-heterodyne receiver is the principle that underlies all these instruments.

Heterodyning is just another name for frequency mixing. If two signals with differing basic frequencies from two separate sources are combined then there is no mutual interference (cf. Section 1.3.2); the signals simply add together. The combined signal will then contain the two original frequencies (ν_1 and ν_2) together with their beat frequencies: ($\nu_1 + \nu_2$) and ($\nu_1 - \nu_2$).

In a super-heterodyne radio receiver, the signal from the source at long wavelengths is usually picked up by a simple half-wave dipole antenna. The half-wave dipole is just two conducting strips, each a quarter of the operating wavelength long and arranged in a line with a small insulating gap at the centre. The incident radio wave induces electric currents in the strips which are fed into the rest of the system by cables. Half-wave dipoles form the active detecting elements in the ubiquitous Yagi TV aerials adorning so many rooftops. The half-wave dipole receives radiation from a large part of its surroundings, but with a low sensitivity. It is therefore usually combined with passive elements that serve to limit the field of view to a smaller region and to increase the signal strength. In the Yagi aerial, these passive elements are the conducting strips placed in front and behind, but insulated from, the dipole. In most radio telescopes, whether terrestrially based or spacecraft borne, the passive elements are parabolic reflecting dishes that have dipoles (or small Yagi aerials) at their prime foci (cf. Figure 1.1a). For spacecraft, the reflectors are small and comparable with satellite TV dishes. Terrestrial radio telescopes, though, can have dishes over 100 m in diameter.* At higher frequencies the detectors may be fed by horn antennas and be SIS devices, Schottky diodes or HEBs.[†]

After the conversion of the radio waves into electrical currents, the super-heterodyne receiver mixes the signal current (Figure 2.1b) with an artificially produced signal which has a similar, but slightly different, frequency. The artificial signal is usually produced by an oscillator that is physically close to the rest of the equipment and so the signal is known as the local oscillator signal (Figure 2.1a). The result of combining the two signals is shown in Figure 2.1c. The lower frequency beat ($\nu_1 - \nu_2$) can be seen easily and it is plotted specifically in Figure 2.1d together with the upper beat frequency ($\nu_1 + \nu_2$).

The lower beat frequency is usually called the intermediate frequency and it can be at a very much lower frequency than those of the original signals – for example in the Ku microwave band (12–18 GHz), a 12.5 GHz original signal might be mixed with an 11.5 GHz

---

* The current largest steerable dish (at Green Bank, West Virginia) has a diameter of 100 m. Static dishes can be larger – Arecibo in Puerto Rico is 305 m across and the Chinese FAST (Five-hundred-meter Aperture Spherical Telescope) instrument is 500 m across. Arrays of smaller dishes can cover even larger areas, culminating in the currently under-construction, Square Kilometre Array.

[†] SIS – Superconductor–insulator–superconductor, HEB – hot electron bolometer. The reader interested in the details of the operation of these devices and of Schottky diodes can find suitable sources of information in Appendix A.

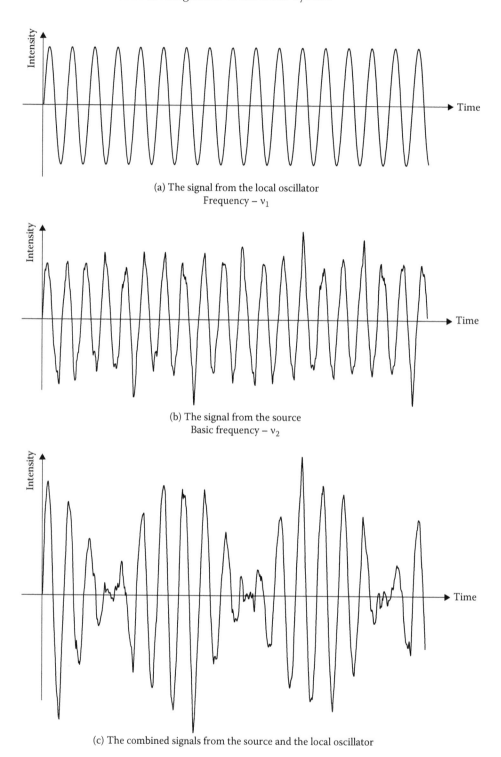

(a) The signal from the local oscillator
Frequency – ν_1

(b) The signal from the source
Basic frequency – ν_2

(c) The combined signals from the source and the local oscillator

FIGURE 2.1 The combination of the signal from the radio source and the local oscillator (frequency mixing, heterodyning).

(Continued)

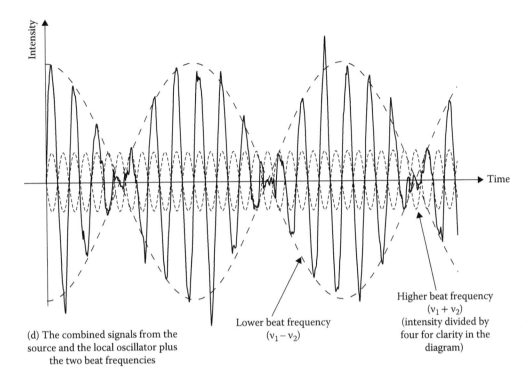

(d) The combined signals from the source and the local oscillator plus the two beat frequencies

Lower beat frequency $(\nu_1 - \nu_2)$

Higher beat frequency $(\nu_1 + \nu_2)$ (intensity divided by four for clarity in the diagram)

FIGURE 2.1 (*Continued*)   The combination of the signal from the radio source and the local oscillator (frequency mixing, heterodyning).

local oscillator to give a 1 GHz intermediate frequency (Figure 2.2). The original and higher beat frequencies are so different from the intermediate frequency that the latter can easily be separated out by electrical filters. If even lower frequencies are desired, the first intermediate frequency can be mixed with a second local oscillator to produce a second and lower intermediate frequency.

It is the signal in its intermediate frequency form that is passed on to the remaining circuits for amplifying, further processing and output. The reasons why the processing is not done directly on the original high-frequency signal are the following: First, the filters to separate the desired signal from unwanted nearby (in frequency terms) signals or to isolate a narrow part of the spectrum of a signal can be made to be of much narrower bandwidths at low frequencies than at high frequencies. Second, the accepted frequencies from the original signal can simply be altered (i.e. the receiver can be tuned) by changing the frequency of the local oscillator and keeping the intermediate frequency at a fixed value. Third, high frequencies require the use of waveguides instead of (the cheaper) coaxial cables and at lower frequencies the amplifiers and other signal processors can also be based on simpler and cheaper circuits.

A given local oscillator will produce data at the same intermediate frequency from parts of the original spectrum that are at the same distances in frequency terms above and below its own frequency. These are called the upper and lower sidebands, respectively. If only one sideband is used then the receiver is known as a single sideband device. However, if the local

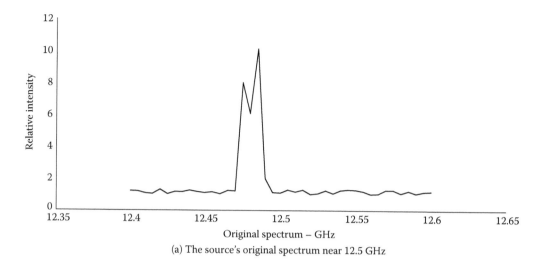

(a) The source's original spectrum near 12.5 GHz

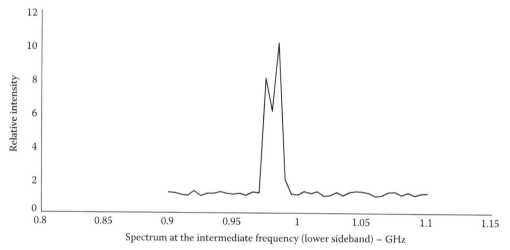

(b) The data moved to a lower frequency after mixing with an 11.5 GHz local oscillator

FIGURE 2.2 Schematic spectra shown at the original frequency and at the intermediate frequency after mixing with the local oscillator and filtering out the higher frequency components.

oscillator is close in frequency to that of the original spectrum, then both sidebands may be used and it is a double-sideband receiver. Thus, a 500 GHz local oscillator with 100 MHz intermediate frequency will obtain spectra near 499.9 and 500.1 GHz. Separating out the two spectra from the data within the intermediate frequency output is accomplished by adjusting the local oscillator frequency slightly. If the local oscillator is increased in frequency then features, such as spectrum lines, within the lower sideband spectrum will increase their intermediate frequency and those in the upper sideband will decrease their intermediate frequency.

The Global Precipitation Measurement Core Observatory's microwave imager has eight frequency channels between 10.6 and 183 GHz. It uses an off-axis 1.2 m diameter mirror that rotates at 32 rpm to map a 900 km swath of Earth's surface with a best horizontal resolution of ~4 km. The mirror has an aluminium coating with a silicon dioxide over the coating to diffuse solar short-wave radiation whose focussed energy could damage the instrument. The microwave beam is directed into four feed horns and then via waveguides to the super heterodyne receivers. It can measure temperatures to around ±0.2 K.

2.3 DIRECT EMISSIONS FROM SOLAR SYSTEM OBJECTS

2.3.1 Introduction

Some of the radio emissions from planets, especially from Jupiter, are easily detectable by terrestrially based instruments. Thus, Jupiter's emissions at 22 MHz were discovered in 1955 by Kenneth Franklin and Bernard Burke using a Mills cross radio telescope. The Mills'cross design, however, would not fit most people's ideas of what a radio telescope should look like – just comprising two orthogonal lines of dipoles. Franklin and Burke's instrument had 66 dipoles in each of its 620 m long arms and was constructed from lengths of wire strung between wooden poles, thus looking like several rather untidy washing lines.

Planetary radio and microwave emissions originate via four main mechanisms:

1. Electrostatic discharges, especially lightning: ≤~100 kHz (≥~3 km). For Saturn, though, the frequencies can reach 100 MHz (3 m). Terrestrial lightning emissions over the 100 Hz to 10 kHz region can be picked up using ground-based very-low-frequency receivers because the emissions are produced below the ionosphere (see Figure 2.3). A search of the Internet will quickly produce several sites that broadcast live terrestrial lightning emissions converted into sound or recordings of earlier such events.

2. Low-frequency cyclotron maser radiation* from electrons within a coherent plasma interacting with the planets' magnetic fields: ~50 kHz to ~40 MHz (~3 km to ~10 m). In Jupiter's case, this radiation is often strongly modulated by interactions with Io. The terrestrial emissions occur at heights around 20,000 km and are linked to the appearances of aurorae which occur at heights of 100–500 km. The emission is thus sometimes called Auroral Kilometric Radiation (AKR) and it is emitted as a narrow fan beam at a tangent to the magnetic field. Aurorae and AKR are also to be found for Jupiter and Saturn. The Internet has several audio conversions of these radio emissions that make for fascinating listening – for example at the time of writing, Cassini's recording of Saturn's emissions is at http://www.nasa.gov/mission_pages/cassini/multimedia/pia07966.html.

3. Synchrotron radiation from charged particle belts: ~100 MHz to ~10 GHz (~3 m to ~30 mm)

4. Thermal radiation: ~100 GHz to ~100 THz (~300 to ~3 µm)

The strongest radiation, by far, results from the second of these mechanisms. It requires a significant magnetic field, so that it is emitted only from Earth and the four gas giants.[†]

---

* Sometimes called synchrotron maser radiation.
[†] Mercury's magnetic field of ~300 nT may be strong enough to generate cyclotron maser radiation, but its very low frequency would mean that it would be absorbed within the plasma formed by the solar wind.

The Earth, observed from a distance of 1 AU, would have a peak intensity near 500 kHz of ~100 kJy*. The gas giants when near opposition and observed from near Earth would have cyclotron maser peak intensities of around

- Jupiter ~5 MJy at 10 MHz

- Saturn ~3 kJy at 100 kHz

- Uranus ~30 Jy at 300 kHz

- Neptune ~6 Jy at 300 kHz

While these peak intensities are potentially easily observable using terrestrial radio telescopes (which can currently observe μJy sources; potentially the Square Kilometre Array (SKA) may observe nJy sources) they are all obscured by Earth's ionosphere (see below) and so require spacecraft-borne instruments for their detection. Some parts of the emissions, though, may observed from Earth through the atmosphere's spectral windows (see below).

The synchrotron emission arises from electrons with energies in the 100 keV to 100 MeV region that are trapped in radiation belts similar to Earth's Van Allen belts by the planet's magnetic field. Jupiter is the main source of this type of radiation. Saturn does not emit it because its rings disturb the region where the radiation belts would form. Weak synchrotron radiation from Neptune has been found near 1.5 GHz, but Uranus' magnetic field appears to be too weak for any significant emissions.

Planetary temperatures range from 60 to 70 K (Uranus and Neptune) to 400 to 450 K (Mercury and Venus). Their peak thermal emissions thus occur over the range from ~3.6 to ~25 THz (~80 to ~12 μm). The thermal emission† is, however, emitted over a broad waveband, so that at 450 K its intensity is still equal to or stronger than 10% of the maximum value over the band from ~4 to ~50 THz (~75 to ~6 μm). While at 60 K, the 10% values occur from ~500 GHz to ~10 THz (~600 to ~30 μm).

Earth's atmosphere is largely transparent to radiation between ~15 MHz and ~30 GHz (~20 m and ~10 mm), but apart from a few narrow transparent gaps (windows) is opaque outside this band (Figure 2.3). Thus, the emissions from Jupiter's radiation belts can be observed using terrestrial radio telescopes in their entirety, together with the higher frequency cyclotron maser radiation. For example ASTRON's‡ Low-Frequency Array for Radio Astronomy (LOFAR) is a recently constructed, Dutch-led project with some 20,000 antennas distributed over much of northern Europe and operating over the 10–230 MHz band. Many of its antennas are simple dipoles, so that it has a close similarity in some ways to Franklin and Burke's Mills'cross. It has recently obtained resolved images of Jupiter's

---

* The jansky (Jy) is a non-SI unit of radiation intensity often used within the microwave and radio regions. 1 Jy = 10^{-26} W·m^{-2}·Hz^{-1}.

† Often alternatively called black body or Planck radiation.

‡ Netherlands Institute for Radio Astronomy.

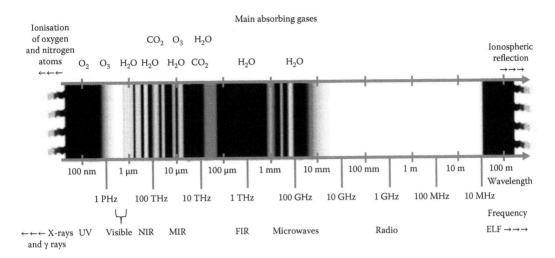

FIGURE 2.3   The transparent and opaque spectral regions of Earth's atmosphere.

radiation belt emissions over the 127–172 MHz band, which result from the magnetic field interactions of 1–30 MeV energy electrons.

Observation of extraterrestrial lightning emissions and of lower-frequency cyclotron maser emissions largely has to depend upon rocket or satellite-borne instruments. They have been detected, though, for Venus, Earth, Mars, Jupiter and especially for Saturn from the highest and driest terrestrial observing sites such as Mauna Kea (Hawaii) and Cerro Tololo, La Silla and Paranal (all in Chile).

Thermal emission, since it can extend into GHz frequencies, can be detected by Earth-based radio telescopes for all the planets and some planetary satellites, asteroids and comets. Only a year after Franklin and Burke's detection of Jupiter at radio wavelengths, for example, the Naval Research Laboratory's 15 m dish was used for the first detections of Mars, Venus and Saturn as well as for further observations of Jupiter at 10 GHz. Thermal radiation from Earth's Moon, meanwhile, was first detected as early as 1945 when Dicke and Beringer used a 900 mm parabolic dish and a microwave radiometer to observe the Moon at 2.4 GHz. Mercury was initially detected in 1960 by William Howard III et al. using the 26 m radio dish at the University of Michigan's Peach Mountain Observatory and a ruby maser radiometer. Uranus' thermal radiation was first detected in 1966 at 2.7 GHz by Ken Kellermann using the Commonwealth Scientific and Industrial Research Organisation's (CSIRO) 64 m radio dish at Parkes in New South Wales. In that same year he and Ivan Pauliny-Toth also made the first detection of Neptune, also with the 64 m dish, at 1.6 GHz.

The first comet to be detected emitting continuum radiation, which is presumed to originate as thermal emission from dust particles, was comet Kohoutek (1973f). Emission at 8 GHz was detected using the three 6 m dishes of the National Radio Astronomy Observatory's (NRAO) Green Bank double-sideband interferometer in 1974. A year earlier Franklin Briggs had used the same instrument to observe the 8 GHz thermal emission from Ceres and from Titan.

Finally, spectrum line emissions lying between ~15 MHz and ~300 GHz from atoms and molecules can be detected from planetary and satellite atmospheres and from the comae and tails of comets by Earth-based instruments (and over a much wider spectral range using spacecraft-borne detectors [Section 2.5]). The Atacama Large Millimeter Array (ALMA), for example, has determined the three-dimensional (3D) structures of the comae of comets C2012 S1 (ISON*) and C2012 F6 (Lemmon). High-resolution direct images of the comae were combined with spectra of hydrogen cyanide, hydrogen isocyanide and formaldehyde in the 339–363 GHz frequency range. The velocities of the molecules obtained from the spectra enabled the 3D structures to be mapped. ALMA has also observed molecular microwave emissions from Titan's atmosphere showing their distributions to be unexpectedly uneven.

The Karl G. Jansky Very Large Array (VLA) has recently been able to map the distribution of ammonia in Jupiter's atmosphere down to 100 km below the visible surface. The observations were made over the 4–18 GHz range and the ammonia appears dark since it absorbs Jupiter's thermal emissions.

2.3.2 Space-Based Observations

The Cluster mission, comprising four identical spacecraft and intended for observations of Earth's magnetosphere, was lost during its launch in June 1996. The replacement, Cluster II, was launched as pairs of spacecraft in July and August 2000. The spacecraft orbit Earth arranged at the corners of a tetrahedron with separations ranging down to 4 km, but more typically in the range 600–20,000 km, so that different parts of the magnetosphere can simultaneously be studied over large regions and triangulation can give the positions of sources. Amongst other instruments, the Wide Band (WBD) receivers make measurements over the 25 Hz to 577 kHz (10,000 km to 520 m) spectral region with three switchable band passes – 25 Hz to 9.5 kHz, 50 Hz to 19 kHz and 1–77 kHz. Observations with these receivers showed the fan-like nature of the AKR emissions in 2008.

The Earth orbiter, Soil Moisture Active Passive (SMAP), carries a synthetic aperture radar (SAR) (active – see Section 2.4 – this failed six months after launch) and a radiometer (passive). The radiometer operates at 1.4 GHz and has a ground resolution of 40 km over a 1000 km swath. Its principle objective is to measure the moisture content of the top few tens of millimetres of the soil, covering the whole Earth every two to three days. The radiometer is fed by a 6 m mesh reflector rotating (like a lasso about a single support arm at its edge) at 14 rpm and it can monitor the soil through clouds and moderate vegetation cover.

The Microwave Instrument for the Rosetta Orbiter (MIRO) was the first spacecraft-borne microwave telescope to be launched. It was based around a 300 mm off-axis paraboloid primary mirror with a hyperboloid secondary mirror (cf. Figure 1.1b). The radiation was fed to two double-sideband super-heterodyne receivers operating at 0.53 mm (562 GHz) and 1.58 mm (190 GHz). The intermediate frequencies after mixing the signal with that from the local oscillator were 5.5–16.5 GHz and ~1 GHz, respectively. The first receiver was also a spectrometer with 4096 channels over a 180 MHz bandwidth (i.e. a spectral resolution

---

* International Scientific Optical Network.

of ~10^7). Their fields of view were about 8′ and 22′, respectively, giving surface resolutions from a distance of 2 km of 5 m and 25 m. Their temperature resolution was about 1 K and the sub-surface temperature could be sensed to a depth of several tens of millimetres. The spectrometer was designed to study the emission lines from water, ammonia, carbon monoxide and methanol and to measure the relative abundances of the oxygen-16, oxygen-17 and oxygen-18 isotopes (Section 2.4).

An early radio astronomy spacecraft was Explorer 38; it was designed to pick up all sufficiently bright celestial radio sources, including those within the solar system. It used two 230 m long V-shaped antennas and a 37 m dipole antenna and operated over the 0.2–9.18 MHz spectral band with a number of different receivers. Four of the receivers were variants of the super-heterodyne design known as Ryle–Vonberg radiometers. Ryle–Vonberg radiometers in turn are a variant of the Dicke radiometer and are sometimes alternatively known as null-balancing Dicke radiometers. The Dicke radiometer improves the stability of a basic super-heterodyne receiver by rapidly switching between the signal from the source and that from a calibration source. The Ryle–Vonberg improves on this further by adjusting the calibration source to have the same intensity as the astronomical source and the settings of the calibration source then form the output from the device.

The two Voyager spacecraft carried detectors to observe radio emissions from the outer planets during their fly-bys, which covered the range from 1.2 kHz to 40 MHz (wavelengths from 250 km to 7.5 m). The two antennas on each spacecraft were 10 m long beryllium-copper tubes set at 90° to each other. Two super-heterodyne receivers were used, the first covering from 1.2 kHz to 1.3 MHz and the second from 1.5 to 40 MHz. The receivers had 198 channels with individual bandwidths between 19 and 300 kHz ($R = 0.5$–130). The radio receivers were turned off in 2008.

The Lunar Reconnaissance Orbiter carries Diviner, a microwave radiometer for studying both the thermal emission from the moon and the reflected solar microwave radiation. It operates with nine spectral bands over the wavelength range from 750 GHz to 860 THz (400 μm to 350 nm). Two 40 mm telescopes with a three-mirror off-axis design feed nine 1 × 21 element linear arrays of thermopiles* through a variety of filters and images are built up by pushbroom scanning. The Chang'e-4 spacecraft is intended to be the first soft lander on the far side of the Moon. A relay spacecraft near the Earth-Moon Lagrange L2 point will provide communications. Since the relay will be largely shielded from mankind's radio and microwave emissions it is intended for it also to carry a low-frequency (< 30 MHz) radio receiver for observing, amongst other sources, the emissions from Jupiter and Saturn.

Juno, a Jupiter orbiter (Figure 2.4), carries six microwave radiometers. The radiometers are centred on 600 MHz and 1.2, 2.4, 4.8, 9.6 and 22 GHz (500–13.6 mm). Only the 22 GHz radiometer requires waveguides; the others are all connected to the receivers by coaxial

---

* A thermopile detector comprises a number of thermocouples. A thermocouple has two dissimilar metal wires connected at their ends. If one of the junctions is held at a different temperature from the other a voltage develops between the junctions (the Seebeck effect) and a current flows around the circuit. By using the incoming radiation to heat the junction, the magnitude of the resulting current gives a measure of the radiation intensity. Thermopiles have a low sensitivity but can operate over a very wide spectral range.

FIGURE 2.4   A computer-generated image of the Juno spacecraft. Three of the microwave radiometer antennas may be seen on the upper surface of the main (hexagonal) body of the spacecraft. The large square at the bottom of the panel is the patch antenna for the 2.4 GHz radiometer. The slightly smaller square and the small square above it are the slot antennas for the 4.8 GHz and 9.6 GHz radiometers. (Courtesy of NASA/JPL-Caltech.)

cables. The two lowest frequency radiometers use 5×5 patch antennas*, the 22 GHZ radiometer uses a feed horn and the other radiometers use 8×8 slot antennas. The receivers are not super-heterodyne, but operate by direct detection at the original frequency, so there is no local oscillator or intermediate frequency involved. The receivers do, however, use switching between the astronomical source and a calibration source to improve their stability (Dicke radiometers). The fields of view of the radiometers range from 12° to 20° and these are swept over the surface of Jupiter by the spacecraft's 2 rpm spin. The temperature profile of Jupiter's upper atmosphere can be measured since the radiation at each frequency is primarily emitted from a different layer of the atmosphere – at 22 GHz it mostly originates from a region with a pressure of ~50 kPa, whereas at 600 MHz it is coming from a region with a pressure around ~5 to ~10 MPa.

Juno is also carrying two almost identical radio instruments. Each instrument has two receivers. One is of a double-sideband heterodyne design and sweeps in 1 MHz bands from 3 to 40 MHz. The other receives directly over the 100 kHz to 3 MHz spectral region.

2.4 RADAR

2.4.1 Introduction[†]

The basic principles of radar (Radio Detection and Ranging) and the closely related lidar (Light Detection and Ranging) are probably already known to most readers of this book. It will therefore suffice to say that the instruments detect objects, measure their distances and sometimes provide information on their velocities and surface structures by emitting

---

[*] Also called a microstrip panel. A patch antenna comprises a number of small strips or squares of a conductor held at a distance of λ/4 (approximately) above, but insulated from, a metal sheet (known as the ground plane). A slot antenna is simply a rectangular hole in a conducting sheet. It is of a similar size to an equivalent half-wave dipole and acts in a similar manner. A horn antenna is simply just that – i.e. a circular or square cross-section flared end to a waveguide. For further information see the sources listed in Appendix A.

[†] Additional material relating to this topic may be found in Sections 7.2.3.4 and 7.2.8.

e-m radiation that bounces off the object and by measuring the time taken between the emission of the signal and the reception of its echo. Often the transmitter and receiver are close together physically (when they are not actually the same instrument) – producing a mono-static radar – or they may be separated by a significant distance – giving a bi-static radar. The radiation may be emitted continuously – continuous wave or CW radar – or in bursts – pulsed radar.

The first equipment to utilise reflecting radio waves from objects was invented around 1903 by Christian Hülsmeyer and intended as an anti-collision system for shipping. It was not a true radar however, since it only indicated the direction of another ship and not its distance (unless two such devices could have been used to triangulate the target). True radar was developed secretly in the five years before the Second World War in a number of countries. It depended upon the development of transmitters capable of emitting short pulses. Oscilloscopes were then used to display the outgoing and incoming pulses and their separation measured to obtain the target's distance. The war years, of course, forced the rapid development of these early radar systems until by 1945 they were recognisably the precursors of today's instruments.

Most investigations in astronomy and astrophysics are pure remote sensing rather than the experimental approaches used in other sciences. That is to say, all the astronomer can do is look to the best of his/her ability – changing the experimental conditions in the way that, for example a chemist might alter the concentration of a solution or its temperature in order to clarify a result, simply cannot be done. The use of radar (or lidar) to investigate an object, however, enables the astronomer to act in some ways as an experimental scientist for once. The power, frequency, pulse repetition rate, distance from the target, etc. of the radar apparatus may all be controllable to a greater or lesser degree in order to improve the observations.

Radar studies of solar system objects take two different forms:

Earth-based investigations using high power transmitters and sensitive receivers combined with the largest available radio telescope dishes

Spacecraft-based instruments that have lower powers and smaller dishes, etc. but nevertheless much higher angular and/or spatial resolutions

2.4.2 Physical Principles of Radar Systems

The performance of a radar system depends upon many factors and these are combined within the radar equation. For targets that are angularly unresolved and where the transmitting dish and receiving radio telescope are the same, the equation takes the form

$$F = \frac{P\alpha A_e^2 v^2}{4\pi c^2 R^4} \tag{2.5}$$

where F is the returned signal strength, P is the power broadcast by the transmitter, α is the radar cross section of the target (defined as the cross–sectional area of a perfectly isotropically scattering sphere which would return the same amount of energy to the receiver as does the target), A_e is the effective area of the transmitting and receiving antenna/dish, v is the operating frequency and R is the separation between the radar and its target.

When the target is angularly resolved (and perhaps also resolved in depth) the target's radar cross section is replaced by an appropriate integral. Thus, a spherical target would have

$$F \approx \frac{PA_e^2 v^2}{4\pi c^2 R^4} \int\limits_0^{\pi/2} 2\pi r \alpha(\varphi) \sin\varphi \left[s\left(\frac{r\sin\varphi}{R} \right) \right]^2 d\varphi \tag{2.6}$$

where r is the radius of the target, ϕ is the angle at the centre of the target to a point on its surface illuminated by the radar beam, $\alpha(\phi)$ is the radar cross section for the surface of the target when the incident and returned beams make an angle ϕ to the normal to the surface and $s(\theta)$ is the sensitivity of the transmitter/receiver at an angle θ to its optical axis.

Whether or not a target can be detected by a radar will depend upon the ratio of F to the noise of the system (N) and this is given by

$$\frac{F}{N} = \frac{P\alpha A_e^2 v^2}{16\pi kc^2 R^4 T_s \Delta v} \tag{2.7}$$

where T_s is the noise temperature of the radar, k is Boltzmann's constant and Δv is the bandwidth of the radar signal. Provided that F/N in Equation 2.7 is significantly larger than unity, then the target should be detectable.

Equations 2.5 and 2.7 enable us to see how the detection of the target can be optimised – the returned signal strength for a given target depends upon R^{-4}, v^2, A_e^2 and P. It may thus be increased in value by decreasing the separation between the radar and the target (halving the distance increases F by ×16) using a higher frequency, using larger telescope dishes and increasing the transmitter power. The noise, N, may be decreased by using a smaller bandwidth and by reducing the system's noise temperature (e.g. mainly by using higher quality components and by reducing their thermal temperatures).

Clearly the fourth-power dependence of the radar signal upon the separation between the radar and its target means that reducing the separation is a very effective way of improving the instrument. It is the reason why small, low power, spacecraft-borne radars with separations from their targets measured in metres to hundreds of kilometres can compete with and beat terrestrially based radars whose powers are measured in megawatts and whose dishes may be hundreds of metres in diameter, but which have separations from their targets of tens of millions to thousands of millions of kilometres.

2.4.3 Terrestrially Based Instruments

The development of radar during the Second World War (Section 2.4.1) meant that by the war's end the direct detection of nearby astronomical objects had become a possibility. In

1946, Project Diana detected the Moon by radar using a 50 kW transmitter operating at 111.5 MHz in New Jersey. The antenna was an 8 × 8 array of half-wave dipoles. The pulses lasted for a quarter of a second and the echo was returned after 2.4 seconds (Figure 2.5).

Apart from the occasional near-Earth asteroid, the next closest solar system object after the Moon is Venus. But the R^{-4} dependence of the radar echo's strength and Venus' hundred-fold greater distance from Earth (even at inferior conjunction) than the Moon meant that other things being equal, the strength of the echo from Venus would be just 0.000001% of that of the Moon's echo seen in Figure 2.5. Clearly, in 1946 Venus was not detectable by radar.

Thus it was not until 1961, after several false claims, that Venus was finally detected using an Earth-based radar. A bi-static CW radar was used for this detection operating at 2.4 GHz (125 mm) with two 26 m parabolic dishes, one as the transmitter, the other as the receiver, both sited at JPL's Goldstone site in the Mojave desert. The transmitter power was boosted for the Venus experiment from 10 to 13 kW and the receiver noise temperature reduced from nearly 1600 K to 64 K by using a maser and a parametric amplifier. The delay for the echo was some six and a half minutes (i.e. a separation between Earth and Venus of ~57 × 10⁶ km when the first observations were made). The results gave a value of 1.49599 ×

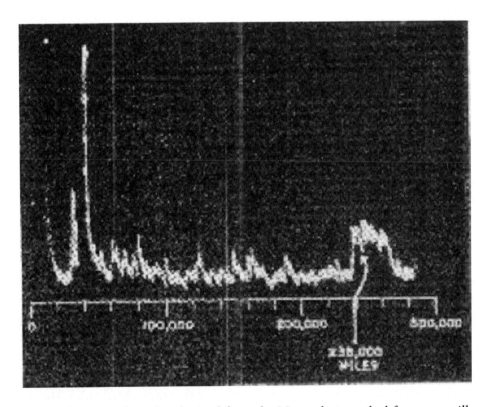

FIGURE 2.5 The first radar echo obtained from the Moon photographed from an oscilloscope display screen. The transmitted pulse is on the left and the (weaker) received pulse on the right. The delay between the pulses is 2.4 seconds, but this has been converted to the Moon's distance in miles on the abscissa.

10^8 km for the astronomical unit – 130,000 km larger than the previously accepted value of 1.49467×10^8 km. When the earlier, claimed, detections were re-analysed using this new value for the AU it was found that they had also detected Venus, but, of course, the credit for the achievement justifiably goes to JPL's Walt Victor and Bob Stevens.

The delay of the radar echo of 6.5 minutes for Victor and Stevens' observations meant that if the pulses were sent out at much shorter intervals (say every few seconds) there could be mis-identifications of the echoes with their initial transmitted pulses. With more distant targets – such as Saturn – the delay can be over two and a half hours. Clearly, sending a single pulse once every 2.5 hours would be highly wasteful of a very expensive telescope's time. Pulses are therefore normally transmitted several times a second. To ensure that an echo is correctly identified with its original pulse, a code is added to the pulses. This may take the form of irregular phase changes within the signal, changes of frequency within a pulse or pulses being sent out at changing intervals of time.

Radar provides information on the direction in space of an object from the pointing direction(s) of the transmitter and receiver. It also measures the distance from the delay between the transmission of the initial pulse and the reception of its echo. By scanning the radar beam over the surface being targeted, a radar image of the surface may be built up. The scan may be made by changing the pointing direction of the radar dish, or, more commonly, especially for spacecraft-borne radars, by the point of view of the radar being tracked over the surface by the spacecraft's orbital motion and/or rotation. Such radars are often then termed imaging radars and they have been especially used for mapping Venus and Titan where there is a solid surface that is heavily obscured by clouds from visual or IR observations.

A comparison of the shape of the pulse and its echo in frequency and time can also give information on the velocity and the depth structure* of the target. Thus, if there is a relative velocity along the line of sight between the radar and its target, the echo will be Doppler shifted when compared with the transmitted pulse. The change in frequency or wavelength is easily converted into the value of the relative velocity via Doppler's equation (using the astronomer's convention that velocities away from Earth are positive):

$$v = -\frac{\Delta v}{v}c = \frac{\Delta \lambda}{\lambda}c \qquad (2.8)$$

where v is the radial velocity.

With pulsed radar the portion of the target nearest to the radar will be the first to return an echo. If other parts of the target are further away, then their echoes will be delayed slightly in comparison. Thus with Venus' spherical shape, the outer edges of the visible disk of the planet are 6050 km further away than the sub-radar spot at the centre of the disk. Echoes from near the edge of the visible disk could thus be delayed by some 40 ms and the echo's duration will be that much longer than that of the transmitted pulse. Thus by analysing cross sections of the echo in time, different segments of the planet may be selected

---

* Sometimes called the range resolution.

FIGURE 2.6   The 70 m antenna at the Goldstone Observatory. (Courtesy of NASA/JPL-Caltech.)

for study. In this way, the Arecibo planetary radar (see below) produced the first maps of Venus' surface with 1–2 km resolution in 1988.

Since 1961 ground-based solar system radar astronomy has been concentrated at two facilities – the 70 m fully steerable antenna at the Goldstone observatory (Figure 2.6) and the fixed 305 m Arecibo telescope in Puerto Rico.* The Goldstone radar principally operates at 8.56 GHz (35 mm) with a transmitter power of 500 kW provided by two klystron[†] amplifiers. The receiver uses HEMT[‡] amplifiers. The instrument can operate as a monostatic radar or as the emitter with other radio telescopes acting as the receivers (bi-static radar). In the latter mode, the receiving telescopes can include other smaller dishes at Goldstone, the 100 m Green Bank Telescope in West Virginia, the 305 m Arecibo telescope and the dishes of the VLA in New Mexico.

The 305 m Arecibo radio telescope can be equipped with transmitters to act as a monostatic radar as well as acting as a bi-static radar with the Goldstone antenna as the transmitter. Funding for this instrument has been very uncertain in recent years and continues to be so at the time of writing. The radar operations have therefore at times had to be suspended, but there is currently an active observational programme on near-Earth asteroids. Transmitters are available at 5–8 MHz (600 kW, CW), 47 MHz (40 kW, pulsed), 430 MHz (2.5 MW, pulsed) and 2.38 GHz (1MW, CW) together with a large number of receivers covering individual bands from 44 MHz to 10 GHz.

---

* Operated by SRI (Stanford Research Institute), USRA (Universities Space Research Association) and UMET (Metropolitan University).
† A klystron vacuum tube generates high power microwaves by converting the energy from a high velocity electron beam into the microwaves whilst it passes between two or more resonant cavities (essentially hollow metal boxes within which the radiation forms standing waves). For further information see Appendix A.
‡ High Electron Mobility Transistors. Further details can be found in A.1.1 and other sources listed in Appendix A.

Earth-based radars to date have been able to provide information on all the planets out to Saturn, many planetary satellites (including Titan), asteroids (Figure 2.7) and comets. Recently, the smallest known asteroid (at 2 m diameter), 2015 TC25, was observed at a

(a)

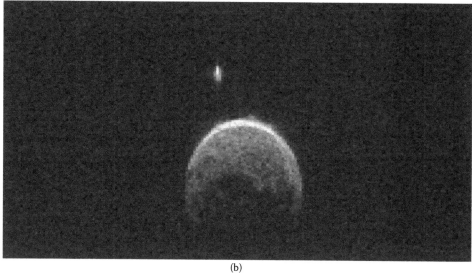

(b)

FIGURE 2.7 Two radar images of Asteroid 2004 BL86 using the Goldstone 70 m dish. The images were obtained in January 2015 when the asteroid was about 1.2 million km away from Earth. The asteroid has a diameter of about 325 m and it has a ~70 m natural satellite seen as a small bright spot above the image of the asteroid. The change in the angular separation between the asteroid and its satellite can clearly be seen between the two images which were taken about one and a half hours apart. The brightest parts of the image are those parts of the asteroid closest to the antenna and are, by convention, almost always shown at the top of the image. (Courtesy of NASA/JPL-Caltech.)

distance of 128,000 km by the Arecibo planetary radar and shown to resemble the very rare achondritic aubrite meteorites.

2.4.4 Spacecraft-Based Instruments

2.4.4.1 Introduction

The earliest spacecraft to carry radar systems did so in order to rendezvous with other spacecraft and/or to dock with them. Thus, Gemini 6A carried a 1.15 kW pulsed radar operating at 1.5 GHz to aid its rendezvous with Gemini 7. The Surveyor lunar landers carried three radar systems. The first started the firing of the landing rockets to make the initial braking of the spacecraft's velocity at a height of 75 km and this was then jettisoned. The actual landing was accomplished with the aid of Doppler and altimeter radars. Although not intended for scientific purposes, the Surveyor radars did provide data on the microwave reflectivity of the lunar surface. The Apollo 11 descent stage carried an altitude radar operating at 9.6 GHz with a power of 90 mW to assist its landing on the Moon's surface.

2.4.4.2 Synthetic Aperture Radar

The first SAR to be carried by a spacecraft was on board Seasat, which was launched in June 1978 for the remote sensing of Earth's oceans, but which failed in October 1978 due to electrical problems. The 1 kW SAR (Seasat also carried a radar altimeter) operated at 1.275 GHz with a height resolution of 25 m and a spatial resolution of 25–40 m. It observed across a 15 × 100 km wide segment of Earth's surface and this area was swept over the surface by the spacecraft's orbital motion (pushbroom scanning) to build up two-dimensional images. Seasat's SAR antenna was a 1024 element microstrip panel (Section 2.3.2) phased array[*], 2.2 × 10.4 m in size. Had a parabolic dish of a similar size been used as a real aperture radar (RAR) operating at the same frequency, then the ground resolution would have been worse than 20 km. The improvement in ground resolution of a factor of ~500 for the SAR shows why the technique is now in widespread use (the improvement could have been even greater – theoretically around a factor of ×4000). SAR was not used earlier because it needs significant computing power to process the data. Nowadays even small home computers would be adequate, but for Seasat the data had to be recorded onto film and processed optically.[†]

SAR essentially relies upon the same process that lies behind arrays of optical or radio telescopes being used as interferometers to improve their angular resolutions and behind the operation of a diffraction grating (Section 1.3.2) in producing a spectrum. In all these

---

[*] A phased array comprises an array of simple antennae (often half-wave dipoles). The signals from each antenna interfere with each other when they are mixed to produce a relatively narrow beam (when acting as a transmitter) and a similarly narrow field of view when acting as a receiver. The pointing direction of the beam can be moved by introducing phase delays between the signals from successive individual antennae. Franklin and Burke's Mills cross (Section 2.3.1) operated on a similar principle.

[†] Images are illuminated by collimated coherent light from a laser. They are placed at one focus of an objective and their Fourier transforms (holograms) are imaged at the back focus of the objective. Manipulation of the Fourier transforms (see footnote in Section 2.4.4.2) and then their conversion back into images accomplishes the required processing.

situations, beams of e-m radiation (or the electrical signals produced by such beams) are mixed together and undergo constructive and destructive interference.

To understand the principles that are involved we may imagine two identical terrestrial radio telescopes acting as an interferometer, separated by a smallish linear distance and observing the same object at the same frequency (Figure 2.8). There will normally be a path difference between the signals received by each telescope. When that path difference is a whole number of wavelengths, there will be constructive interference between the two signals and the combined output will be at a maximum. When the path difference is a whole number of wavelengths plus half a wavelength, there will be destructive interference between the two signals and the combined output will be zero. The Earth's rotation, of course, moves the source across the sky and thus changes the angle between the direction of the source and the interferometer's baseline. The path difference will thus also change with time and so the interferometer output will vary between its maximum and minimum values (Figure 2.9a).

If we now imagine replacing the single source with two sources, equally bright and separated by a small angle along the line of the interferometer's base line, then each will produce an individual output similar to that in Figure 2.9a. These will simply add together, since the two sources are mutually incoherent, to produce the overall output of the interferometer. If the angle between the two sources is such that the maximum of one individual output is superimposed upon the minimum of the other (a phase difference of $2n\pi + \pi/2$ – where n is an integer), then they will simply fill each other in to produce a nearly constant output (Figure 2.9b). If one maximum coincides with the other (a phase difference of $2n\pi$), then the output will be a sine wave with double the amplitude of either of the individual outputs

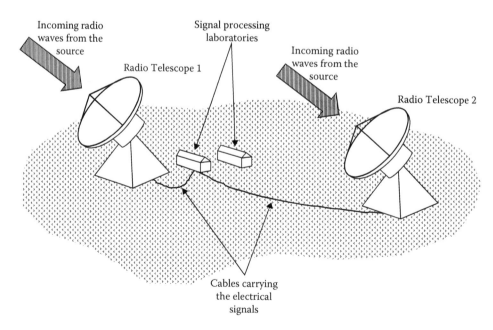

FIGURE 2.8   Two radio telescopes acting as an interferometer

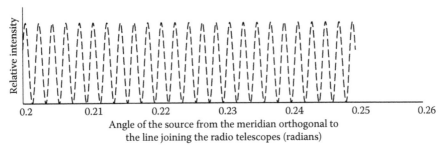

(a) The output from a two-element radio interferometer observing a single unresolved source at 1.5 GHz (200 mm wavelength) using a 100 m baseline.

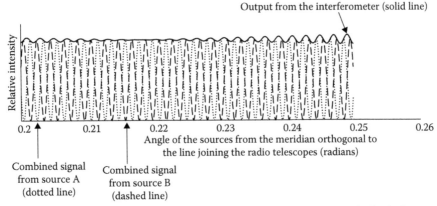

(b) The output from a two-element radio interferometer observing a pair of individually unresolved separated by 0.005 radians (17′) along the line joining the two telescopes at 1.5 GHz (200 mm wavelength) using a 100 m baseline showing the time of minimum variability of the output.

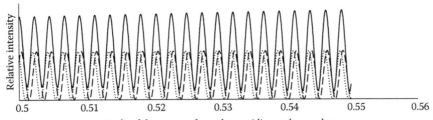

(c) The output from a two-element radio interferometer observing a pair of individually unresolved sources separated by 0.005 radians (17′) along the line joining the two telescopes at 1.5 GHz (200 mm wavelength) using a 100 m baseline showing a time near maximum variability of the output.

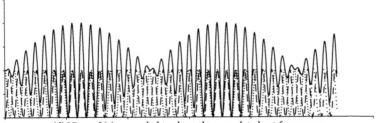

(d) View of (c) expanded to show the complete beat frequency.

FIGURE 2.9   Outputs from a two-element interferometer (cf. Figures 1.9 and 1.10)

(Figure 2.9c). Now, as the source moves across the sky, the phase difference between the two individual outputs will also change and they will sometimes be in phase ($2n\pi$) and sometimes out of phase ($2n\pi + \pi/2$) with each other. The interferometer output during a day's observing will thus alternate between that seen in Figure 2.9b and c and, when looking over a long time period, we shall see a beat frequency (Figure 2.9d).

Comparing the output of the interferometer for a single source (Figure 2.9a) with that for a double source (Figure 2.9b, c, and d), we clearly see that they are quite different in their natures. Understanding how an interferometer operates, it would be quite easy to interpret the first output as coming from a single source and the second from a double source. Furthermore, by measuring the beat frequency in the second case and knowing the interferometer's baseline and operating frequency it would be straightforward to estimate the separation of the two components of the double source (17′ for the situation illustrated in Figure 2.9).

In reality (unless measuring the diameter of a uniform disk and regarding the two halves of the disk as the separate sources), the two sources may be of differing intensities or may be of a more complex shape altogether. Also, in reality, most interferometers have more than two telescopes and in many cases the telescopes can be moved to change the length of the base line. Two-dimensional imaging of sources is undertaken by Earth rotation synthesis – by observing over a 12-hour interval, Earth's rotation will have rotated the baseline through an angle of 180°, allowing measurements of the source to be made in two orthogonal directions. In these more complex cases, the nature of the source cannot be obtained by a simple inspection of the output, but is given by the Fourier transform* of the output.

Returning now to the SAR, it might not seem to have much in common with the system(s) just described. However, the SAR transmitter and receiver are usually carried on board an aircraft or spacecraft that is moving over the ground. If the SAR makes one observation *and records it*, then a second observation obtained a little later will be with the instrument in a new position with respect to its target. The recorded observation may then be combined with the new one to give the equivalent of the observations fed into an interferometer from its two (static) telescopes. Furthermore, so long as the target remains within the SAR's field of view, further observations may be made at intervals to give the equivalent to a multi-element interferometer.

This mode of operation of an SAR gives rise to a property that, to most scientists who have some familiarity with telescopes, seems entirely counter-intuitive: if not degraded in other ways, the ground resolution of an SAR equals half the size of its antenna, i.e. to

---

* The Fourier transform, F, of a function, f, is given by

$$F(s) = \int_{-\infty}^{\infty} f(x) e^{-2\pi i \times s}\, dx$$

and the inverse Fourier transform F^{-1} by

$$f(x) = \int_{-\infty}^{\infty} F(s) e^{2\pi i \times s}\, ds$$

For further information, see sources in Appendix A.

improve the angular resolution of an SAR you *decrease* the size of its transmitting/receiving dish.

Thus as mentioned earlier, Seasat's SAR with a 10 m wide antenna could have had a ground resolution of 5 m – compared with the 20 km resolution for a 10 m dish operating at a 235 mm wavelength and used conventionally. This last comment, however, gives the clue to why an SAR has such comparatively high resolution – the 20 km resolution of the 10 m dish means that the target would be visible to the SAR whilst its spacecraft moved a distance of 20 km around its orbit. The synthesised instrument would thus have an effective width of 20 km. Now the angular resolution of an interferometer is given by (cf. Equation 1.12)

$$\alpha = \frac{\lambda}{2d} \text{ radians} \qquad (2.9)$$

where d is the maximum separation of its elements.

Thus for Seasat with $d \approx 20$ km and $\lambda = 235$ mm, we have $\alpha = \sim 1.2''$; so for a space craft altitude of ~800 km, the theoretical ground resolution is ~5 m.

The theoretical resolution may not be achieved for several reasons. One of the main problems is the changing distance of the target from the SAR as the spacecraft moves (the curvature of the orbit generally has a negligible effect). Thus for Seasat, at an altitude of 800 km, the target is about 630 mm further away when it enters the field of view than when it is directly below the spacecraft – almost three wavelengths, when about the maximum possible for useable results is a phase shift of $\sim\lambda/4$. The changing distance may be compensated by phase-shifting the observations, producing a focussed SAR, but this was not possible with Seasat. Seasat's SAR is thus termed an unfocussed SAR and its actual ground resolution was ~25 to ~40 m.

A second problem is the A_e^2 dependence of the SRA's signal-to-noise ratio (Equation 2.7). Most antennas for SARs are parabolic dishes, sometimes circular in shape, at other times rectangular (like Seasat's antenna). Either way, increasing A_e means increasing the size of the dish and thus decreasing the SAR's ground resolution. Practical instruments thus have to be a compromise between these two conflicting requirements.

The Sentinel-1A and Sentinel-B spacecraft (see also Section 1.4.2) of ESA's Copernicus Earth resources programme aim to monitor the arctic sea ice and land movements and to map forests, water and soil using their 5.4 GHz (C-band) SARs. The spacecraft have the same orbit but are separated by 180°. Their antennas are 821 mm × 12.3 m in size and their ground resolutions can range from 5 m to 25 m over swathes ranging from 80 to 400 km across.

2.4.4.3 Imaging Radar Spacecraft Missions

The first spacecraft-borne radar images of another planet were obtained of Venus' surface by the USSR's Venera 9 orbiter. The spacecraft carried the transmitter for a bi-static radar with the echoes being picked up by terrestrial radio telescopes. The system operated at 940 MHz and, along with the similar instrument on the Venera 10 orbiter, provided images with resolutions down to 5 km – comparable with the then resolutions of Earth-based planetary radars.

The Pioneer Venus Orbiter Spacecraft carried a 20 W mono-static radar altimeter operating at 1.76 GHz that had a height resolution of 200 m and a surface resolution of 7 km at best. It produced topographic maps covering 93% of Venus' surface.

Veneras 15 and 16 were the first spacecraft designed to map Venus using radar. They used 1.4×6 m segments of parabolic dishes for their synthetic aperture imaging radars and 1 m circular parabolic dishes for their radar altimeters. The SARs operated at 3.75 GHz, transmitting pulses for 200 μs and then switching to receiving the echoes for 3.9 ms. About a quarter of Venus' surface was mapped to a resolution of 1–2 km. The radar altimeter operated at the same frequency as the SARs and shared their processing electronics. It had a height resolution of about 50 m. The radars were switched between the electronic systems every 300 ms.

The Magellan spacecraft (Figure 2.10a) spent four years mapping Venus' surface and producing maps and images of 98% of the planet's surface using its SAR. Magellan's 325 W SAR operated at 2.385 GHz and had a 3.7 m parabolic dish as its antenna. Use of the dish was shared with an altimetry radar whose vertical resolution was 30 m and a passive thermal radiometer that measured the surface temperature to an accuracy of 2°C. The SAR images have a resolution of about 100 m and can appear photographic in quality although it should be remembered that they are not visible light images, but 126 mm radio wavelength images (Figure 2.10b).

The SAR carried by the Cassini spacecraft was designed for observing Titan although it could also be used for some of Saturn's other satellites. Like the Magellan SAR, it shared the use of the 4 m, Cassegrain design, high-gain antenna with a radar altimeter and a passive radiometer, but this antenna was also used for communications back to Earth. The SAR operated at 13.78 GHz with a ground resolution, at best, of around 350 m.

For the future, Russia has plans for a Venus orbiter/lander called Venera-D that could be launched in 2025 at the earliest. It is expected to be similar to Veneras 15 and 16 but to carry a more powerful SAR that will map potential landing sites for later missions.

2.4.4.4 Altitude, Doppler, Terrain and Other Radars

Conventional radars and lidars (i.e. not SARs), as already discussed, have been widely used to assist close approaches and dockings between spacecraft and landings on the surfaces of solar system objects. The lidars used on the Hayabusa asteroid sample-return spacecraft, for example, have been mentioned in Section 1.2.1.2. Another similar example was the lidar carried by the NEAR Shoemaker spacecraft for its Eros mission. This operated at a 1.064 nm wavelength with a range resolution of 6 m over a distance of up to 50 km. The first successful Mars landers – NASA's Vikings 1 and 2 – carried pulsed radar systems that started operating at a height of about 1400 km and were used to initiate the deployment of their parachutes. The final descent was controlled by four CW Doppler radars that measured the landers' horizontal velocities to ± 1 m·s^{-1}.

A Doppler radar is planned as the only instrument for the ADM Aeolus (Advanced Dynamics Mission Aeolus) mission. To be called ALADIN*, it will monitor winds in

---

* Atmospheric Laser Doppler Instrument.

(a)

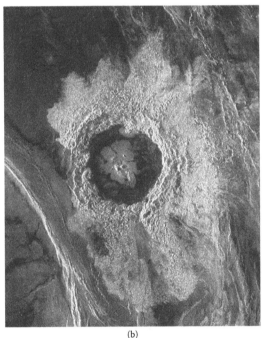

(b)

FIGURE 2.10 (a) The Magellan spacecraft during its deployment from the space shuttle 'Atlantis'. The 3.7 m radar antenna can clearly be seen at the left of the image. (Courtesy of NASA/JPL.). (b) A Magellan SAR image of the 69 km diameter Dickinson* impact crater on Venus. (Courtesy of NASA/JPL.)

atmosphere using a lidar based upon an Nd:YAG 150 mJ laser. The laser is frequency tripled to operate at 355 nm and will emit a hundred pulses per second. A second, identical laser is available as a backup. The laser beam will illuminate Earth's atmosphere at a 35° slant angle and observe up to heights of 30 km with about a 1 km vertical resolution. Photons

---

* Named for the American poet Emily Dickinson (1830–1886).

backscattered towards the spacecraft by molecules (Rayleigh scattering) and small dust and aerosol particles (Mie scattering) will be collected by a 1.5 m Cassegrain telescope with silicon carbide mirrors. Two separate etalon-based spectrometers (Section 1.3.5) will then be fed the Rayleigh and Mie photons* via a polarising beam splitter in order to measure their Doppler shifts and hence obtain the line-of-sight wind velocities. Both spectrometers will use CCDs as their detectors. Wind velocities up to 150 m·s^{-1} should be measureable to a precision of about ±1 m·s^{-1}.

If a range radar is used from a spacecraft whose orbit is well known, then its measurements can be used to map the height of the target's surface features and it is then commonly called an altitude radar. The lidar on NEAR Shoemaker, for example, was used in this way to determine the shape of Eros. The data sets from the radar altimeters on Veneras 15 and 16, Magellan and Cassini (see above) were used to produce topographic maps of Venus and Titan, in the latter two cases in conjunction with the data from their SARs. The Magellan altitude radar operated at 2.385 GHz with a vertical resolution of 50 m, while the Cassini radar has been discussed in the preceding section.

A lunar topographic map was produced by the lidar carried by the Clementine spacecraft. This was based upon a 180 mJ Nd:YAG laser operating at 1.064 μm. The spacecraft's high-resolution telescope picked up the returning pulse and a dichroic beam splitter then sent it to a silicon avalanche photodiode detector (Section 1.4.1.4). The instrument's height resolution was about 40 m. Similarly, Mercury's topography was mapped by the laser altimeter on board MESSENGER. This operated at 1.064 μm with eight 20 mJ pulses per second and received the reflected pulse through four 115 mm diameter refractive telescopes. The instrument had a ground horizontal resolution between 15 and 100 m and a height resolution of 300 mm.

The Jason 2 and 3 spacecraft monitor ocean heights on Earth using a microwave radar. They operate at 5.3 and 13.6 GHz with single measurement accuracies of better than 30 mm. The two spacecraft are in the same orbit and separated by 80 seconds. Combining and averaging their measurements reduces the errors to only a few millimetres.

Earth's icecaps and sea ice are currently being monitored by an SAR altimeter on board Cryosat-2 (the first Cryosat's launch failed). SIRAL (SAR and Interferometric Radar Altimeter) operates at 13.575 GHz and is a development of the Jason radars. Its best ground resolution is about 250 m. The spacecraft monitors its position to a precision of about 40 mm by measuring Doppler shifts of the signals from radio beacons on Earth and also, from Earth, the position is monitored by using laser pulses reflected from a small array of retroreflectors mounted on the spacecraft (Section 7.2.8).

A radar can also be used to measure the roughness of the surface that it is observing – then the instrument is usually called a scatterometer. In this mode, the radar cross section normalised to the area being illuminated by the radar is measured over a range of angles of incidence. The Cassini radar could operate in this mode and has shown average slopes on Titan to lie between about 5° and 13°. The data from the scatterometer could also be used to estimate

---

* Since the thermal motions of the molecules are much greater than those of the dust (etc.) particles, the original spectrally narrow 355 nm laser line will be broadened much more by Rayleigh scattering than by the Mie scattering.

the dielectric constants of the surface materials and so suggest possible compositions for the materials. For Titan these have been found to lie between 1.9 and 3.6 – liquid hydrocarbons have constants between 1.6 and 1.9 and water ice and solid simple hydrocarbons have constants between 2 and 3, while for water-ammonia ice the value can be up to 4.5.

A dual frequency radar is used by the Global Precipitation Measurement Core Observatory to monitor precipitation (rain and snow) over the whole Earth. At 16.6 GHz (Ku band) moderate to heavy rainfall is measured, whilst at 35.5 GHz (Ka band) snow and light rainfall can be monitored. Both radars have ground resolutions of 5 km and cover overlapping swathes of 245 km (Ku band) and 120 km (Ka band). The radars are pulsed, emitting about 4000 pulses per second with pulse lengths of 1 or 2 μs.

The CALIPSO mission (Cloud Aerosol Lidar and Infrared Pathfinder Satellite Observations), as its name suggests, used lidar to study clouds and droplets in Earth's atmosphere. It had two identical Nd:YAG lasers for redundancy, but also so that a planned replacement of the primary laser could be made after three years of operation. The lasers emitted 110 mJ at each of their 532 nm and 1.06 μm wavelengths (the shorter wavelength being obtained by frequency doubling of the longer basic emission). Their pulse emission rate was about 20 per second with a pulse length of about 20 ns and the beam diameter was 70 m at Earth's surface. A 1 m Cassegrain telescope with beryllium mirrors received the returned pulses and directed them, via beam splitters, to a photomultiplier tube (PMT) detector (Section 1.4.3.3) for the 532 nm radiation and an avalanche photodiode detector for the 1.06 μm radiation.

The four spacecraft making up the Cluster II mission each carry identical instruments called WHISPER (Waves of High Frequency and Sounder for Probing of Electron Density by Relaxation Experiment). The instruments act in a manner similar to a radar but the 'echoes' are indirect. A brief radio pulse is transmitted in the 2–80 kHz band (150–3.75 km) from two 50 m wire antennas. These pulses trigger oscillations in the plasma within which the spacecraft are embedded and these oscillations are then detected and the plasma density measured.

2.4.4.5 Ground-Penetrating Radar

These instruments are additionally known as sounding radars (see also Section 7.3.1). Ground-penetrating radars operate at very low frequencies where the radiation is able to penetrate some distance into a solid material. For a given material, the depth of penetration depends upon the operating frequency, with the lower the frequency, the greater the depth reached. By scanning the frequency, the depth structure of the ground can thus be determined. In dry sand, typical penetration depths are the following:

~20 m at 100 MHz (3 m wavelength)

~10 m at 250 MHz (1.2 m wavelength)

~5 m at 500 MHz (600 mm wavelength)

~2 m at 1 GHz (300 mm wavelength)

<1 m at 2 GHz (150 mm wavelength)

However, the depths attained can be much less than this in wet, salty materials.

Ground-penetrating radars have been carried on several spacecraft, particularly on missions to Mars. Thus, MARSIS (Mars Advanced Radar for Subsurface and Ionospheric Sounding*) on board the Mars Express spacecraft is based around two 20 m long antennas and operates between 1.3 and 5.5 MHz. It is designed to detect liquid water and ice and it can observe down to several kilometres below the surface. The Mars Reconnaissance Orbiter carries SHARAD (shallow sub-surface radar) that is intended to complement the MARSIS observations. It operates over the 15–25 MHz waveband to search for liquid water and ice down to a maximum depth of 1 km. It has a best vertical resolution of 7 m and a best horizontal resolution of 300 m.

MARS 2020, a possible future Mars lander/rover mission, is currently being studied by NASA. A ground-penetrating radar mounted on the rover is being designed and called RIMFAX (Radar Imager for Mars Subsurface Exploration). It will operate from 150 MHz to 1.2 GHz with a depth penetration in the region of 10 m and a vertical resolution of some 140 mm and its aim will be to provide data on the immediate sub-surface layers as the rover moves over the ground.

Another possible proposal in the early planning stages is the Europa Multiple Flyby mission, which is expected to carry the Radar for Europa Assessment and Sounding: Ocean to Near-surface (REASON). This is planned to operate at 9 and 60 MHz to allow penetration of Europa's icy surface to a depth of ~200 to ~300 m.

The terrestrial radio telescopes at Arecibo and Green Bank (Section 2.4.3) have been used as a ground-penetrating radar for the Moon. Working together as a bi-static radar at 430 MHz and with Arecibo as the transmitter, the lunar surface can be probed to a depth of ~15 m.

2.4.4.6 Occultations

Whilst not strictly a form of radar, the occultation of artificial radio sources is best mentioned here since it does involve a radio transmitter and receiver and the interaction of the radio waves with the target. Any spacecraft in orbit around a solar system object is almost certain on some occasions to pass behind that object as seen from Earth, or in some cases as seen from another spacecraft. Some fly-by spacecraft may also briefly pass behind their object.

If the spacecraft carries a radio transmitter, then both before and after the occultation the radio waves will be passing close to the object's surface. The effects on that radiation as it passes through the object's atmosphere, magnetic field, radiation belts, etc. may then be used to infer something about the properties of that atmosphere, magnetic field or radiation belt. In some cases, the spacecraft's communications system provides the transmitter, in others a purpose-built transmitter may be used. A transmission from a terrestrial radio source may also be used in a similar, but reverse, fashion by being received on the spacecraft.

---

* Sounding of the ionosphere occurs when the spacecraft's altitude is over 800 km and uses frequencies from 100 kHz to 5.5 MHz.

A related technique, which does not require an occultation, is to use the Doppler shifts of the radio transmitter to determine the orbit of the spacecraft very precisely and then to model the gravity field of the object involved (see Section 8.3).

The Mars fly-by, Mariner 4, was the first spacecraft to be used successfully for radio occultations using its S band (2–4 GHz) transmitter. The Martian atmospheric pressure was found to lie between 400 and 700 Pa (~0.5% of Earth's surface pressure). Mariners 6, 7 and 9 were also used for radio occultation observations of Mars, while Mariners 5 and 10 were similarly used for Venus.

The telecommunications systems on both the Voyager 1 and 2 spacecraft, which operated over the S and X bands (2–4 GHz and 8–12 GHz, respectively), were used during occultations to provide information on the atmospheres and ionospheres of Jupiter, Saturn, Titan, Uranus and Neptune, the nature of the material forming Saturn's rings and the masses of the planets and some of their satellites.

The electron density in the region around Comet Halley's nucleus was studied by several of the spacecraft missions to the comet via its effects upon the radio waves propagating through it. Thus with the USSR's Venus and Halley fly-bys, Vega 1 and Vega 2, phase changes in the L band (1.67 GHz) communications downlinks showed that the electron density was 2×10^9–3×10^9 m^{-3} at a distance of 8500 km from the nucleus and that the density varied with the inverse fourth power of the distance near to the comet and as the inverse square further out. Giotto, with transmitters operating at 2.3 GHz and 8.4 GHz, measured a deceleration due to dust impacts of 1.7×10^{-3} $m \cdot s^{-2}$ implying that some 100 mg to 1 g of cometary material had hit the spacecraft.

Cassini transmitted at 2.3 GHz, 8.3 GHz and 32 GHz with all three frequencies available for occultation measurements and the first and third reserved for them. Saturn's ionosphere and rings and Titan's atmosphere have all been studied by their radio occultations and in 2006 a fortuitous occultation of the water plume emitted by Enceladus was observed.

Rosetta's CONSERT (Comet Nucleus Sounding Experiment by Radiowave Transmission), although unfortunately unable to function because of the failure of the Philae lander on Comet 67P/Churyumov-Gerasimenko, nonetheless was a type of radio occultation experiment. Unlike those just discussed, however, it was intended to probe the interior of the comet's nucleus. The orbiter would have transmitted a 90 MHz signal to the lander when the former was on the far side of the nucleus from the lander. The signal would have passed through the nucleus to be received on the lander which, after some processing, would have transmitted the results back to the orbiter along the same path through the interior of the nucleus.

The interior of comet 67P/Churyumov-Gerasimenko was however probed by Rosetta in another way. Precision measurements of the Doppler shift in its telecommunications signals allowed the spacecraft's orbit to be determined extremely precisely. A mass of just less than 10^{13} kg was found for the comet giving it a mean density of 530 $kg \cdot m^{-3}$. The results also ruled out the possibility of any large (kilometre-sized) caverns inside the nucleus.

The gravity field of Mercury, its interior structure and the shape of the Sun will be probed by the MORE (Mercury Orbiter Radio Science Experiment) instrument when Bepi-Colombo is launched. The spacecraft's orbit will be measured to an accuracy of ~150 mm and its velocity to ~2 × 10⁻⁶ m·s⁻¹ using its Ka band (32.5 and 34 GHz) telecommunications equipment.

Radio occultation observations can be made using natural sources when the object happens to pass in front of them. Thus in January 1974, Comet Kohoutek passed over the extragalactic radio source PKS* 2025–15 and was observed at 327 MHz using the Indian National Centre for Radio Astrophysics' (NCRA) Ooty radio telescope. Although the observed intensity fluctuated strongly it was not possible to relate the fluctuations to features within the comet.

2.5 SPECTROSCOPY

Radio spectroscopes can be as simple as the tuning knob on your domestic radio. They operate through scanning over a band of frequencies by changing the local oscillator frequency (Section 2.2), whilst keeping the intermediate frequency fixed. More commonly a receiver will have a number of channels centred on slightly different frequencies and observe them all simultaneously. Both these methods give the spectrum intensity at each frequency directly.

A quite different approach, however, is also in widespread use. This uses a receiver that is sensitive to radiation over a range of frequencies. At a given instant of time the output will be the sum of the intensities of all the constituent frequencies at that time. The output will change with time as the various frequencies fall in and out of step with each other. For example if there are just two narrow emission lines with little in the way of continuum emission, then there will be a beat frequency between the two (cf. Figure 2.9d). More complex spectra will produce other patterns of outputs. Measurements of the output are made at twice the rate of the highest frequency component within the spectrum[†] – i.e. to obtain a spectrum over the 199–200 MHz spectrum range, measurements would have to be made at 2.5 ns intervals (= 400 MHz). The spectrum is obtained from the Fourier transform (see footnote in Section 2.4.4.2) of the autocorrelation[‡] of the output signal.

Fourier transforms are also involved in a design of spectroscope used at microwave frequencies. This is known as the Fourier transform spectroscope (FTS) or Michelson interferometer spectroscope (Figure 2.11). Its design is based upon the interferometer used by Albert Michelson and Edward Morley in 1887 to try and detect Earth's absolute motion in space by measuring the speed of light in different directions.[§] The incoming radiation

---

* Parkes Radio Catalogue.

[†] The sampling theorem states that the nature of a signal whose highest frequency is f is completely determined by measuring it at a frequency of 2f. This latter frequency is usually known as the Nyquist frequency.

[‡] Autocorrelation is the cross correlation of a function with itself and the cross correlation formula for two functions, f and g, is

$$f \star g = \int_{-\infty}^{\infty} f(u)g(u-x)du$$

where ⋆ denotes the operation of cross correlation.

[§] One of the most famous scientific failures in history – their finding that the velocity of light was constant in all directions, no matter how Earth was moving, led directly to the Special Theory of Relativity.

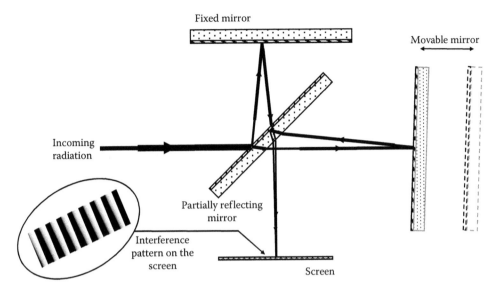

FIGURE 2.11 The optical layout of Fourier transform spectroscope. The mirrors are shown as slightly angled so that the ray paths are clear. In many modern instruments the main mirrors (i.e. not the beam splitter mirror) are corner-cube reflectors (Section 7.2.8).

is split into two beams and then sent on two orthogonal paths before being recombined to produce an interference pattern (Figure 2.11). If the screen is replaced with an appropriate single element detector, then the detector will register the intensity of that part of the interference pattern that is falling onto it. However, one of the mirrors in the instrument is movable so that the path difference between the two rays being recombined can be changed. Thus by slowly moving the mirror, the interference pattern can be scanned over the detector to produce the instrument's output. The system is essentially a two-element interferometer and so if the incoming radiation is monochromatic a simple sine-wave interference pattern is produced (cf. Figures 1.8, 1.9a, 1.10b and 2.9a). If the incoming radiation spectrum has two monochromatic emission lines at different frequencies, then a beat frequency will be seen (cf. Figure 2.9d). As with the autocorrelation spectroscope discussed above, these two patterns can be seen by inspection to result from spectra containing a single monochromatic emission line and two monochromatic emission lines at different frequencies, respectively. More complex spectra are obtained from the real part of the Fourier transform of the output.*

For mapping the thermal emissions from Saturn's and Titan's atmospheres, from the rings and from the surfaces of the icy satellites, the Composite Infrared Spectrometer (CIRS) carried by the Cassini spacecraft had two FTSs that shared the same telescope and mirror scanning mechanism. Its design owed much to the Voyager FTSs (see below). The CIRS's telescope was a 508 mm Cassegrain with beryllium mirrors (Figure 1.2 below the NAC). A pair of 45° mirrors split its field of view, sending the beams separately into the two spectrometers.

---

* The derivation of this relationship may be found in reference A.1.1 – see Appendix A.

The MIR spectrometer was a conventional FTS covering the spectral band from 18 to 42 THz with a spectral resolution at best of 15 GHz. Two separate 1 × 10 HgCdTe arrays were used for the detectors, one covering the range 18–33 THz, the other from 33 to 42 THz, and their field of view per pixel was 1′.

The FIR spectrometer operated from 300 GHz to 18 THz with a spectral resolution of 15 GHz and an angular resolution of 13′. Its optical design is known as a Polarizing Modulation FTS, which first of all linearly polarizes the beam of radiation. Then a polarizing beam splitter, its axis set at 45° to the first, produces two orthogonally linearly polarized beams, one reflected and one transmitted. These follow conventional FTS paths to retroreflectors before being recombined. The recombined beam is now elliptically polarized and the degree of ellipticity varies as the moving mirror is scanned. A third polarizer reflects one linear component of the radiation on to a thermopile detector (see Section 2.3.2) and transmits the orthogonal component to a second thermopile detector. The outputs from the two detectors as the mirror is moved are thus modulated and the modulation provides the outputs for Fourier transforming into the spectrum. The polarizers are grids of fine metal wires spaced at intervals of a tenth or less of the wavelength of the radiation. Radiation whose electric vector is parallel to the wires is reflected, and radiation with an orthogonal electric vector is transmitted*.

The FTSs on board the Voyager spacecraft[†] used 500 mm diameter Cassegrain mirrors. Dichroic beam splitters transmitted the visible and near-infrared (NIR) radiation to be detected by radiometers and reflected the radiation between 5.4 and 7.5 THz into the spectrometers. Both the radiometers and the FTSs used thermopiles as their detectors. The spectral resolution was 130 GHz and the angular resolution 15′. The Voyager Infrared Radiometer Interferometer and Spectrometer (IRIS) instrument's design was developed from an instrument of the same name carried by Mariner 9 to observe Mars' thermal emission. The Mariner IRIS observed from 6 to 60 THz with a spectral resolution of 70 GHz and was, in its turn, developed from the IRIS instruments on board Earth-observation spacecraft Nimbus 3 and 4.

The Venus Express and Mars Express orbiters had FTSs that were very similar in their designs. Known as double pendulum interferometers, the main mirrors (Figure 2.11) are corner-cube reflectors that are rigidly mounted together on a solid support. The transmitted and reflected beams from the partially reflecting mirror are sent, either directly or via secondary mirrors, into the corner-cube reflectors. After reflection, the beams retrace their paths back to the partially reflecting mirror and after transmission/reflection again are recombined to form the interferogram. The path lengths of the beams are changed by rotating the support through a small angle[‡] about a pivot lying on the symmetry axis of the instrument.

---

* This may seem the wrong way round (think about pushing a long stick through the bannisters on a stair case). However when the electric vector is parallel to the wires it induces electric currents in them leading to reflection just as if it were a solid sheet of metal. With the electric vector at right angles to the wires, the wires are so thin that electrons cannot move far and so the currents are very weak, enabling the radiation to pass through.

[†] Infrared Radiometer Interferometer and Spectrometer (IRIS).

[‡] That is, a movement similar to that made by the bob of a pendulum.

The Planetary Fourier Spectrometer (PFS) for Venus Express used two pendulum FTSs (short- and long-wave instruments) fed by a dichroic beam splitter. The long-wave instrument covers from 6.7 to 60 THz and the short-wave instrument from 60 to 330 THz, both with spectral resolutions of 45 GHz. The ground resolution when the spacecraft is at pericytherion is 13 and 7 km for the long- and short-wave instruments, respectively. The detectors are a $LiTaO_3$ pyroelectric* bolometer (long wave) and a PbSe/PbS photoconductor (short wave).

The ExoMars Trace Gas Orbiter's Atmospheric Chemistry Suite comprises three spectrometers. The TIRVIM[†] is a Fourier transform instrument operating over the 1.7–17 μm spectral range (see Section 1.3.2 for a discussion of the other two instruments). The TIRVIM is also of a double pendulum design with a KBr beam splitter. It has a spectral resolution of ~30,000 and uses HgCdTe detectors cooled by a Stirling engine.

Radio and microwave line emissions originating from the top layers of planetary atmospheres and from the comae and tails of comets can be observed using terrestrial-based instruments. Thus, the first generally accepted detection of the radio emission from a comet was of the 1.67 GHz OH lines from Comet Kohoutek. These were detected in 1973 using the Nançay radio telescope. This telescope uses a flat tiltable metal mesh mirror 40 m high by 200 m long feeding a 300 m long reflector that is curved in the horizontal direction. The radio waves are focussed by the second reflector onto two horn antennas located between the two mirrors. The receivers are super heterodyne and the spectra obtained by autocorrelation.

The presence of ethyl alcohol together with seven other organic[‡] molecules was detected for the first time in a comet in 2015 when comet C/2014 Q2 (Lovejoy) was observed using the 30 m IRAM[§] dish. A super heterodyne receiver, the Eight Mixer Receiver (EMIR), was coupled to an FTS to obtain spectra over the 210–272 GHz waveband with a spectral resolution of 200 kHz. An alternative processor, WILMA (Wideband Line Multiple Autocorrelator), on IRAM obtains spectra by autocorrelation.

---

* A detector that responds to a change in its temperature by developing a temporary voltage across its sensitive element.

[†] Thermal Infrared V-shape Interferometer Mounting. Also based upon the initials of Vasilii Ivanovich Moroz who pioneered the use of FTSs at the IKI (Institut Komicheskih Issledovanyi – Space Research Institute).

[‡] That is molecules formed from hydrogen, carbon and oxygen. The term does NOT imply that they originated from a living entity.

[§] Institut de Radioastronomie Millimétrique. Operated by the CNRS (Centre National de la Recherche Scientifique), IGN (Instituto Geográfico Nacional) and the MPG (Max Planck Gesellschaft).

CHAPTER 3

X-Ray and γ Ray Regions

3.1 INTRODUCTION

It is the usual practice (and it is also convenient) to characterise x- and γ rays by their photon energies expressed in electron volts (see also Table 1.1). There are no sharp boundaries between the various regions of the e-m spectrum, but at the high energy end some widely used definitions are

Extreme ultraviolet (EUV):	~10 to ~100 eV
Soft x-rays:	~100 eV to ~10 keV
Hard x-rays:	~10 to ~100 keV
Gamma rays:	~100 keV to ∞*

or in wavelength terms (these are not direct conversions of the energy limits just listed):

Extreme ultraviolet (EUV):	~100 to ~10 nm
Soft x-rays:	~10 nm to ~100 pm
Hard x-rays:	~100 to ~10 pm
Gamma rays:	~10 pm to 0

The use of frequency over this region is very rare, but Equations 1.1 through 1.6 give the conversion formulae if needed.

Earth's atmosphere protects us and other life-forms from x- and γ radiation, which is fortunate for our survival, but hampering when we are trying to observe celestial sources of that radiation. Instruments have therefore to be lifted up to heights of 25–100 km or more before they can detect x- and/or γ rays. This, of course, means that the instruments must be carried on high-altitude balloons, rockets or spacecraft.[†]

---

* A slight exaggeration here – the highest energy γ ray observed to date had an energy of 16 TeV.

† Very high energy (GeV to TeV) γ rays can be detected by Earth-based instruments from their fluorescent emissions as they pass through the upper atmosphere. Such energetic γ rays, however, do not originate within the solar system.

3.1.1 Early Observations

3.1.1.1 X-Rays

The first detection of celestial x-rays (from the Sun) was made in January 1949 when an ex-WWII V2 rocket carried an x-ray detector to a height of 60 km.

X-rays in the energy range 10–100 keV from Earth's aurora were postulated by James van Allen as the explanation of the detections made by Geiger counters flown on sounding rockets in 1953. The results were not announced though, until 1957, when the alternative explanations (protons and electrons) had been eliminated.

In 1962, a US Air Force (USAF) Aerobee 150 sounding rocket was launched to look for scattered solar x-rays from the Moon. It failed in its objective but picked up an enormously bright source about 25° away from the Moon. That source is now known as Sco X-1; it is the second brightest x-ray source in the sky after the Sun and is a neutron star in a binary system with a small normal star. Scattered x-rays from the illuminated half of the Moon were eventually detected in 1990 using the ROSAT spacecraft. The third solar system source of x-rays was found in 1979 when NASA's Einstein x-ray observatory detected Jupiter's emissions. At the time of writing, x-rays have been detected from Venus, Earth, Moon, Mars, Jupiter, Saturn, several natural satellites and several comets.

3.1.1.2 Gamma Rays

Gamma rays have energies higher than 100 keV. Those with energies in excess of 1 GeV originate from outside the solar system, but γ rays in the keV and MeV regions can be produced by solar system objects. Thus in March 1958, Larry Peterson and John Winkler made the first detection of γ rays produced within the solar system when they detected γ rays from a solar flare using an ionisation chamber and a Geiger counter (Section 3.2 and Chapter 4) flown on a high-altitude balloon.

The first γ ray spacecraft, Explorer 11, was launched in April 1961 carrying a scintillator detector and a Čerenkov detector (Section 3.2 and Chapter 4). Of its 22,000 detections, just 22 were due to γ rays (the rest were charged particles) – and none of these could be attributed to anything within the solar system.

Perhaps surprisingly the Moon is brighter* than the quiet Sun in γ rays. The lunar γ rays originate from high-energy cosmic rays interacting with the lunar surface layers (see below). The Energetic Gamma Ray Experiment Telescope (EGRET) carried by the Compton Gamma Ray Observatory spacecraft first detected this lunar emission in 1991. Fermi's large area telescope has also, more recently, observed these lunar emissions. Gamma rays with a similar origin and in some cases from natural radioactive decays (which can also produce x-rays) have additionally been detected from Mercury, Mars and Eros.

---

* This statement does not allow for the shorter distance to the Moon – i.e. in visual astronomical terminology it is the apparent magnitudes that are being compared here.

3.1.2 X- and γ Ray Sources

We may distinguish between x- and γ ray sources that are artificially induced by human intervention and those that occur naturally. The former result from techniques such as neutron activation analysis and are discussed in Section 7.2.3.5. Here, we are concerned with the natural sources occurring within and around solar system objects.

The high-energy photon 'emissions' from many solar system objects are actually scattered solar radiation. The main mechanism is fluorescence x-ray scattering. This is thought to be the process behind emissions from Venus', Mars', Jupiter's and Saturn's atmospheres, Saturn's rings and, at times of large solar flares, the Earth's atmosphere. It comprises the absorption of the original solar x-ray by an atom in the object and its rapid re-emission. The re-emission can be in any direction (i.e. scattering) and the energy of the photon may be the same as that of the original x-ray, or at a somewhat different energy characteristic of the absorbing atom (fluorescence – Section 7.2.2.5). Thus for Venus and Mars, the re-emission is found mainly at 280 eV and 530 eV – lines originating from carbon and oxygen atoms, respectively.

The OSIRIS-REx (Origins, Spectral Interpretation, Resource Identification, Security, Regolith Explorer) mission to the asteroid Bennu is carrying REXIS (Regolith X-ray Imaging Spectrometer) that will use 300 eV–7.5 keV fluorescence x-rays produced by solar x-rays to image the asteroid's surface. It will use a coded mask x-ray telescope (XRT) (Section 3.3) and five broadband and 11 narrow band filters to measure chemical element abundances.

Thomson scattering* of soft solar x-rays was observed coming from the Earth's atmosphere as early as 1982 by the HEAO-1 A-2[†] experiment. Thomson scattering is elastic so that the x-ray's energies are unchanged during the interaction. The soft x-rays can be detected even during quiet Sun interludes – higher energy x-rays scattered from the Earth's atmosphere are only to be found during major solar flares.

The solar wind charge exchange (SWCX) emission mechanism is thought to be behind at least some of the x-ray emissions from Venus' and Mars' x-ray halos, the geocorona, Jupiter's atmosphere and comets. This process occurs when highly charged, medium mass ions in the solar wind – such as C^{+6}, O^{+7} and Ne^{+9} – collide with neutral atoms in the 'target'. Electrons from the neutral atoms are lost to the highly charged solar wind ions (charge exchange). Most of the resulting ions will then be highly excited. As these ions return to the ground state, they emit the x-rays. The ions will also recombine with electrons and the subsequent cascade of the electron down to the ground state may emit x-ray as well as lower energy photons.

Most of the other x-ray emissions result from free–free (bremsstrahlung – Section 2.1 and Chapter 4) radiation as high-energy electrons interact with magnetic fields. They are known as auroral emissions since, although the γ ray radiation originates many thousands of kilometres above the planet's surface, aurorae (at heights of ~100 km to several × 100 km on Earth) often occur at the same time and arise from the same bursts of high-energy electrons. Thus in 2015, the European Space Agency's International Gamma Ray Astrophysics

---

* Scattering by a free charged particle such as an electron or proton.
[†] The A-2 experiment – also known as the Cosmic X-ray Experiment (CXE), covered the 2–60 keV x-ray photon range using gas proportional counters (Section 3.2 and Chapter 4).

Laboratory (INTEGRAL), in a serendipitous observation,* imaged 17–60 keV x-rays from high above the Earth's north polar region which coincided with a strong auroral display at lower heights over Siberia and Japan, then later over Canada and Greenland (see also Figure 3.1).

At one time, scattering of x-rays by dust particles was suggested for the origin of the x-rays from comets and from Mars. The x-ray emission from Mars, however, shows no correlation with the presence or absence of dust storms, so this process is now generally discounted.

X-rays arising from other (i.e. non-free-free) processes are mostly lumped together as 'non-auroral emissions' and are also sometimes called disk emissions since they are not normally to be found in the polar regions. Jupiter's auroral emissions, though, are thought to originate mainly from charge exchange with a small proportion coming from free–free interactions.

Gamma rays emissions from solar system objects mostly originate from cosmic-ray interactions and natural radioactivity.

Cosmic rays are able to penetrate into the solid surface layers of an object when it has no, or only a very thin, atmosphere. Protons are the main particles in galactic cosmic rays and during their collisions with nuclei in the surface layers, neutrons may be ejected from the surface nuclei. These neutrons, in turn, collide with and excite other surface nuclei and the latter, upon returning to the ground state, emit γ rays. Since the object's atmosphere is thin enough to allow the cosmic rays to penetrate through to the surface, the γ rays (and also some of the neutrons – Chapter 6) can emerge back out into space to be detected by orbiting spacecraft.

Radioactivity, as a phenomenon, is likely to be familiar to readers of this book, Little, therefore, need be added to its production of γ rays (and also x-rays) save, firstly, to note that if the photons are to emerge from the surface and travel up to where spacecraft are orbiting, then only those originating in the immediate surface layers will be able to do so without being absorbed first. Second, note that only the radioactive elements and isotopes that emit x and γ radiation are detectable through this process.

Terrestrial Gamma Flashes (TGFs) were detected in 1994 by the Burst and Transient Source Experiment (BATSE) on board the Compton Gamma Ray Observatory and have since been detected by other spacecraft. They occur at a rate of at least 50 per day – perhaps many more if the emissions are tightly beamed. They are closely associated with large lightning discharges, but not every such flash produces a TGF. The mechanism for their production is still very uncertain but may involve free–free radiation from electrons accelerated by the electrical fields associated with the discharge. The photon energies in TGFs can be up to 50 MeV. The flashes last typically for 100 µs and such short flashes can currently only be detected by instruments carried by the Fermi, Reuven Ramaty High-Energy Solar Spectroscopic Imager (RHESSI) and Astro-Rivelatore Gamma a Immagini Leggero (AGILE) spacecraft (see below).

The Solar Dynamics Observatory (SDO) observed x-rays and extreme ultraviolet (EUV) emissions from Venus during its 2012 transit of the Sun. The origin of the radiation is

---

* INTEGRAL was trying to determine the diffuse deep space x-ray background by observing it directly and then comparing that observation with another obtained when the Earth was obscuring some of the radiation.

uncertain, but it may arise by scattering of solar e-m radiation within Venus' magneto-tail. Venus was observed using the SDO's Atmospheric Imaging Assembly (AIA) which uses four f/21, 200 mm diameter, multi-layer (silicon and molybdenum), normal incidence, Ritchey–Chrétien telescopes (Sections 1.2.1.3 and 3.3.1) to image the whole solar disk in the light of eight multiply ionised atoms at wavelengths from 9.4 to 160 nm plus broadband images at 170 and 450 nm. The telescopes have 4096 × 4096 pixel charge-coupled device (CCD) arrays as their detectors.

Gamma ray emissions can occur through many other processes than those just mentioned – the inverse Compton effect, deuterium formation and electron–positron mutual annihilations (the latter two producing spectrum lines at 511 keV and 2.2 MeV, respectively) – but these are not significant in a solar system context.

3.2 DETECTORS

3.2.1 Introduction

Detectors for x- and γ rays are to a large extent the same as those for energetic charged particles (Section 4.2) with minor adaptations to optimise them for a slightly differing situation. The reason for the overlap is that both energetic charged particles and x- and γ rays cause ionisations or excitations within materials when they pass through them or are absorbed within them. So any of the detectors discussed in Section 4.2 that operate through the production of ion–electron pairs or through the excitation of electrons within atoms or through photo-emission can be used to detect x- and γ rays. Additionally, some extended optical region (EOR) detectors, such as CCDs, avalanche photodiode (APDs) and photomultiplier tube (PMTs) can be adapted to have sensitivity to high-energy photons (Section 1.4).

Here we shall discuss the adaptations needed for the detectors for use with high-energy radiation, with references being made to the appropriate section(s) in Chapters 1 and 4 for the details of the physics underlying the detectors.

3.2.2 Proportional Counters and Related Instruments

See Sections 4.2.1 and 4.2.3 for additional details.

Proportional counters are used for detecting x-rays and soft γ rays. At the lower energy end of the range, the counter will need a window that the photons can penetrate. Typical windows are formed from very thin sheets of plastic or mica, but the softest of x-rays are still absorbed within them. The lower energy limit for detection of x-rays by a proportional counter is thus ~100 to ~300 eV.

An upper limit is reached when the photons are so energetic that they pass completely through the detector and this occurs at ~50 keV and above. Higher energy x- and γ rays will still be detected, but not all their energy will be deposited inside the detector and so their total energy will not be determined.

The energy resolution of a proportional counter is determined by the energy needed to produce an ion–electron pair and this is about 30 eV. Photons of 500 eV will thus produce around 17 ion–electron pairs and 50 keV photons around 1700 pairs. For a Poisson distribution, this will give standard deviations of ~4 and ~40 ion–electron pairs, respectively and so to the 2.5σ significance level the energy resolutions will be ± ~300 and ± ~3000 eV, respectively.

Proportional counters can also act as polarimeters. The initial track of the electron released in the ionisation will be driven by the electric vector of the x-ray and so it will (roughly) be in the same direction. If the beam of x-rays is polarised then there will be a preferred direction for the electron tracks. If it is unpolarised, then the tracks will be random.

The imaging position-sensitive proportional counter (Sections 4.2.1 and 4.2.3) is used as the detector for a number of XRTs (Section 3.3). The Einstein x-ray observatory (Section 3.1.1.1) had three proportional counters on board – two Imaging Proportional Counter (IPCs) and an Monitor Proportional Counter (MPC). The IPCs were identical except for the entrance windows, these being a dual polypropylene/polycarbonate (Lexan™) film, 2.4 µm thick, for detector 'A' and a 3 µm thick polyethylene terephthalate (Mylar™) film for detector 'B'. The lower energy detection limit for detector 'A' was thus about 100 eV, while for detector 'B' it was about 150 eV. The IPCs used an Ar/Xe/CO_2 gas mix in the ratio 84%/6%/10% at a pressure of 100 kPa. The pressure was maintained by using a controlled slow leak. Two planes of cathode wires sandwiched a single plane of anode wires at 3 mm spacings and they were maintained at 900 and 3600 V, respectively. They operated over the 400 eV to 4 keV x-ray energy range with an angular resolution of 1′. The IPCs were fed by a 600 mm glancing incidence telescope (Section 3.3 – also known as grazing incidence telescopes) with four nested reflectors of the Wolter type 1 design.

Einstein's MPC monitored the 1–20 keV x-rays from the sources being observed by the IPCs, but not using the main glancing incidence telescope. Instead a simple collimator (Section 3.3) limited its field of view to 1.5°. The proportional counter used a 1.5 µm beryllium window and the filling gas was Ar/CO_2. Four anodes collected the electrons and it had an energy resolution of ~1.2 keV at 6 keV.

The Joint European X-ray Monitor (JEM-X) on board ESA's INTEGRAL spacecraft uses two imaging proportional counters as its detectors. The instrument uses a coded mask (Section 3.3) as its imager giving it a field of view of 4.8° and an angular resolution of 3′. It operates over the 3–35 keV energy band with a spectral resolution of 1.3 keV at 10 keV. The filling gas of the counters is a xenon (90%)/methane (10%) mix at a pressure of 150 kPa.

The Polar spacecraft was a sister mission to the Wind spacecraft (Section 5.2.4.5). Launched in February 1996 into an orbit taking it close to the Earth's poles, it continued its operations until April 2008. Amongst other instruments, it carried Polar Ionospheric X-ray Imaging Experiment (PIXIE) to obtain x-ray images of the Earth over the 2–60 keV photon energy range. It used two pinhole cameras with four sizes of holes available and four position-sensitive proportional counters. Two proportional counters were stacked within a single gas chamber with the top counter responding to 2–12 keV x-rays and the bottom counter responding to 10–60 keV x-rays. In March 1996, PIXIE obtained the first x-ray image of the Earth (Figure 3.1) showing emissions arising from high-energy electrons interacting with atoms in the Earth's ionosphere; at lower altitudes aurorae are produced (see also Section 3.1.2).

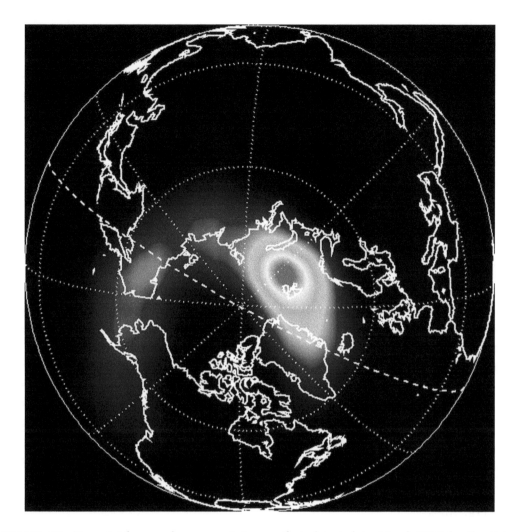

FIGURE 3.1 Terrestrial auroral x-ray emissions – first detected in March 1996 by the Polar Ionospheric X-ray Imaging Experiment (PIXIE) instrument on the Polar spacecraft. (Courtesy of NASA.)

As well as detecting γ rays from the Moon, ROSAT picked up x-rays from comet Hyakutake and from the July 1994 comet Shoemaker–Levy collision with Jupiter. It carried two (for redundancy) position-sensitive proportional counters that could be positioned at the focus of the XRT (Section 3.3). Each had two separate multi-wire grids; the first was the x-ray imager, the second acted as an active anti-coincidence (AC) shield. The gas mix was Ar (65%), Xe (20%) and methane (15%). The counters had polypropylene windows and operated over the 100 eV–2.4 keV energy range with an energy resolution of ~450 eV at 1 keV. The counters' angular resolution was about 25″ at best and their fields of view were about 2°.

The NEAR Shoemaker spacecraft carried three gas proportional counters to detect x-rays from Eros and also a scintillation detector for the γ rays (Section 3.2.3). The x-ray/gamma ray spectrometer (XGRS – Figure 3.2) was designed to determine the elemental

X-ray/gamma ray spectrometer

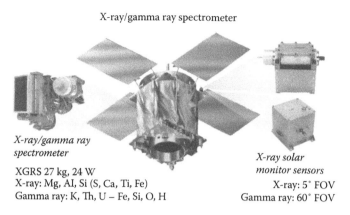

X-ray/gamma ray spectrometer

XGRS 27 kg, 24 W
X-ray: Mg, Al, Si (S, Ca, Ti, Fe)
Gamma ray: K, Th, U – Fe, Si, O, H

X-ray solar monitor sensors

X-ray: 5° FOV
Gamma ray: 60° FOV

FIGURE 3.2 The x-ray/gamma ray spectrometer (XGRS) carried by the NEAR Shoemaker spacecraft. (Courtesy of NASA/John Hopkins University Applied Physics Laboratory [JHUAPL].)

composition of the asteroid's surface and near-surface layers. Eros' x-rays originate by fluorescence in the asteroid's surface layers induced by solar x-rays. The detectors were single wire proportional counters determining x-ray spectra over the 500 eV to 19 keV energy range with a resolution of 850 eV at 5.9 keV. A simple honeycomb collimator restricted their fields of view to 5°. The counters had 25 μm thick beryllium windows and in addition, one had an 8.5 mm thick magnesium filter and a second, an 8.5 mm thick aluminium filter in order to separate out the lines from aluminium, magnesium and silicon. The gas fill was argon (90%) and methane (10%) at a pressure around 100 kPa. The signals arising from x-rays originating from Eros were distinguished from those arising from cosmic-ray interactions by utilising the longer rise times of the latter signals. The observations were typically made from heights between 35 and 100 km above the surface.

MESSENGER carried the x-ray spectrometer (XRS) developed from NEAR Shoemaker's XRGS to detect x-ray fluorescence emissions from Mercury's surface induced by solar x-rays. Like the XRGS it used three proportional counters but these had shields of anti-coincidence wires added around them. The counters operated over the 1–10 keV photon energy region with an energy resolution of 850 eV at 5.9 keV. XRS had a 12° field of view provided by a honeycomb collimator (Section 3.3) giving a best ground resolution of 40 km. The gas fill was an argon (90%)/methane (10%) mix. The surface abundances of elements such as aluminium, calcium, iron, magnesium, silicon and titanium were measured and the majority of Mercury's surface was mapped. A separate instrument monitored solar x-rays in order to calibrate the XRS's measurements.

In September 2015, Astrosat was launched on a mission for spectroscopy and timing studies of both solar system and more distant objects at wavelengths ranging from the visible to hard x-rays. The Large Area Xenon Proportional Counter (LAXPC) is a non-imaging detector for the 3–100 keV photon energy region. It has an energy resolution of 2 keV at 22 keV and a 1° field of view produced by a honeycomb collimator (Section 3.3). Each unit has an aluminised 360 mm × 1000 mm, 50 μ thick, Mylar entrance window to exclude photons under about 3 keV in energy and is 150 mm in depth with 60 anode wires at 30 mm centres throughout its volume. Its total collecting area is thus 1.08 m² – almost 10 times that of NuSTAR.* The counter units are filled with a xenon (90%)/methane (10%) mix at a pressure of 200 kPa.

Gas-based proportional counters are little used in space applications for detecting γ rays because of their low stopping powers. Analogous detectors using solid materials are discussed in Section 3.2.5.

3.2.3 Scintillation Detectors

See Section 4.2.4 for additional details and also Sections 3.2.5.3 and 4.2.8.

Scintillation counters, especially those based upon dense liquids or solids, can be used to detect both x- and γ rays to well over 10 MeV. However, spectral features due to nuclei largely die out above 10 MeV so that only continuum spectra then tend to be observed. Below 10 MeV, instruments giving high spectral resolutions will be encountered, whilst only low spectral resolutions are generally needed above that limit. Scintillation detectors based upon NaI and CsI crystals are thus often to be found in the higher energy instruments.

Like most of the detectors discussed in this section, scintillators are sensitive to both high-energy photons and high-energy charged particles (Section 4.2). It is often possible, however, to distinguish between photon and particle interactions since the photons are absorbed quickly and produce a sharp pulse, while the particles take longer to lose their energy and so produce longer pulses.

Scintillation counters are often used as active† shields since they can be available in the form of large sheets of plastic which may be used to enclose the primary detectors completely. They can be used in several modes such as:

1. Rejecting primary detections when two simultaneous detections are made in the shield (showing that the photon passed completely through the instrument and did not deposit all its energy in the primary detector).

2. Discriminating between high-energy photons and high-energy charged particles (see above) and rejecting primary detections of the latter.

3. Acting as a collimator – when made into a cylinder with one closed and one open end and with the primary detector at the bottom of the cylinder, rejecting primary detections that are simultaneous with one or more in the shield will limit the acceptance angle of the primary detector to just those photons coming through the cylinder's open end.

---

* A spacecraft launched in June 2012 and still operating at the time of writing. Its mission is to make x-ray observations of black holes.
† Also known as anti-coincidence (AC) shields. Passive shields are simply pieces of material – usually formed from or containing heavy elements such as lead – which just absorb the unwanted radiation or particles. Passive shields are generally of high mass and so are little used for balloon, rocket or spacecraft-borne instruments, although the Jupiter orbiter, Juno, carries a 170 kg titanium vault containing its most sensitive electronics which reduces the radiation levels by a factor of 800.

Examples of the use of scintillation counters for solar system missions include the Lunar Prospector, NEAR Shoemaker, AGILE, the Compton Gamma Ray Observatory and the balloon-borne Balloon Array for Radiation Belt Relativistic Electron Losses (BARREL) instrument.

The Lunar Prospector orbited the Moon between January 1998 and July 1999 carrying a gamma ray spectrometer (GRS). The primary detector was a cylindrical Bismuth/germanium/oxygen (BGO) crystal mounted on the end of a 2.5 m boom to isolate it from the main spacecraft. The BGO crystal was surrounded by a plastic scintillator that acted as an active shield and which could also detect fast neutrons. Twenty-five to thirty passes of the spacecraft over a portion of the lunar surface were needed to build up enough detections to be able to estimate element abundances. Abundance estimates then had uncertainties between 7% and 30%, depending on the individual element being assessed. The surface resolution was about 150 km. Amongst other discoveries, Lunar Prospector, from the distribution it measured for thorium, found a volcanic eruption on the lunar far side whose diameter was nearly 300 km (similar to the size of Hawaii's Mauna Kea at its base at the bottom of the Pacific).

In addition to the proportional counter x-ray detectors making up NEAR Shoemaker's XGRS (Section 3.2.2, Figure 3.2), the suite of instruments included scintillator-based γ ray detectors. These used BGO and NaI(Tl)* as the scintillator materials and PMTs as the optical detectors. The NaI(Tl) scintillator formed the primary detector and it was surrounded by the BGO as both an active and a passive shield. The detector operated from 100 keV to 10 MeV with an energy resolution of 60 keV at 660 keV. The penetrating nature of these high-energy photons meant that the surface could be probed to a depth of 100 mm. The γ ray detector was designed particularly for the detection of hydrogen, iron, oxygen and silicon and for the radioactive emissions from potassium, thorium and uranium.

The AGILE spacecraft has a primary aim of detecting x- and γ rays from outside the solar system. However, it was soon found that it could pick up TGFs as well. The minicalorimeter (MCAL – Section 3.2.5.3) instrument detects photons over the energy range from 350 keV to 50 MeV and has a time resolution of 3 μs, thus it is ideally suited to the detection of TGFs and in its first few years of operation detected about three per day. The three main detectors on AGILE, however, are completely surrounded by 6 mm thick plastic scintillator panels that act as an active shield and it was realised that the delay time imposed on the MCAL's response by the shield was leading to many short TGFs being missed. Thus in 2015, the software controlling the MCAL was re-programmed to disable the shield's delay, resulting in the number of TGFs detected rising to ~30 per day.

BARREL is a 76 mm sized NaI scintillator operating over the 20 keV to 10 MeV range with a 45 keV spectral resolution at 660 keV. Multiple (5–10) identical instruments are carried to heights up to 35 km by separate balloons near the North and South poles for periods of 5–10 days at a time. The mission's aim is to monitor electrons from the van Allen belts by observing the bremsstrahlung x-rays (Section 2.1) produced by the electrons colliding with neutral atoms in the Earth's upper atmosphere.

---

* Sodium iodide activated with thallium.

As has already been mentioned (Section 3.1.2) the BATSE instrument on board the Compton Gamma Ray Observatory discovered TGFs in 1994. BATSE comprised eight identical modules oriented like the faces of an octahedron so that the whole sky could be covered. Each main detector was based around a 12.7 mm thick disk of NaI which was 505 mm in diameter. This was supported on a quartz disk and had a plastic scintillator sheet as an active shield placed before it. Three large PMTs detected the scintillation flashes. It detected photons with energies from 20 keV to 1.9 MeV. Each BATSE module also contained a smaller scintillation counter optimised for spectroscopy. These were based upon NaI disks 76 mm thick and 127 mm in diameter and used a single PMT as the optical detector. The spectroscopic detectors covered the 10 keV to 100 MeV energy range.

Numerous TGFs have also been picked up by the gamma ray burst monitor (GBM) on board the Fermi spacecraft. The GBM uses four groups of three NaI scintillators placed around the spacecraft in order to monitor the whole sky and to determine the positions of bursts to within a few degrees. The NaI detectors are 127 mm wide disks and 12.7 mm thick, each coupled directly to an individual PMT. The BGO detectors have the same diameter but are 127 mm long and have two PMTs each. The NaI scintillators also provide spectroscopic measurements between about 8 keV and 1 MeV and are supplemented by two BGO scintillators to extend the range up to 40 MeV.

3.2.4 Microchannel Plates and Related Instruments

See Section 4.2.5 for additional details and there is some background material in Section 1.4.3.

Microchannel plates (MCPs) are widely used for EUV and x-ray imaging with telescopes or collimators (Section 3.3) providing the images. There is almost no difference in the design or operation of an MCP used for high-energy photon detection from one used for high-energy charged particle detection (Section 4.2.5) except for the electron emitting coating. High-energy photon detecting MCPs generally use CsI as the photoemitter.

The closely related PMT (in terms of its principle of operation – Section 1.4.3.3) is now rarely used as a primary detector, but is still widely used as the optical detector for scintillation counters (Section 3.2.3).

MCPs will usually be able to detect photons over a wide range of energies. They generally therefore need screening from unwanted radiation and also from charged particles. Thus an MCP used for detecting x-rays will often be screened from UV and EUV photons by a filter comprising a thin layer of aluminium coated onto a plastic film. The use of large area plastic scintillators or other detectors as active shields (Section 3.2.3) may also be required.

The wide field camera (WFC) on the ROSAT spacecraft used a pair of chevron MCPs as its detectors. The photoemitter was CsI and they had 12.5 μm channels at a 15 μm pitch over a 45 mm diameter area. Their spatial resolution was about a factor of two better than that of the image produced by the telescope. The MCPs were curved to match the focal plane of the telescope. Six filters could be used, composed variously of aluminium, beryllium, carbon and tin coated on Lexan™ films, enabling various energy bands from 17 to 24 eV at the lowest energies to 90–210 eV at the highest to be selected.

Seven MCPs were used as detectors on the Extreme Ultraviolet Explorer (EUVE) spacecraft which *inter alia* observed the collision of comet Shoemaker–Levy with Jupiter and may have detected helium on Mars. The instruments were four glancing incidence imaging telescopes (Section 3.3) and three spectroscopes (Section 3.4). For the highest energy EUV and soft x-ray photons, two of the telescopes fed separate MCPs and covered the photon energy band from 34 to 280 eV.

The Einstein spacecraft's high-resolution imager (HRI) utilised two MCPs and was fed by glancing incidence mirrors (Section 3.3) with a 2″ angular resolution. The first MCP used a magnesium-fluoride photoemitter and fed the second MCP. Some 5×10^7 electrons were then collected by a wire grid anode for each incident x-ray. The detectors were protected from UV photons and ions by an aluminium coated polymer film. The instrument had no intrinsic spectral resolution, but some could be provided within the 150 eV–3 keV detection range of the MCPs by the use of interchangeable filters and objective gratings.[*]

JAXA's Nozomi Martian orbiter, although it failed to reach Mars, was carrying the XUV instrument to detect the neutral and ionised helium emissions from the Martian atmosphere and ionosphere within the 10–60 eV photon energy range. The detector was a three-stage MCP fed by a normal incidence 60 mm f1.2 multi-coated mirror (Section 3.3). The first stage used a CsI photoemitter and the overall gain was about 10^7. Aluminium and carbon coated polymer films shielded the detectors from lower energy photons.

The Chandra X-Ray Observatory, although primarily concerned with observing objects outside the solar system, did, in July 2000< observe the x-rays coming from comet C/1999 S4 (LINEAR[†]) and more recently from comets C/2012 S1 (ISON) and C/2011 L4 (PanSTARRS[‡]). The x-rays were coming from nitrogen and oxygen ions via the SWCX mechanism (Section 3.1.2). Rather surprisingly, in 2015, Chandra also detected a few 300–600 eV x-ray photons coming from Pluto. The origin of these x-rays is uncertain but may also be from nitrogen via the SWCX mechanism. Two MCPs are used as the detectors for the eight glancing incidence mirrors of the HRI giving it an angular resolution of 0.5″. Each MCP is 100 mm square and has 6.9×10^7 10 μ wide channels. The gain is 3×10^7 and the electrons are collected by a wire grid. The MCPs are stacked in a chevron formation.

3.2.5 Solid State Detectors and Related Instruments

See also Sections 1.4 and 4.2.6.

Solid state detectors can be used for both x-ray and γ ray detection. Their operational principles are based either upon detecting the electron–positron pairs produced by γ rays interacting with nuclei of (for example) tungsten or the detection of the electrons from electron–hole pairs produced by x- and γ rays in other materials (especially silicon and germanium). Like all the detectors in this section, they can mostly also be used for the detection of high-energy charged particles and in some cases for the detection of optical

---

[*] A very coarse grating placed in front of the telescope's objective. Each point source image is replaced by two short spectra (the first order interference fringes – see Section 1.3).

[†] Lincoln Near Earth Asteroid Research.

[‡] Panoramic survey telescope and rapid response system.

photons. A review of the band structure of solids and of semiconductors is given in Section 4.2.6.1 which may also be helpful to some readers for this section as well.

3.2.5.1 Charge-Coupled Devices

CCDs (Section 1.4.1.2) are directly sensitive to x-rays in the same manner that they detect EOR photons, since the x-rays can penetrate into their silicon substrates. The higher energy of x-ray photons when compared with EOR photons, however, means that each photon will usually produce many electron–hole pairs.

NASA's Swift spacecraft has detected EUV and x-ray emissions from several comets. The XRT was used in 2009 to observe comet C/2007 N3 (Lulin) and in 2005 to observe the x-ray emissions from comet 9P/Tempel-1 following the collision of the Deep Impact spacecraft with that comet's nucleus. The images, produced by twelve nested glancing incidence telescopes (Section 3.3), are detected by a 600 × 602 pixel frame-transfer CCD which operates over the 200 eV–10 keV region. The energy resolution ranges from 60 eV at 200 eV to 200 eV at 10 keV. An aluminium coated polyimide film shields the CCD from UV and visible light. The pixel size of 40 μm corresponds to an angle of 2.4″. Swift's ultraviolet and optical telescope (UVOT)< which uses a 300 mm Ritchey–Chrétien telescope feeding an MCP-intensified CCD array, is aligned with the XRT and has also been used for cometary imaging, detecting water via the emissions from OH molecules.

The Hitomi spacecraft's primary mission was* to study the wider universe in x-rays and soft γ rays but doubtless it would also have been used for solar system studies on an opportunistic basis. It carried four glancing incidence telescopes (Section 3.3) and four detectors. The soft x-ray imager (SXI) used four large CCDs and was fed by the 450 mm, f/12.4 SXT-I soft x-ray telescope. The other instruments also used solid state detectors: the hard x-ray imager and soft gamma ray detector (Section 3.2.5.2) and the soft x-ray spectrometer (Section 3.2.5.4). The SXI operated over the 400 eV to 12 keV photon energy range with an energy resolution of <200 eV at 6 keV and an angular resolution of <1.3′. Its field of view was 38′ across.

The advanced CCD imaging spectrometer (ACIS) was used for Chandra's observations of comet C/1999 S4 LINEAR. ACIS can acquire high angular resolution images and moderate spectral resolution spectra simultaneously using ten 1024 × 1024 CCD arrays. Four of the CCDs are used exclusively for imaging; the other six can be used for imaging or spectroscopy. Two of the latter CCDs have been thinned and are back illuminated giving them responses to lower energy photons and better energy resolutions than the front illuminated CCDs. The front-illuminated CCDs respond to photons with energies in the range 700 eV and 11 keV, while the back-illuminated CCDs respond between 400 eV and 10 keV. The energy resolution varies considerably over the CCDs: from ~80 to ~200 eV at 1.5 keV and from ~150 to ~500 eV at 5.9 keV. The angular resolution is 2″ at 1.5 keV and 2.5″ at 6.4 keV. Aluminised polymer films are used to shield the detectors from lower energy photons.

The Soft X-ray imaging Telescope (SXT) on Astrosat has a glancing incidence telescope with 41 concentric shells (Section 3.3) to form images within its 300 eV to 8 keV region. Its field of view is ~41′ and its angular resolution ~3′. Images are obtained by a 600 × 600 pixel

---

* The spacecraft broke up 5 weeks after launch.

frame-transfer CCD (Section 1.4.1.2) that registers single photons so that the energy of each x-ray can be measured.

A non-imaging variant of the CCD, known as a swept charge device (SCD), has recently found application in space missions. SCDs have the advantages of high sensitivity, rapid read-out, good spectral resolution and not needing active cooling. The SCD differs from an imaging CCD in the way in which the accumulated charges are read out. The charge transfer channels run parallel to two of the sides of the device, meeting at a diagonal central charge transfer line (Figure 3.3a). The elements* of the SCD are formed by subdivisions running parallel to the other diagonal. The detection and charge transfer processes are the same as for a CCD. The output is obtained as illustrated in the sequence of drawings in Figure 3.3. We may see how this is performed by following the track of an individual charge within one element (Figure 3.3a). A single standard CCD charge transfer cycle moves the charge one element closer to the diagonal central charge transfer channel (Figure 3.3b). The charge movement is parallel to the sides of the device – i.e. downwards in the example shown in Figure 3.3b or to the left for elements in the bottom right half of the device (not shown). A second clock cycle transfers the charge down to the next element (Figure 3.3c). The third clock cycle (in this example) transfers the charge to the central diagonal charge transfer channel (Figure 3.3d). Finally, two more clock cycles transfer the charge one element down the central diagonal charge transfer channel and then out to the read-out electrode (Figure 3.3e and f).

Thus far, the device probably seems similar to a CCD, except for the read-out geometry. However, as well as the element just considered being read out in five clock cycles, many other elements will also be read out in five cycles and their charges will simply combine together. The 'five-cycle' elements lie on the same diagonal† as the one first considered (Figure 3.3g).‡ Other elements may be read out in two, three, four, . . ., many clock cycles and lie on similar diagonals (Figure 3.3i). Thus the first charge emerging from the read-out electrode will be formed from the charges in two of the original elements, the second, from the charges in four original elements, the third, from the charges in six original elements – and so on. There is no way to identify the measurement made by a single element, so these emerging charges are then, in their turn, added together to give a single measurement for the intensity of the exposure over the whole area of the SCD.

The first SCD to be used for a space mission was the D-CIXS§ on the SMART-1 lunar orbiter. This detected the presence of aluminium, magnesium and silicon in the lunar surface via x-ray fluorescence induced by solar x-rays over the 500 eV to 10 keV spectral range. The D-CIXS made the first x-ray fluorescence measurements of the lunar surface since the Apollo 15 and 16 missions and the first with a global coverage.

The detectors were 24 SCDs with a spectral resolution of 200 eV. The detectors were grouped in three sets of eight and used miniature gold/copper honeycomb collimators

---

* These are not called pixels because an image is not generated by the device.
† If you try checking this – remember that the first movement is parallel to one of the sides of the device and then down the diagonal.
‡ The name of the device originates from the way in which the charges from numerous elements are thus 'swept' together.
§ Demonstration of a compact imaging x-ray spectrometer.

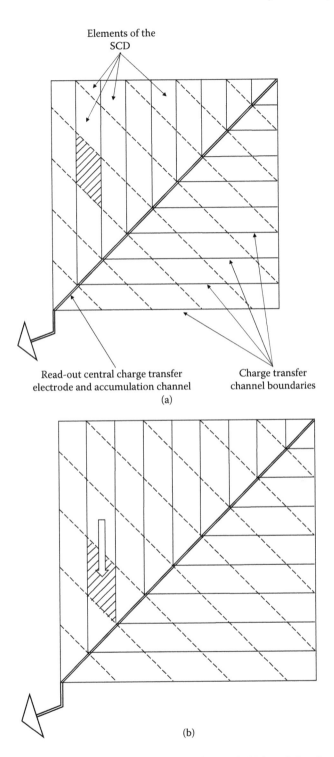

FIGURE 3.3 The read-out sequence of a swept charge device (SCD) and the elements contributing to each output charge package. (a) Charge stored in shaded SCD element. Clock cycle = 0. (b) Charge stored in originally shaded SCD element has transferred down to the next element. Clock cycle = 1.

(Continued)

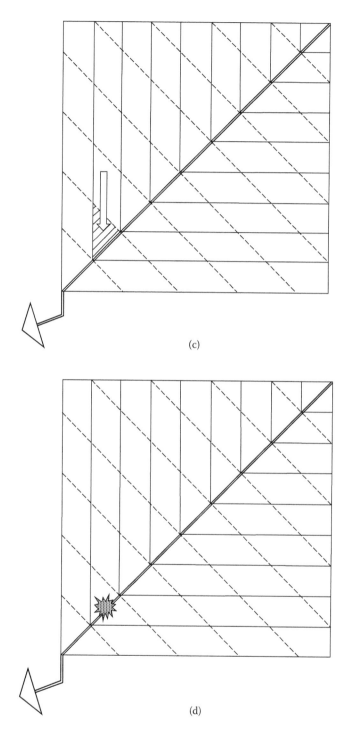

(c)

(d)

FIGURE 3.3 (*Continued*) The read-out sequence of a swept charge device (SCD) and the elements contributing to each output charge package. (c) Charge stored in originally shaded SCD element has transferred down one more element. Clock cycle = 2. (d) Charge stored in originally shaded SCD element has transferred to the central charge transfer and accumulation channel. Clock cycle = 3.

(Continued)

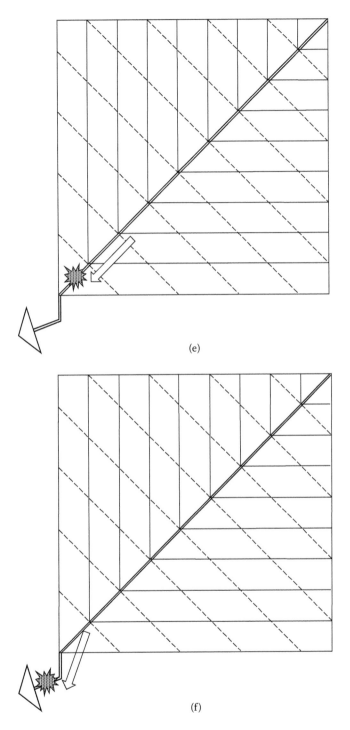

(e)

(f)

FIGURE 3.3 (*Continued*) The read-out sequence of a swept charge device (SCD) and the elements contributing to each output charge package. (e) Charge stored in originally shaded SCD element has transferred one stage down the central charge transfer and accumulation channel. Clock cycle = 4. (f) Charge stored in originally shaded SCD element has transferred to the read-out electrode. Clock cycle = 5.

(*Continued*)

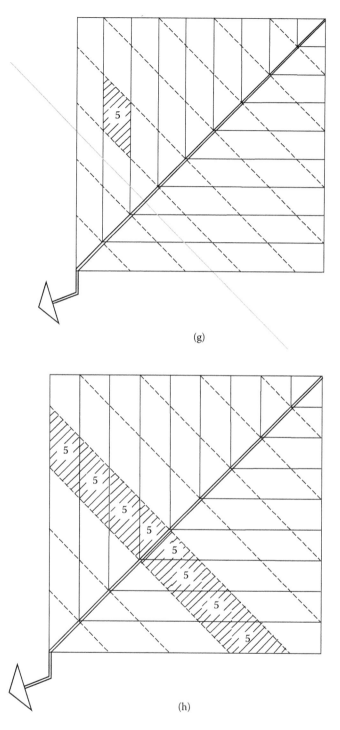

(g)

(h)

FIGURE 3.3 *(Continued)* The read-out sequence of a swept charge device (SCD) and the elements contributing to each output charge package. (g) Charge stored in originally shaded SCD element has taken five clock cycles to transfer to the read-out electrode. (h) The charges in all the elements shown along the shaded diagonal will reach the read-out electrode simultaneously after five clock cycles.

(Continued)

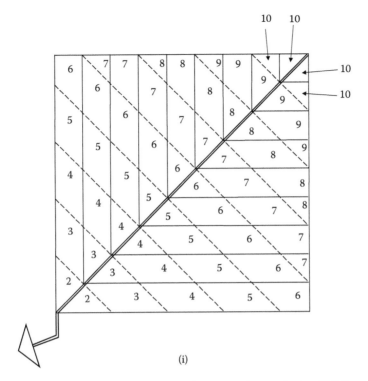

(i)

FIGURE 3.3 *(Continued)* The read-out sequence of a swept charge device (SCD) and the elements contributing to each output charge package. (i) The charges in all the elements along each diagonal will reach the read-out electrode simultaneously and ranging from two clock cycles (bottom left) to ten clock cycles (top right).

(Section 3.3) to give fields of view of 8° and 12° thus giving a best ground resolution of ~40 km. Each detector comprised 575 three-phase polysilicon electrodes and had a total area of 100 mm². A separate instrument was used to monitor the solar x-ray intensity and the D-CIXS measurements were calibrated* by using surface samples returned by the Luna-20 and 24 missions.

The Chandrayaan-1 spacecraft was India's first mission to the Moon. It carried an array of 24 SCDs, known as the Chandrayaan-1 X-ray Spectrometer (C1XS), also to map element abundances on the lunar surface by x-ray fluorescence. It covered the spectral range, 800 eV to 20 keV, with a spectral resolution of ~150 eV at 5.9 keV. Passive cooling kept the devices' temperatures to within the range from −10 °C to +5 °C (cf. a typical operating temperature for a 'normal' CCD would be −70 °C to −50 °C). The best ground resolution was about 25 km.

The Indian Space Research Organisation's (ISRO's) Chandrayaan-2 mission is planned, at the time of writing, for a 2018 launch into a polar orbit around the Moon and will comprise a lunar orbiter, lander and rover. The orbiter will carry the Chandrayaan Large Area

---

* Also known as ground truth samples (Section 7.2.7).

Soft x-ray Spectrometer (CLASS) based upon SCDs; again the objective of the instrument will be mapping the lunar surface elemental composition. CLASS is a development of the C1XS instrument using sixteen 20 mm × 20 mm passively cooled SCDs. The SCDs will be formed into four 2 × 2 arrays. Miniature honeycomb collimators will restrict the field of view to 14° – a ground resolution of ~40 km from an orbital height of 200 km. A quarter of the array, however, will be equipped with a finer collimator and provide a best ground resolution of 12 km. The SCDs will cover the spectral range from 800 eV to 6.5 keV with a resolution better than 200 eV at 6 keV. The detectors will be shielded from high-energy particles by beryllium foils and be covered by a door when the spacecraft passes through the Earth's radiation (van Allen) belts.

3.2.5.2 Semiconductor-Based Detectors

CCDs and SCDs could also have been included in this section, but were deemed of sufficient significance to be dealt with separately (Section 3.2.5.1).

3.2.5.2.1 Germanium-Based Detectors   The MESSENGER spacecraft detected γ rays from Mercury's surface using what was essentially a solid ion chamber (Section 4.2.2). The gamma ray and neutron spectrometer (GRNS) comprised a 50 mm × 50 mm cylinder of very pure n-type germanium kept by a Stirling engine to cooler than about 90 K. A triple layer thermal shield was also needed to help maintain that temperature at the spacecraft's distance of about 58 million kilometres from the Sun. A potential difference of 3.2 kV between the cathode and anode caused the electrons to drift towards the latter, where they were detected. The instrument was intrinsically sensitive to the γ rays' energies since the number of electron–hole pairs produced within the germanium block by a photon was proportional to the energy of that photon. It detected γ rays in the 100 keV to 10 MeV energy range produced by cosmic-ray interactions and natural radioactivity in the topmost layers of Mercury's surface (Section 3.1.2) in order to map element abundances. The energy resolution ranged from 3.3 keV at 1.2 keV to 4.8 keV at 6.1 MeV. The detector had an active shield of a boron-infused plastic scintillator viewed by a PMT and that shield could also be used to detect neutrons (Chapter 6).

The Mars Odyssey orbiter also carries a germanium-based γ ray detector similar to, but larger than MESSENGER's GRNS, for the purpose of mapping surface element abundances on Mars. The Odyssey GRS is based around a 1.2 kg cylinder of germanium with a 3 kV potential difference between cathode and anode. Gamma rays between 300 keV and 8 MeV energy are detectable with an energy resolution ranging between ~4 and ~10 keV. The instrument was deployed, once in Mars orbit, at the end of a 6 m long boom in order to reduce the possibility of false detections arising from radioactive emissions from the spacecraft. The instrument has a spatial resolution of about 300 km and its results suggest the presence of iron sulphates containing chemically bound water in the top 0.5 m of the surface.

The planned mission to the metal-rich asteroid, 16 Psyche, would include a germanium GRS with a boron plastic AC shield. It would have an energy resolution of 4 keV at 1.3 MeV and aim to measure the abundances of aluminium, calcium, iron, nickel, potassium, silicon, sulphur, thorium and uranium. The design of this instrument will also be an updated version of MESSENGER's GRNS.

3.2.5.2.2 Silicon Strip Detectors   See also Section 4.2.6.2. As mentioned in Section 3.2.3, the AGILE spacecraft had its programming changed in 2015 to enhance its detection of TGFs. In addition to the MCAL mentioned there and in Section 3.2.5.3, the spacecraft carries a hard x-ray imaging detector (Super-AGILE) and a gamma ray imaging detector (GRID). All the instruments are protected by the active scintillator shield discussed in Section 3.2.3.

Super-AGILE covers the photon energy range 18–60 keV with an energy resolution of ~4 keV and an angular resolution of 6′. It uses four tungsten one-dimensional (1D) coded masks (Section 3.3) placed 140 mm above the detector plane to obtain its images. A honeycomb collimator (Section 3.3) lies between the coded masks and the detector plane. Sixteen silicon microstrip detectors (Section 4.2.6.2) form the detector plane formed into four 2 × 2 arrays and electron–hole pair production by the x-rays is the basis of the detection mechanism.

GRID operates between 30 MeV and 50 GeV with a best energy resolution of ~8 keV and an angular resolution of between 1.2° and 3.5°. Its main purpose, however, is imaging rather than spectroscopy. Its detectors are silicon microstrip particle detectors (Section 4.2.6.2) and these pick up the electron–positron pairs produced by the γ rays within layers of tungsten. The microstrip detectors are stacked in 12 layers with the first 10 being sandwiched between 245 μm thick tungsten sheets. Each layer in the stack contains sixteen 95 mm square individual detector arrays with the strips at a pitch of 121 μm.

3.2.5.2.3 Cadmium-Zinc-Telluride Detectors   Cadmium-zinc-telluride (CZT, also CsZnTe) detectors are formed by alloying cadmium telluride and zinc telluride. CZT is a semiconductor with a band gap of between 1.4 and 2.2 eV depending upon the ratio of the two compounds (cf. HgCdTe detectors – Section 1.4.1.2). They operate like CCDs (Section 3.2.5.1 and Figure 1.19), with a similar physical structure, but without the charge-coupling. They thus store the accumulated electrons from the electron–hole pairs produced by the high-energy photons below the anodes that form the pixels and these are then read out individually. They have a high sensitivity to x- and γ rays because both cadmium and tellurium have high atomic numbers. They can also be operated without cooling and have a good energy resolution over their main spectral range from 10 keV to 1 MeV.

The Burst Alert Telescope on the Swift spacecraft uses 32,768 CZT pixels in 256 modules of 128 pixels, allied to a 52,000 tile coded mask (Section 3.3) to make detections within the 15–150 keV energy band. It observes nearly 25% of the whole sky at a time and pinpoints sources to an accuracy of better than about 4′.

Astrosat's Cadmium-Zinc-Telluride Imager (CZTI) instrument operates over the 10–150 keV energy band using a tantalum coded mask (Section 3.3) to obtain its images. Its energy resolution is about 1.2 keV at 60 keV. Each CZT pixel is 2.5 mm^2 and they are formed into 16×16 modules. Sixteen modules are then grouped into a 4×4 quadrant and finally the four quadrants made up into a square some 300 mm across. The CZTI can also act as a polarimeter for photon energies over 100 keV. The instrument detects high-energy charged particles and these signals are separable from those produced by x-rays the multiple simultaneous detections of the former and the single detection events of the latter. The CZTI has a field of view ranging from 6° to 17° for photons with energies below and above 100 keV, respectively and an angular resolution of 8′.

3.2.5.3 Calorimeter
See also Section 4.2.8.

A calorimeter is an instrument designed to measure the total energy of a high-energy photon (or charged particle). It does this by ensuring that the photon and its interaction products are completely absorbed within the detector and then measuring the total energy deposited into the detector. Usually, this is achieved by rejecting any detections in which the photon and/or some of its interaction products escape from the detector. The loss of some of the photon's energy by escape from the detector is detected using an active shield that surrounds the detector.

Mostly for high-energy photon detection, calorimeters are based upon heavy-element absorbers (converters) that first convert the photons' energies into electron–positron pairs and then scintillators that detect the resulting free–free radiation from those charged particles.

Some calorimeters, however, operate by detecting the small change in temperature of the detector arising from the energy input from the photon. Since, even for very low mass absorbers, the temperature change is small, the absorber has to be cooled to well below 1 K, so that the change is detectable. Such low temperatures are difficult to attain on board spacecraft, nonetheless, the Susaku spacecraft* carried a micro-calorimeter comprising 32 mercury telluride absorbers with silicon thermistors to detect the temperature changes and cooled to 60 mK. The instrument briefly carried out spectroscopy over the 300 eV to 12 keV energy band with a resolution of ~7 eV. ESA's Advanced Telescope for High-Energy Astrophysics (ATHENA[†]), currently planned for a 2028 launch, at one time was expected to use a micro-calorimeter array for x-ray detection but this has now (probably) been superseded.

γ rays from the Moon were first detected (Section 3.1.1.2) by the EGRET instrument on board the Compton Gamma Ray Observatory. EGRET covered the 30 MeV to 30 GeV energy range with an energy resolution of 150 MeV at 1 GeV and an angular resolution of 1.7° at the same energy. The instrument was in two sections. The first section was a spark

---

* A problem with the cooling system caused the reservoir of liquid helium to boil off by August 2005. The x-ray spectrometer, for which the micro-calorimeter was the detector, therefore ceased to function after this date. Susaku is Japanese for 'Vermillion (or red) Bird'; it is a sky area in oriental astrology divided into seven mansions that (very roughly) correspond to the modern constellations of Gemini, Cancer, Hydra, Crater and Corvus.

† Developed from the now cancelled Constellation-X, XEUS (ESA – X-ray Evolving Universe Spectroscopy) and IXO (NASA – International X-ray Observatory) missions.

chamber* with a stack of 27 thin layers of tantalum in which the γ rays were converted to electron–positron pairs. Between the tantalum layers were wire grids with 992 wires in each of the *x* (cathode) and *y* (anode) directions. The sparks flowed between the wires and were detected by PMTs. Triggering of the spark chamber was accomplished by the detection of the passage of a photon through two 4 × 4 arrays of plastic scintillator tiles. The chamber had a neon/argon/ethane gas fill which could be replenished occasionally. EGRET's second section was the calorimeter and this was a 200 mm thick, 760 mm diameter disk of NaI (Tl) scintillator crystals. The optical detectors for the scintillators were an array of 16 PMTs. A second, much simpler, spark chamber was used to eliminate those photons entering the instrument from below and the whole thing was enclosed in a dome of plastic scintillator as a shield.

The MCAL (Section 3.2.3) on the AGILE spacecraft covers the energy range from 350 keV to 50 MeV with a resolution of 170 keV at 1.3 MeV and it works in tandem with GRID (Section 3.2.5.2.2) to detect TGFs. The positrons and electrons emerging from GRID deposit their energies within MCAL to add spectroscopic information about the incoming γ rays to GRID's positional data. MCAL can also operate as a stand-alone non-imaging detector over the 300 keV to 100 MeV photon energy range. MCAL uses thirty 375 mm long CsI bars arranged in two orthogonal layers. Each bar has photodiodes at its ends to detect the particles' scintillations.

3.3 IMAGERS

The penetrating nature of x- and γ radiation means that the conventional reflective telescope and camera designs discussed in Section 1.2 have to be extensively modified for use with high-energy photons and that refractive designs cannot be used at all. For the EUV and soft x-ray regions, reflective surfaces can be formed from multiple layers of elements such as molybdenum and silicon and relatively conventional telescope designs utilised. X-rays up to about 100 keV in energy can also be reflected, but only if their angles of incidence to the reflecting surface are very shallow – glancing or grazing incidence. A somewhat incomplete analogy would be the way stones may be skipped (reflected) if thrown at a water surface from a low angle, whereas they would fall straight to the bottom if simply dropped in. At energies above about 100 keV, images have to be produced by scanning with a restricted the field of view (collimating/pin-hole camera – Section 3.3.3) and/or by obscuring a part of the field of view in a more complex (coded mask – Section 3.3.4) way, the latter also usually combined with scanning and/or changing the masks.

Images, of a sort, can be built up with any high-energy photon detector by observing the target being occulted by a nearer object. Point sources will disappear (or reappear)

---

* A spark chamber is closely related to the Geiger counter (Section 4.2.1) except that, usually, there are many electrodes in the form of a stack of flat sheets or wire grids with small separations and the series of sparks induced between the layers by the passage of the photon or particle allows its path to be followed. Positional detection of the sparks is by direct observation of the Townsend avalanches (sparks) by photomultipliers, etc. Spark chambers are now mostly just of historical interest, but their low power requirements, since the electrodes are only charged to their operating voltage immediately after the passage of a photon or particle has occurred, mean that they may still have uses for spacecraft-borne instrumentation.

almost instantaneously.* Extended sources will disappear over a period of time and the changes in their intensity with time may sometimes be used to form a map of the target's structure. Occultations by the Earth as seen by the Imager on Board the Integral Satellite (IBIS) instrument on board the INTEGRAL have been used in this way to determine the structure of extra-solar system sources.

3.3.1 Multi-Layer Reflectors

Mirrors reflecting at near-normal incidence and based upon Bragg reflection (Section 1.3.5, Equation 1.22) can operate with reasonable (~50%) efficiency up to photon energies of about 15 eV. The layers are made alternatively from heavy and light elements, or their compounds, that have slightly differing refractive indices. One layer acts as the spacer, the other as the Bragg scatterer/reflector. Commonly used combinations are Si/Mo, Si/Mo$_2$C, Y/Mo and SiC/Mg.

The thicknesses of the layers are determined by the Bragg condition (Equation 1.22) so that the reflectivity drops off rapidly away from the optimised wavelength, since for near-normal incidence ($\phi \approx 90°$) in Equation 1.22, the layer thickness for first-order Bragg reflection is half the wavelength that it is desired to reflect (44 nm at 15 eV). Thus a wider reflected bandwidth can be obtained by varying the thicknesses of the layers, with the thicker layers at the top, reflecting longer wavelengths, and the thinner layers lower down, reflecting the shorter wavelengths. Hundreds of layers are usually required for the final mirror. A combination of multi-layer coatings and glancing incidence (Section 3.3.2) may in the future enable photons up to 300 keV or so to be reflected.

Multi-layer reflective coatings could be used in any of the reflective telescope and camera designs discussed in Section 1.2; however, their low reflectivity (by optical standards) favours prime focus designs using just a single reflection.

So far, multi-layer reflective instruments have all been targeted at the Sun. The Hinode spacecraft, for example, carries the extreme UV imaging spectrometer (EIS) which operates between 4 and 7 eV. It uses a 150 mm off-axis paraboloidal Si/Mo multi-layer coated mirror feeding a toroidal multi-layer coated reflection grating and has a CCD detector.

3.3.2 Glancing Incidence Reflectors

As mentioned in the introduction to this section, x-ray photons with energies up to about 100 keV may be reflected if they hit a surface at a very shallow angle. The angle does, however, have to be *very* shallow – about a third of a degree to the plane of the reflecting surface for 12 keV photons incident onto a gold-plated surface, for example. Nonetheless XRTs based upon glancing incidence have been designed and built. Several designs, including the one widely used in practice, are due to Hans Wolter. In 1952, Wolter produced three two-mirror designs for glancing incidence telescopes that had reasonably wide fields of view due to their being free from spherical aberration. The initial designs, however, still suffered from coma, but this could be eliminated by small (<1 μm) corrections to the mirror surfaces, resulting in what are now called the Wolter-Schwarzschild designs, although curvature of the focal 'plane' still remains.

---

* At optical wavelengths the Fresnel diffraction fringes around the shadow of the object causes the disappearance of point sources to last up to a second, but the fringes will not be observable for these much shorter wavelength photons.

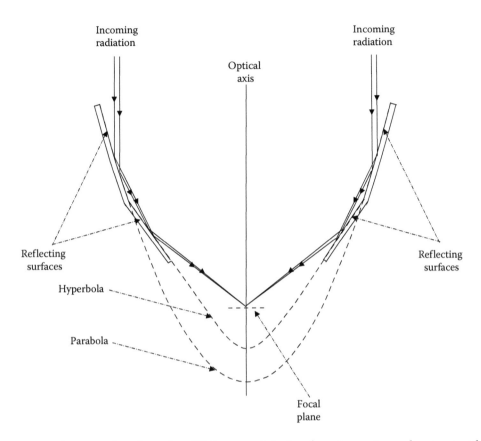

FIGURE 3.4 Cross section through a Wolter type-I design for a two-mirror glancing incidence x-ray telescope. NB: for clarity, the angles of the ray paths to the mirror surfaces are shown as much larger than they would be in practice.

3.3.2.1 Macroscopic Systems

The first of the designs* – Wolter type-I – has a thin concave parabolic primary mirror and a thin concave hyperbolic secondary mirror (Figure 3.4). Both mirrors are in the form of annular rings taken far from the central axes of their conic sections and thus resemble cylinders or the basal segments of narrow angle cones. It is the most widely used design for spacecraft instruments because of the rugged supports possible for the mirrors and the ease of nesting many confocal such telescopes, one inside another, in order to build up the collecting area.

The mirrors are thin shells constructed in some cases by evaporating firstly the support layer of nickel onto a mandrel whose shape is the inverse of that required for the final mirror and then the reflecting layer of gold, platinum, etc. onto the nickel. In other cases, the support may be a thin ceramic or glass shell figured to the required shape. The reflecting surface has to be extremely smooth if the image quality is not to be degraded and typically irregularities are kept to 300 pm or less by the

---

* The other designs are to be encountered in the laboratory. Interested readers can find further details in sources in Appendix A.

manufacturing process (cf. surface irregularities of ~50,000 pm or so for a good quality optical telescope). At the low end of the energy range, or if the highest quality images are not needed, the mirror shells can be thin aluminium foils and/or just made in the shape of cones, not conic sections.

The Einstein observatory was the first spacecraft to obtain high-resolution x-ray images and it had two orders of magnitude higher sensitivity than previous missions. It used a glancing incidence telescope based on the Wolter type-I design that covered the 100 eV to 4 keV photon energy region. The telescope had four nested sets of mirrors with the outermost one having a diameter of 600 mm. The mirrors were made from fused quartz coated with a layer of nickel and had a 3.3 m focal length and an effective collecting area of 0.04 m². The angular resolution was 10″ to 15″ over a 1° field of view. Four instruments could be placed at the focus – a position-sensitive proportional counter (400 eV to 4 keV – Section 3.2.2), a chevron MCP (150 eV–3 keV, 2″ angular resolution – Section 3.2.4), an SSD based upon cooled Si/Li (500 eV to 4.5 keV – Section 3.2.5) and a Bragg crystal spectrometer (420 eV to 2.6 keV, energy resolution ~400 eV – Section 3.4).

ROSAT's XRT used four nested Wolter type-I pairs of mirrors with a diameter for the outer pair of 840 mm and a focal length of 2.4 m (Figure 3.5). The supporting shells were 20 mm thick Zerodur* with a gold reflective coating. It operated over the 100 eV to 2 keV energy range. Two (for redundancy) position sensitive proportional counters (Section 3.2.2) and a high-resolution chevron MCP imager (Section 3.2.4) could be moved into the focal plane of the telescope. The former had an angular resolution of 25″ at 1 keV over a 2° field of view and a spectral resolution of about 400 eV at the same energy. The latter had a field of view of 38′ and an angular resolution of 2″.

ROSAT also carried an EUV/soft x-ray WFC based upon three nested Wolter-Schwarzschild type-I mirror pairs. The diameter of the largest of the mirror sets was 576 mm and the instrument's focal length was 525 mm. The mirrors were formed from a nickel-coated aluminium substrate with a gold reflective coating. The field of view was 5° with a best angular resolution of 2.3′. The effective collecting area was 0.046 m². Two identical curved (to match the curved focal plane) chevron MCPs with CsI photoemitters were used as the detectors with eight interchangeable filters and it operated over the 20–200 eV energy band. A Geiger counter and a channel electron multiplier (Section 4.2.5.1) were used to eliminate charged particle detections and to switch off the instrument during passage through the worst regions of the Earth's radiation belts.

The XRT on the Swift spacecraft utilises 12 Wolter type-I shells which were flight spares for the JET-X instrument on the cancelled Spektr-RG† mission and have an effective collecting area of 0.01 m². Its angular resolution is 15″ at 1.5 keV over a 24′ field of view and it covers the 200 eV to 10 keV energy range with a resolution of 140 eV at 5.9 keV. Its detector is a CCD (see Section 3.2.5.1). The shells have diameters of ranging from 191 to 300 mm and their focal lengths are 3.5 m.

---

* A low-expansion ceramic material often used for optical telescope mirrors.
† Based upon the Russian words for Spectrum, Röntgen and Gamma. This mission was cancelled in 2002. Not to be confused with a second Spektr-RG spacecraft currently under construction with a possible launch date of September 2017 (Section 3.3.2.2.4). JET-X stands for Joint European Telescope for X-ray astronomy.

FIGURE 3.5 The entrance aperture of the ROSAT spacecraft showing the tops of the four nested Wolter type-1 glancing incidence mirrors. The diameter of the largest mirror is 840 mm. (Courtesy of the ROSAT mission and the Max-Planck-Institute für extraterrestrishe Physik.)

ESA's X-ray Multi-Mirror (XMM)-Newton spacecraft has observed x-ray emissions from comets (Figure 3.6) and from the Deep Impact collision with comet P/Tempel-1. It has three XRTs each with 58 Wolter type-I nested nickel shells. The shells are 600 mm deep, 0.4–1.2 mm thick and have gold reflective coatings. The focal length is 7.5 m, the maximum shell diameter is 700 mm and the minimum shell diameter is 306 mm. The effective collection area is 0.15 m². The surface roughness is about 500 pm and the angular resolution about 12″.

The telescopes each feed separate cameras. All the cameras are cooled to ~150 K by radiators, and have six-position filter wheels and overall fields of view of 30′ with angular resolutions between 4.5″ and 6.6″.

EPIC-1[*] is a detector sensitive to the photon energy range 150 eV to 15 keV. The detector comprises twelve 64 × 189 pixel individual rear-illuminated CCD arrays arranged in a 378 × 384 overall array.

---

[*] European Photon-Imaging Camera.

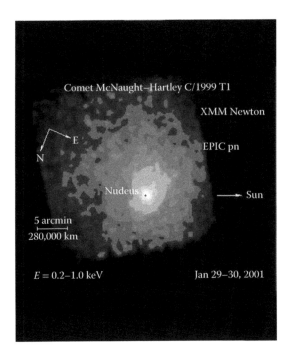

FIGURE 3.6   An x-ray image of comet C/1999 T1 (McNaught–Hartley) obtained over the 200 eV to 1 keV photon energy region by the XMM-Spacecraft in 2001. (European Photon-Imaging Camera [EPIC]. Courtesy of ESA.)

EPIC-2 and EPIC-3 use seven 600×600 pixel frame-transfer metal-oxide semiconductor CCDs (MOS-CCDs) arranged as a single central array surrounded by the other six arrays. They have good detection efficiency over the 200 eV to 10 keV energy range. They are front-illuminated and have a partially open structure to aid the detection of the lower energy photons. The telescopes feeding these two detectors also have objective gratings (Section 3.4) and the spectra are detected at separate foci by separate 384×1024 pixel MOS-CCD arrays. Forty percent of the photons are diverted into the spectra and 50% go to the direct image. The spectrometers cover the energy range from 350 eV to 2.5 keV with a resolution ranging from 0.4 eV at 350–4 eV at 1 keV.

3.3.2.2 Microscopic Systems

3.3.2.2.1 Lobster-Eye Imagers   Images are produced within the eyes of most species of life by refraction. However, a few examples are known whereby the images result from reflection – most notably for lobsters and related species. The lobster's eye is made up from many tiny square tubes arranged like the squares on a sheet of graph paper and curved into a portion of a sphere. Light entering a tube either passes straight through or is reflected one or more times off the (shiny) walls of the tube. With suitable geometry for the structure, many of the light beams emerging from the tubes will be concentrated towards a single point (i.e. brought to a focus) at half the radius of curvature of the sphere. A number of beams, however, will go in other directions and so there is also a bright, unfocussed background to the images.

In 1979, Roger Angel realised that the structure of the lobster's eye would also work for imaging EUV photons and x-rays since the reflections are at low angles. The x-ray lobster-eye imager (or collimator) closely resembles the MCP (Sections 1.4.3.2 and 4.2.5). Essentially an MCP disk is manufactured with square holes in a rectangular grid. No photoemitter or electrodes are used, but the walls of the tubes sometimes have a reflective coating added. The disk is then heated and bent (slumped) over a mandrel to take up the required spherical shape. A spherically shaped detector with half the radius of the tubes' sphere will then pick up the images. Potentially, very wide field instruments can be based upon lobster-eye imagers.

The first lobster-eye x-ray imager was carried on a sub-orbital mission by a Black Brandt IX sounding rocket in December 2012 to demonstrate wide angle x-ray imaging. The tubes' grid had a radius of curvature of 750 mm and the detector was a chevron MCP. Developments of that instrument may be flown in the near future as 'CubeSats' with the aim of studying SWCX emissions.

3.3.2.2.2 Micropore Imagers   Significant improvements on the lobster-eye imager can be made by doubling it up so that it has the Wolter type-I optical design (or a close approximation to it). Two lobster-eye sets of tubes are stacked with the second one having a third of the radius of curvature than the first (Figure 3.7). The focus is then a quarter of its radius of curvature inside the outermost set of tubes.

Ideally, the telescope's individual holes would have a parabolic longitudinal shape in the first section and a hyperbolic longitudinal shape in the second section. They should be increasing angled away from the centre of the device so that all the holes feed the same focal point and be arranged in a symmetrical radial fashion to provide two–dimensional (2D) imaging. Mostly, in practice, the profile of the holes will be a simple straight tube or be the base of a narrow-angle cone. Nickel or iridium reflective coatings may sometimes be added to the insides of the tubes. Typically, the holes are a few microns wide, a millimetre or so long and the substrate is glass.

With a micropore imager, XRTs are similar in their layout to optical refractors. The micropore disk replaces the objective lens and an appropriate x-ray detector takes the place of the camera or eye. This is a much more robust system than the complex nests of large, thin and delicate shells discussed in Section 3.3.2.1 and generally more suited to withstanding the rigours of being launched into space.

Bepi-Colombo, ESA's orbiter mission to Mercury, is due for launch in October 2018. Its mercury imaging x-ray spectrometer (MIXS), intended to study x-ray fluorescence, comprises an imaging telescope and a non-imaging collimator. The latter (MIXS-C) is a lobster-eye imager, but with the tube lengths optimised for the straight-through radiation, rather than the lower energy focussed radiation.

The telescope (MIXS-T) will be the first high-resolution XRT intended specifically for planetary observations. It will use a conical approximation to the Wolter type-I design with a 210 mm diameter and a 1 m focal length. The micropore optics will be in the form of 36 square arrays in each set. The outer ('parabolic') set will have a radius of 4 m and the inner ('hyperbolic') set, a radius of 1.33 m. It will have an angular resolution of better than 9′ over a 1.1° field of view and cover the energy range 500 eV to 7.5 keV with a resolution of 100 eV at 1 keV.

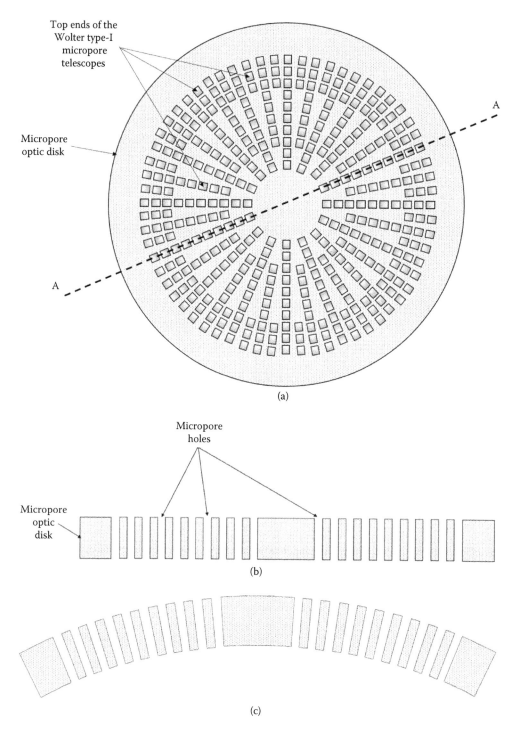

FIGURE 3.7 Micropore x-ray imager – (a) view from above the entrance apertures. (NB – There would be millions of tubes in a practice.) (b) Cross section through one of the micropore disks before slumping. (c) Cross section through one of the micropore disks after slumping.

(*Continued*)

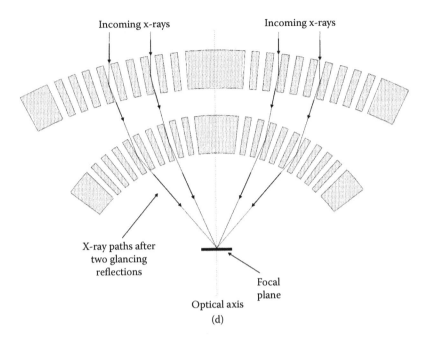

Optical axis

(d)

FIGURE 3.7 *(Continued)* Micropore x-ray imager – (d) Two slumped micropore disks acting as Wolter type-1 imagers.

3.3.2.2.3 Silicon-Pore Imagers   Silicon pore optics are essentially identical to micropore optics except that they are fabricated by large scale integrated circuit techniques and so are of superior optical quality (i.e. the holes can have parabolic and hyperbolic longitudinal profiles). Clearly, the substrate is silicon, not glass, which at the moment means that they are, like-for-like, more massive than micropore optics. This latter drawback is likely to disappear as fabrication methods improve.

The basic element of silicon-pore imagers is a square wafer of silicon a few tens of millimetres across. Channels a few hundred microns wide and a hundred or so microns deep are etched across the wafer. The channels are the equivalent of the tubes in the micropores devices. Reflective coatings of gold, iridium, platinum or tungsten are applied to the wafers. Several such wafers are stacked on top of each other and bonded together thus closing the tops of the channels and making a very rigid and robust structure. During this process, the wafers are deformed to the required parabolic or hyperbolic shapes. The final telescope will comprise hundreds or thousands of these stacks and millions of the reflecting channels.

The ATHENA spacecraft will have two telescopes and these will be silicon pore-based Wolter type-I instruments with a 12 m focal length and an effective area of ~2 m². The optics will each contain around 1½ million pores, each a millimetre or so in size, assembled into stacks 60 mm to each side.

The hard x-ray imager (HXI) will cover the 10–40 keV energy band with a resolution of 1 keV at 40 keV. The field of view will be 12′ and the pixels, at 220 μm across, correspond to an angular resolution of 4″.

The wide field imager (WFI) will cover the 100 eV to 15 keV energy band with a resolution of 70 eV at 1 keV. The field of view will be 18′ and the pixels, at 100 μm across, correspond to an angular resolution of 1.7″.

3.3.2.2.4 Polycapillary Optics*   Almost identical in construction to the micropore and silicon-pore imagers, polycapillary imagers concentrate x-rays by multiple glancing incidence reflections inside very narrow circular glass tubes. The tubes may be curved and/or tapered so that they individually focus the x-rays and then many (sometimes thousands) can be stacked so that they all concentrate their outputs into the same spot. The changes in the shapes of the tubes must be sufficiently gentle that the reflection angles remain less than the critical angle. For borosilicate glass the critical angle, θ, is given by

$$\theta \approx \frac{2\,\text{keV}}{E}\,\text{degrees} \tag{3.1}$$

for the photon energy, E, in keV. Thus θ is about 2° for 1 keV x-rays.

To date, polycapillary optics have not been used for spacecraft-borne instruments but they were considered for the Astronomical Röentgen Telescope—X-ray Concentrator (ART-XC) instrument on the Spektr-RG mission currently scheduled for an April 2018 launch. The instrument would have covered the 5–80 keV x-ray region with an energy resolution of about 2%, however the instrument has been redesigned and is now based upon conventional glancing incidence mirrors.

3.3.3 Collimators

Reflecting optics, whether of normal incidence or glancing incidence, currently enable x-ray photons up to an energy of about 50 keV to be imaged. Possibly, x-rays up to 300 keV may be imageable using both glancing reflection and multi-layer coatings together. But for energies above that (i.e. for hard x-rays and γ rays) then, for the foreseeable future, other approaches must be adopted if images are wanted.

The other approaches all involve restricting the field of view of the detector in some way, without producing genuine (focussed) images.

3.3.3.1 Pin-Hole Cameras

The simplest approach is the pin-hole camera or its equivalent. The detector is placed inside an enclosure that is opaque to the radiation being detected. The enclosure has a small hole in one side, through which the radiation can fall on to the detector. The device is scanned over the target to build up the image or it may be observed directly if a position-sensitive detector is available.

---

* Also known as Kumakhov lenses after Muradin Kumakhov who made the first such working lens in 1985. The lenses may also be used to concentrate neutrons and other neutral particles (Chapter 6).

If making the enclosure opaque by constructing it from (say) thick lead sheets would result in a device too massive for use on rockets or spacecraft, then an alternative is to make it from an active shield material, Plastic scintillator (Sections 3.2.3 and 4.2.4) is commonly used as the shield in this type of design. The scintillator takes the form of a deep tube, closed with more scintillator at one end and open at the other. Detections occurring within the main detector that are simultaneous with detections in the scintillator shield are rejected. Only those detections in the main detector which do not have corresponding detections in the shield are accepted and these are the ones that have come down the entrance aperture (open) end of the tube and been absorbed completely within the main detector.

Much the same result may be obtained by using a small(ish) scintillator in combination with a small primary detector. In this design, only simultaneous detections in the two detectors are used.

Restricting the view of a pin-hole camera to a narrow field of view, so that fine details (high resolution) of the target may be observed, requires the entrance aperture to be as far away from the detector as possible and for it to be as small as possible. It also requires an imaging detector with as high a spatial resolution as possible, or a non-imaging detector that is as small as possible. However, most hard-x-ray and γ ray sources are quite faint. Thus whilst a high resolution might be physically realisable, it will be of little practical use if only one of the required photons per day (say) makes it through to the detector. Thus angular resolution has to be traded off against the speed with which an observation may be obtained.

As mentioned in Section 3.2.2, the Polar spacecraft's PIXIE comprised two pin-hole cameras with four interchangeable sizes for the holes. The angular resolution of the cameras was ~45′, giving a ground resolution from apogee (where the exposures could be longest – height 55,000 km) of ~700 km.

The Compton Gamma Ray Observatory's BATSE instrument (Section 3.2.3) used NaI scintillator detectors in combination with plastic sheet scintillators to obtain a restricted field of view.

3.3.3.2 Honeycomb Collimators

Honeycomb collimators are essentially lots of pin-hole cameras stacked together. For a given field of view, more radiation gets through to the detector(s), enabling shorter observations to be made, or higher resolutions to be achieved for a given exposure time.

The collimators are narrow tubes of hexagonal, square or circular cross section stacked in linear, rectangular (Figure 3.8), radially symmetric (Figure 3.7) or other geometrical arrangements. They simply act to limit the field of view in the same way as a pin-hole camera does when they are placed in front of the detector(s). Images are usually built up by scanning over the target. The usefulness of honeycomb collimators is limited at high photon energies by some photons being able to penetrate the walls of the tubes and at low photon energies by some photons reflecting off the walls of the tubes. In both cases, the result is the detection of photons from directions outside the desired field of view.

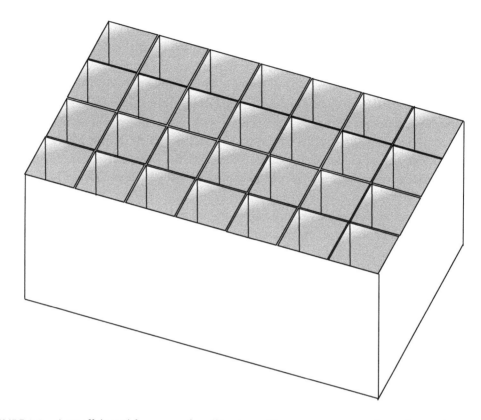

FIGURE 3.8 A small (4 × 7) honeycomb collimator with square cross-section tubes and rectangular stacking.

Unless physically large, honeycomb collimators are fabricated in much the same way as micropore optics (Section 3.3.2.2.2) and can clearly be seen to be identical to lobster-eye collimators, except that they have not been slumped into a spherical shape.

Honeycomb collimators have been used within instruments on a number of x- and γ ray missions. For example, NEAR Shoemaker XGRS detectors (Section 3.2.2) had a simple honeycomb collimator which restricted their fields of view to 5°, MESSENGER's XRS (Section 3.2.2) had a 12° field of view provided by a honeycomb collimator, the D-CIXS on SMART-1 (Section 3.2.5.1) had copper/gold collimators with fields of view of 8° and 12° and Chandrayaan-2 (Section 3.2.5.1) is expected to employ honeycomb collimators with fields of view ranging from 4° to 14°.

3.3.4 Coded Masks

Coded mask imaging differs little in its basic physical structure from a pin-hole camera. The extraction of the image from its data, though, is rather more complex. The coded mask itself comprises an unfilled 1D or 2D array of square tiles that are opaque to the radiation of interest. They are arranged in a pseudo-random pattern so that there are roughly as many areas obscured by a tile as there are clear (tile-less) areas (think of a crossword puzzle's black and white squares). The complete mask is placed in front of the position-sensitive detector(s) and shields or collimators placed in front of the mask to restrict the field of view. If a single point source lies within the field of view then a shadow which is a replica of the mask will be cast onto the detectors. In this situation, it will be simple to interpret the outputs from the detectors as being due to a single point source. If there are two point sources, they will each cast shadows and, although rather more complex, the interpretation of the detector outputs will again be interpretable in terms of two point sources together with their relative intensities and positions.

If there are many point sources and/or extended sources then it will no longer be simple to convert the outputs of the detectors to an image. However, if you have read Chapter 2 in this book, you may be getting a feeling of déjà vu at this point. This is because we have already encountered similar situations in quite different circumstances several times:

- A radio interferometer's output (Section 2.4.4.2, Figure 2.9) is easily and directly interpretable for one or two point sources. More complex targets require the Fourier transform of the output to obtain the image

- A radio spectroscope's output (Section 2.5) is easily and directly interpretable for one or two monochromatic sources. More complex spectra require the Fourier transform of the autocorrelation of the output to obtain the image

- A Fourier transform spectrometer's output (Section 2.5) is easily and directly interpretable for one or two monochromatic sources. More complex spectra require the real part of the Fourier transform of the output to obtain the image

It will thus probably not come as a surprise that to obtain the image of a complex target from a coded mask observation, the cross correlation (see footnote on page 90) of the output with the coded mask's pattern is used.

An additional complication is that the masks generally have to be physically larger than the detectors so that objects towards the edge of the field of view are still completed covered by them. This, however, means that each object projects only a part of the mask (equal to the area covered by the detectors) onto the detectors – and that it is a different part of the mask for every object in the field of view.

Mask patterns may be of various types (Figure 3.9):

- Random – which tends to have a high background noise level.

- Uniformly redundant array (URA) and modified uniformly redundant array (MURA) with segments of the array repeated across the array and sometimes different-sized tiles. This has a reduced background noise level, but also has ghost images.

- Fresnel – like a Fresnel lens.

These provide different levels of ease of manufacture, angular resolution, simplicity of computation, etc.

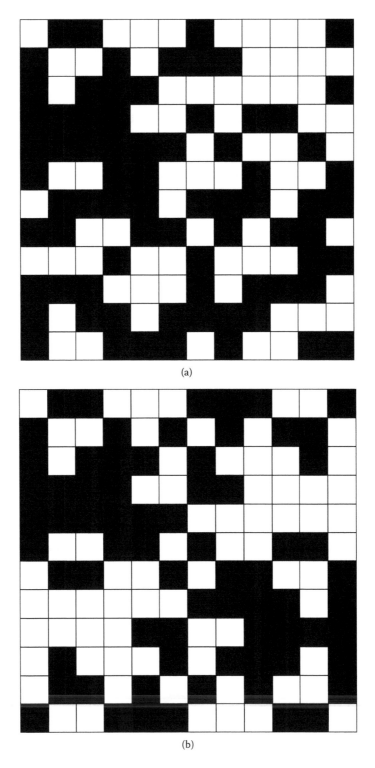

(a)

(b)

FIGURE 3.9 Schematic designs for random and uniformly redundant array (URA) coded masks. (a) A coded mask with a schematic random pattern. (b) A coded mask with a schematic uniformly redundant pattern.

Spacecraft carrying coded mask instruments include, amongst many others, INTEGRAL (IBIS, JEM-X and SPI – see also Sections 3.2.2 and 3.3), AGILE (SuperAGILE – see also Sections 3.2.5.2.2 and 3.2.5.3) and Astrosat (CZTI – see also Sections 3.2.5.2.3 and 4.2.6).

INTEGRAL's IBIS uses a MURA-type coded mask of 95 × 95 rectangular tungsten tiles, 16 mm thick and placed 3.2 m before its detectors. It has an array of 128 × 128 SSD CdTe detectors for the energy range from 15 keV to 1 MeV and an array of 64 × 64 CsI scintillators detectors for 1–10 MeV. Its angular resolution is 12′ over an 8° field of view and its spectral resolution varies from 8 keV at 100 keV to 100 keV at 1 MeV. Passive (lead) and active (BGO) shields exclude unwanted detections.

JEM-X comprises two instruments using coded masks placed 3.4 m above the detectors and operates over the 3–35 keV spectral band. The masks are identical but oriented at 180° to each other. The masks use 500 μm thick hexagonally shaped tungsten tiles in a URA pattern. Its field of view is 6.6° across with an angular resolution of 3′.

The spectrometer for INTEGRAL (SPI) also uses a URA pattern coded mask with hexagonal tungsten tiles. The mask has 127 cells, of which 63 are 68 mm thick and thus opaque to the x and γ photons. Actively cooled germanium crystals are used for its detector. It has an angular resolution of 2° over an 8° field of view and it covers the spectral range from 20 keV to 8 MeV with a resolution of 2 keV.

SuperAGILE uses 1D coded masks in a 2D array for its imaging. The masks are placed 140 mm above the detectors and formed from 117 μm thick tungsten sheets with 242 μm slots cut across them. Its field of view is in excess of one steradian (~8% of the whole sky) with an angular resolution of about 6′.

The coded mask used for the CZTI instrument on Astrosat is formed from a 500 μm thick plate of tantalum with holes cut through it to form the clear parts of the mask. Each quadrant of the whole mask is an identical 64 × 64 array of tiles and holes, rotated by 90° for each successive quadrant. The mask is followed by a square cross-section honeycomb collimator to restrict the field of view to 6° and then the CZT detectors some 480 mm below the mask. Its angular resolution is ~8′. It covers the 10–150 keV photon energy band with a resolution of 1.2 keV at 60 keV.

3.4 SPECTROSCOPES

Many of the detectors used for x- and γ rays have some intrinsic sensitivity to the energies of the photons with which they are interacting (Section 3.2). Furthermore, as we have seen (Section 3.3) designing reflective optical components for high-energy photons is no trivial matter and they are then often only 50% efficient at best. Spectroscope designs such as those considered in Sections 1.3 and 2.5, where there may be up to 10 reflections, are thus out of the question.

Some additional spectral resolution, over and above that intrinsic to the detector, may be obtained by the use of filters. These are simply disks of various materials and thicknesses that absorb photons below some specified energy, while allowing some, at least, of the higher energy photons to pass through and be detected. Alternatively, several layers of absorbers and detectors may be stacked, such as in the GRID instrument of the AGILE spacecraft (Section 3.2.5.2.2), so that only the most energetic photons reach the lowermost layers of detectors.

There are, however, more 'active' types of spectroscopes that can be used for high-energy photons based upon Bragg reflection and diffraction gratings.

3.4.1 Grating-Based Spectroscopes

The use of objective gratings in connection with obtaining spectra of high-energy photons has already been mentioned in connection with the Einstein spacecraft's HRI (Section 3.2.4) and the XMM – Newton Wolter type-I telescopes (Section 3.3.2.1). The optical principles involved are covered in Section 1.3.2.

3.4.1.1 Transmission Gratings

Transmission gratings were used in some early missions, such as the Einstein observatory and EXOSAT. These were typically fabricated from gold wires forming a parallel grid. The Einstein spacecraft had two such grids with 100 and 500 wires per millimetre that could be moved into the beam of x-rays from the XRT. In the former case, the wires were 5 μm wide and 200 pm thick at 10 μm centres and in the latter were 1 μm wide and 200 pm thick at 2 μm centres. The two first order sets of fringes formed the spectra and were imaged by the HRI. Aberrations limited the spectral resolution to less than 50. EXOSAT* also had two gratings with 1000 and 500 lines per millimetre, but on two separate Wolter type-I telescopes. The spectra covered the spectral region from 50 eV to 2 keV and the detectors were channel electron multipliers (Section 4.2.5.1).

The Chandra spacecraft carries several sets of transmission gratings. For high-energy photons there are gold gratings with 5000 and 2500 lines per millimetre that cover from 400 eV to 10 keV. The 'wires' of the gratings are actually thin strips deposited onto a plastic film. The gratings are used simultaneously with the first being fed x-rays from the two inner shells of Chandra's Wolter type-I telescope and the second being fed by the two outer shells. Confusion between the spectra is avoided by mounting the gratings with their lines at different angles so that the spectra are also at different angles. At lower energies (70 eV to 7 keV) a single grating is used that reaches a spectral resolution of $R \approx 2000$ at 80 eV and has 1000 lines per millimetre.

3.4.1.2 Reflection Gratings

Glancing incidence reflection gratings were planned to be used on board the now cancelled Constellation-X mission, so their main application to date has been for the Reflection Grating Spectrometers (RGS) on the XMM-Newton spacecraft. As with the glancing incidence telescopes, the glancing incidence reflection gratings require many nested sub-units. For XMM-Newton, the gratings are placed immediately after two of the three telescopes' Wolter type-I 'objectives' on the spacecraft. They allow about 50%

---

* Its mission was to use lunar occultations to obtain precise positions of x-ray sources.

FIGURE 3.10 The glancing incidence reflective grating assembly of the XMM-Newton's RGS spectrometers. (Courtesy of ESA.)

of the radiation to pass straight through, to produce the direct images, whilst around 40% is reflected to one side, by about an angle of 7°, towards the spectroscopes' detectors. The gratings are arranged in 28 co-aligned modules covering the whole output beams from the telescopes (Figure 3.10). Individual gratings are 100 mm × 200 mm in size with 645 lines per millimetre and formed from silicon carbide with a gold reflective coating. The whole assembly requires 182 individual gratings. The spectrometers have resolutions ranging from 0.4 eV at 350 eV to 4 eV at 1 keV and cover the energy range from 350 eV to 2.5 keV.

3.4.2 Bragg Diffraction-Based Spectrometers

If a parallel beam of mixed energy x-rays illuminates a crystal, then only those wavelengths undergoing constructive interference between reflections from different layers within the crystal will emerge (Section 1.3.5, Equation 1.22). Thus the outgoing x-rays from such a reflection will essentially be monochromatic because only the first order interferences will be of significant intensity. To make a spectrometer, the angle of incidence of the beam onto the crystal may be scanned through a range of angles (usually by rotating the crystal), thus (Equation 1.22) changing the wavelength of the reflected beam, or the crystal may be distorted (bent) so that different wavelengths are reflected from different points along its length. Crystals in common use for this purpose include potassium acid phthalate (KAP), lithium fluoride and lithium hydride.

The Einstein observatory used a bent Bragg crystal spectrometer to obtain spectra over the photon energy range 180 eV to 3.4 keV with a spectral resolution ranging from $R = 100$ to $R = 1000$. Six crystals could be used as the reflectors:

PET (pentaerythritol – 1.9–3.4 keV)

ADP (ammonium dihydrogen phosphate – 1.2–2.1 keV)

TAP (thalium acid phthalate – 700 eV–1.2 keV eV)

RAP (rubidium acid phthalate – 500–800 eV)

PbL (lead laurate – 260–450 eV)

PbS (lead stearate – 180–320 eV)

The instrument had an angular resolution of ~1′ above an energy of 1.5 keV. The detector was a position-sensitive proportional counter.

Bragg diffraction spectrometers are also used to analyse powdered rock samples obtained by solar system spacecraft's landers and rovers (Section 7.2.2.5) and they are known as x-ray powder diffractometers. In a powder sample, the individual particles have random orientations to an illuminating monochromatic parallel beam of x-rays. When a particle's crystal layers happen to be correctly oriented for the wavelength, Bragg reflection will occur. If a scintillating screen is placed behind the sample, then a flash will be seen on it at the Bragg angle to the direction of the beam of x-rays. Since there will normally be millions of particles in the sample and there is rotational symmetry around the axis of the beam of x-rays, then there will be many other crystals oriented at this correct angle all around the x-ray beam. A bright circle will thus be seen on the screen centred on the x-ray beam, with its radius given by the Bragg angle.

Now crystals have regular atomic structures and we tend to think of there just being a single set of parallel layers off which the x-rays are reflected. However, even in cubic crystals there are many more reflecting layers than just those forming the faces of the cubes. For example, if we place the crystal so that the cubes' tops and bottoms are level and their sides vertical then we have the three obvious layers parallel to those tops, bottoms and sides. But there are three more layers to be obtained by looking along the diagonal planes running through the edges of the cubes. Yet more layers occur if we take the cubes in pairs and look along the top left-hand edge of the first cube to the bottom right-hand edge of the second cube – and so on.* In the powder sample, there will be crystals oriented so that Bragg reflection can occur off all of these planes as well as off the basic plane. Since the separation of these planes will be different from those of the basic planes, their Bragg angles will be different and the screen will show a series of concentric circles. The radii of the circles are related to the layers' separations and so to the crystal structure of the mineral – in many cases leading to the immediate identifications of the mineral or at least to

---

* This structure can be labelled by the Miller indices – the first cases would be 100, 010 and 001, the second, 110, 101 and 011, etc. The interested reader is referred to appropriate mineralogical sources for further information.

restricting it to a small range of possibilities. The complete screen with its concentric circles may be recorded by suitable x-ray detectors, or, more usually, a radial section detected giving an output that looks like a normal emission line spectrum except the abscissa gives the spacing of the layers, not wavelength.

The CheMin instrument of MSL's Curiosity Mars rover is an x-ray powder diffractometer and can also act as an x-ray fluorimeter. It is the first such instrument used off the Earth. It can analyse an unlimited number of samples because the sample cells can be emptied and reused. The x-ray source is cobalt-59 bombarded by high-energy electrons and emitting at 6.9 keV. A 532 × 600 pixel x-ray sensitive CCD used in photon-counting mode is the detector. A single set of measurements takes around 10 hours.

II

The Detection and Investigation of Sub-Atomic, Atomic and Molecular Particles

CHAPTER 4

Detectors

4.1 INTRODUCTION

Charged particles – electrons, positrons, protons, ions, and so on – together with uncharged (neutral) particles* have been the objects of measurement and study using instruments on board spacecraft since the third successful artificial satellite, Explorer 1, was launched on 1 February 1958. That spacecraft carried a scientific payload which included a Geiger counter for detecting the high-energy charged particles called cosmic rays. The counter picked up so many hits that it was saturated most of the time. The experiment was repeated with Explorer 3 and the high particle flux attributed to a torus-shaped region surrounding the Earth in which electrons and ions from the solar wind and cosmic rays were trapped by the Earth's magnetic field.[†] Subsequent spacecraft mapped out the torus – now called the inner Van Allen[‡] belt – in more detail and also detected the outer Van Allen belt.[§] Two spacecraft launched in 2012, the Van Allen Probes, are currently studying the belts in great detail with much more modern instruments than Geiger counters.

In fact, the possibility of the Earth's magnetic field trapping charged particles had been suggested by Carl Størmer 60 years earlier (1907) as a result of his work in investigating the aurora borealis. He showed that charged particles would be reflected by the converging magnetic field lines in the crescent-shaped regions of magnetic field around a dipole magnet (Figure 4.1). Sadly, Størmer died in August 1957 – just five months before he would have seen his prediction confirmed by the Explorer 1 results.

Charged particles are also studied in terrestrial laboratories and there is frequently a need to detect them and measure their energies and fluxes in civilian, medical and military

---

* Charged and neutral dust grains and micrometeorites are considered in other sections of this book.

[†] A significant increase in radiation levels was detected by a Geiger counter on board Sputnik 2, but the implication of this measurement was not realised at the time.

[‡] Named after James Van Allen who designed and built the instrumental packages on Explorers 1 and 3.

[§] The belts' positions and shapes can change radically with events occurring on the Sun. They can merge into a single belt or, as briefly in 2013, a third belt can form. The inner belt is relatively stable and typically occupies a region from about 600 to 10,000 km above the surface of Earth. It is composed mostly of protons. The outer belt lies roughly between 13,000 and 60,000 km above Earth's surface and is mostly electrons. The particle densities in the belts range downwards from a few tens of millions of particles per cubic metre.

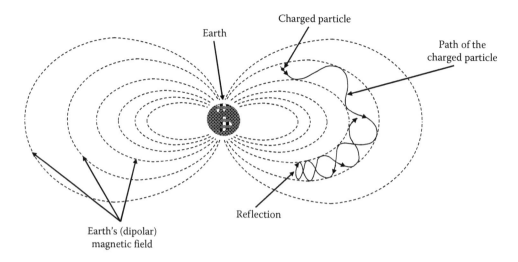

FIGURE 4.1   The trapping of charged particles within the Earth's dipole magnetic field.

contexts. The particle detectors designed for these applications often appear quite different from those on board spacecraft. In particular, laboratory instruments may be very large and massive. The ATLAS detector for the Large Hadron Collider, for example, is 46 m long and weighs 7000 tonnes. Conversely, radiation detectors for use by (say) the civil defence forces after a radioactive material spillage need to be small, lightweight, self-contained, reliable and easy to use. However, the operations of all these detectors are based upon the same few physical principles irrespective of their gross design features. We therefore first look at the main types of charged particle detectors before considering those most appropriate for space applications.*

4.2 DETECTORS

The interactions of high-energy charged particles are currently detected via seven main physical processes:

1. The excitation of an electron within an atom or ion to a higher energy level – also called electron–hole pair production.

2. The emission of one or more electrons from the surface of a material following the deposition of energy into that material of energy from a colliding charged particle (called secondary electron emission and essentially the same process as photo-emission).

3. The ionisation of an atom (also called ion pair production) – or the further ionisation of an ion.

---

* Almost all charged particle detectors can also be used to detect x-rays and γ rays and some will detect ultraviolet photons as well. This means that charged particle detectors will often need screening from x-rays and γ rays if their measurements are not to be contaminated. Likewise x-ray and γ ray detectors will usually need to be screened from charged particles (Chapter 3).

4. Via bremsstrahlung radiation from the particle.

5. Via transition radiation from the particle.

6. Via Čerenkov radiation from the particle.

7. Following its interaction with a nucleus via the strong nuclear force.

The inverse Compton effect also is potentially usable to enable detections to be made, but is not at present so used – however, the Compton effect is the operational basis of some γ ray detectors.

The interactions can take place in a gas, a liquid, at the surface of a solid or within the bulk of a solid. The various combinations and the detector types in current use are summarised in Table 4.1. Detectors for neutrons (Section 6.2) are also included here for completeness.

Frequently, several different charged particle detectors will be carried as a spacecraft's payload. This is normally to provide measurements of several different aspects of the particle (e.g. velocity, charge, mass, position) simultaneously, since the same high energy particle will often pass through many of the detectors before being stopped or escaping back into space. Multiple measurements may also be needed when a detector is sensitive to only a restricted range and when information on particles outside that range is needed. Additionally, obtaining measurements of a number of parameters may enable other properties of the particle to be calculated, when these have not been measured directly.

4.2.1 Geiger Counters

Geiger counter detectors were used on several early space-craft, such as Sputnik 2, Explorers 1, 2, 3 and 4 and Luna-9 but have long since been superseded,* although their descendants, such as proportional counters, are still in widespread use. Their basic principle is that of detecting the ingress of a charged particle into the detector from the ions and electrons (ion pairs) that it produces amongst the neutral atoms and molecules of the gas contained within the detector. Typically an energy of 20–50 eV is needed to produce an ion pair in argon.† So a 10 MeV (say) charged particle from a solar flare could potentially produce around a hundred thousand to half a million ion pairs. However, such a particle will normally pass completely through the detector, losing only a small fraction of its energy whilst doing so. Thus the actual number of ion pairs produced in the detector by a charged particle will range from one (or zero) to a few thousands.

The detection is made by applying a high voltage between two electrodes immersed within the gas. The initial few ions and electrons produced by the charged particle are attracted towards the cathode and anode, respectively. The electrons, however, will quickly reach velocities a thousand or more times those of the ions and in colliding with the

---

* Geiger counters are still in use in the laboratory and for military and medical applications, especially for monitoring exposure to x-rays.

† The gas often used to fill these detectors and chosen because, as a monatomic gas, energy cannot be lost into unwanted reactions such as dissociation, rotation or vibration as would be the case with a molecular gas.

TABLE 4.1 Possible Configurations for Charged Particle Detectors

	Solid (interaction within the bulk of the material)	Solid (surface interaction only)	Liquid	Gas
Bremsstrahlung radiation	Calorimeter			
Čerenkov radiation			Čerenkov detector	Čerenkov detector
Elastic scattering	Elastic scattering neutron detector			
Electron excitation (electron–hole pair production)	Scintillation counter Silicon microstrip detector (solid state detector)		Scintillation counter	Scintillation counter
Interaction with a nucleus	Calorimeter Radioactive decay (prompt or delayed) neutron detectors			Čerenkov detector[a]
Ionisation (ion pair production)			Proportional counter	Drift chamber (MicroMegas/time projection chamber) Geiger counter Imaging position-sensitive proportional counter (multi-wire proportional chamber) Ion chamber Position-sensitive proportional counter Proportional counter Streamer chamber
Secondary electron emission		Channel electron multiplier (continuous channel multiplier/electron multiplier) Microchannel plate		
Transition radiation		Transition radiation detector		

[a] Sometimes the incoming particle will produce Čerenkov radiation directly; at other times it will be produced by the secondary particles produced during the incoming particle's interaction with a nucleus.

remaining neutral atoms and molecules will produce more ion pairs. The new electrons from these pairs will then be accelerated in their turn and produce yet more charged particles. In this way, one electron produced initially by the incoming charged particle will result in millions of electrons arriving at the anode. The multiplication of the free electrons in this way is called a Townsend* avalanche or an electron cascade.

---

* After Sir John Townsend (1868–1957).

A commonly used design for a Geiger counter is to have the cathode as a cylindrical tube and the anode as a wire running down the centre of the tube. There are several advantages to this design, but the main one is that the electric field increases very rapidly close to the anode. Its strength at a distance, r, from the anode, $E(r)$, is given by

$$E(r) = \frac{V}{r \ln(b/a)} \qquad (4.1)$$

where V is the voltage difference between anode and cathode, a is the radius of the anode wire and b is the inner radius of the cathode cylinder.

Thus for $a = 100$ μm, $b = 10$ mm and $V = 2000$ V, we have, at the inner surface of the cathode cylinder ($r = 10$ mm), $E_{(10\,mm)} = 4 \times 10^4$ V·m^{-1}, whilst at the surface of the anode wire ($r = 100$ μm), we have $E_{(100\,μm)} = 4 \times 10^6$ V·m^{-1}.

The strength of the electric field has to be above the Townsend threshold before an electron cascade can develop. For argon the threshold is around 6×10^5 V·m^{-1}. In the above example, such a value will only be reached when an electron is 600 μm away from the anode wire (i.e. $r = 700$ μm). Wheresoever the originating charged particle passes through the tube, therefore, the Townsend avalanche develops mostly very close to the anode wire. The avalanche thus is confined to a small length of the anode wire and its position, when needed, along the wire can be accurately determined (Figure 4.2).

The electron cascades from the various ion pairs resulting from the passage of a charged particle effectively all arrive at the anode wire simultaneously (the electron flight time, even from close to the cathode cylinder, is a few nanoseconds) depositing their charges as a single pulse. If suitably connected, this will produce a current flowing around the rest of the circuit which may be detected in numerous ways. Frequently the pulse is just sent to a small loudspeaker, so that the ubiquitous 'beep' of a radiation detector as depicted on the TV and in the movies is produced.

One problem with the basic Geiger counter is that the electron cascade is usually saturated by the time that it reaches the anode (i.e. all available atoms and molecules have been ionised) and so the strength of the electron pulse is independent of the energy of the incoming charged particle. A detector in which the pulse strength depends to at least some degree on the energy of the original particle – a proportional counter – may, though, be obtained by reducing the voltage between the anode and cathode of the standard Geiger counter. The reduced voltage means that fewer ions and electrons are produced, but that the cascade no longer saturates and its strength depends upon the number of ions and electrons produced by the original charged particle (i.e. usually upon its energy). The reduced voltage also helps with a second problem of the basic Geiger counter: its dead time. This is an interval of typically 100 μs after the detection of one particle when the positive ions from the Townsend avalanches that are close to the anode reduce the electric field strength to below the Townsend threshold and it becomes insufficient to cause another electron cascade. A second charged particle following very closely behind the first will thus not be detected.

The arrival direction of the charged particle may be inferred by using two (or more) aligned Geiger counters and only registering those particles which hit both counters.

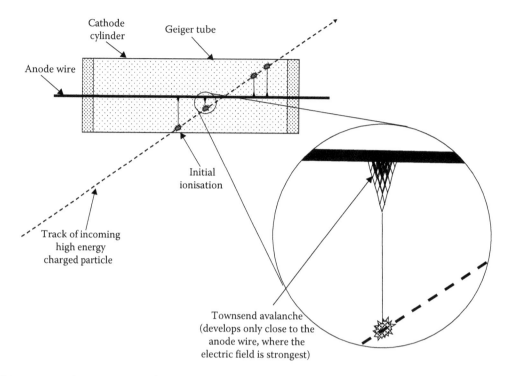

FIGURE 4.2   A cross section through a Geiger counter showing Townsend avalanches. NB: Many more initial ionisations than the four shown are likely to occur in a real situation.

A better system uses a resistive anode and detects the electron pulse from both of its ends. The pulse strength reduces as it travels along the anode and so the position along the anode of the electron cascade may be inferred from the ratio of the strengths of the pulses abstracted from each end. Although not used in actual Geiger counters, this system, together with the reduced voltage, is the basic operating principle of the position-sensitive proportional counter. Finally, by making the gas chamber a box with the cathode forming the top and by using many parallel wires as the anodes we obtain the imaging position-sensitive proportional counter,* although this is used more for the detection of x-ray images than for charged particles.

An important part of the design of Geiger counters and their relatives is to deal with the ions which are produced during the original interaction. The velocity of the ions is so low compared with that of the electrons that they are effectively stationary during the production of the electron pulse. However they *will* reach the cathode eventually and be absorbed. When the ion recombines with an electron from the cathode, sufficient excess energy may be released for a second electron to be ejected from the cathode. This will be a free electron and it will be accelerated towards the anode in exactly the same way as an electron produced in an ion pair by the passage of a charged particle through the counter, so it will produce a second delayed and unwanted pulse.

---

* Often called the multi-wire proportional chamber.

The ions have therefore to be neutralised (quenched) before they reach the cathode. This is accomplished by adding a small proportion (5%–10%) of a second gas, typically ethyl alcohol, to the argon. The ions recombine with electrons from the quench gas, leaving the latter with a positive charge. The quench gas molecules are thus the ones which arrive at the cathode and are neutralised. However, the complex structure of the quench gas molecules means that any excess energy from the neutralisation is channelled into dissociating the molecules and not into the production of new electrons. Unfortunately organic quench gases are gradually consumed, limiting the life of the detector to around 10^9 counts in total. Recently, the use of a halogen (often around 0.1% chlorine in neon) as the quench gas has become popular since the chlorine atoms will recombine after dissociation thus replenishing the quench gas.

A Geiger counter operated with a pulsed voltage and parallel plate electrodes has been used by particle physicists in the laboratory in the past. The electron avalanche does not develop because the pulse of the applied voltage is too short. Instead the ions and electrons recombine along the track of the incoming charged particle emitting light and the track can be seen from these emissions, which appear as faint streamers between the electrodes. The device is thus known as a streamer chamber. It is not suitable for use on spacecraft and has also generally been superseded within the laboratory by other detectors.

The operating principles of Geiger counters are recognisably similar to those of many, more sophisticated, charged particle detectors. Almost all such detectors have three main segments. First, there is a process whereby some or all of the energy of the charged particle is converted into one or more free electrons (ion pair production in the argon gas for Geiger counters). Second, there is an electron multiplying stage whereby the primary electron(s) generate many more secondary electrons (acceleration towards the anode wire for Geiger counters and the consequent Townsend avalanche), whilst the last stage collects and registers the electrons (the charge pulse from the anode wire in Geiger counters).

4.2.2 Ion Chambers

An even simpler charged particle detector than the Geiger counter is the ion chamber. It is essentially a Geiger counter run at a very low voltage. The voltage is too low (~100 V) for any secondary ion pairs to be produced and the detection is that of the primary electrons and ions as they are collected when they have drifted to the anode and cathode, respectively. Ion chambers thus lack the electron multiplying stage of other particle detectors.

The output from an ion chamber is a continuous current whose strength varies with the intensity of the incoming radiation. However, the current is very small (~10^{-15} to ~10^{-10} A) and so the devices must employ very sensitive ammeters to show the radiation intensity. The ion chamber may be identical in its physical design to a Geiger counter (i.e. a central anode wire and a surrounding cathode cylinder) or it may use a pair of parallel plate electrodes charged to differing voltages. The output from an ion chamber varies little with the voltage difference between the anode and cathode (Figure 4.3) provided that this remains within its prescribed operating range.

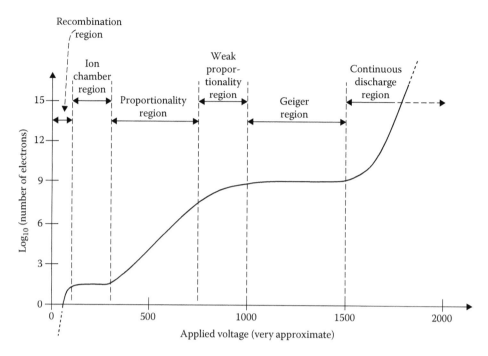

FIGURE 4.3 Electron production in response to the applied voltage. NB: The voltage shown on the graph is only intended as a rough guide – the actual applied voltage required to produce each of the regions will depend in detail upon the design of the detector. The six regions arise as follows:

1. Recombination region – the applied voltage is too low to keep the electrons and ions apart and so they recombine.
2. Ion chamber region – the applied voltage is sufficient to prevent the electrons and ions recombining and it pulls them towards the anode and cathode, respectively, but at velocities too low to cause the production of secondary ion pairs. This is the region over which ion chambers operate.
3. Proportionality region – the applied voltage accelerates the electrons to velocities sufficient to produce secondary ion pairs, but the current does not saturate. Only one avalanche is produced for each ion pair originating from the incoming charged particle (i.e. the number of avalanches is proportional to the original number of ion pairs). This is the region over which proportional counters operate. NB: Some confusion may arise from the apparent linear (i.e. proportional) response shown in the figure over this region. However, note that the number of electrons is plotted logarithmically, whilst the voltage is plotted linearly, so the actual response over this region is exponential.
4. Weak proportionality region – the applied voltage is sufficient to lead to the production of secondary avalanches of electrons that arise from ultraviolet (UV) photons originating within the initial avalanches. The number of avalanches is thus more than the number of original ion pairs.
5. Geiger region – the applied voltage is sufficient to cause saturation of the discharge current. This is the region over which Geiger counters operate.
6. Continuous discharge region – the applied voltage is sufficient to cause continuous discharges within the gas whether or not a charged particle has passed through the detector (i.e. the voltage is above the breakdown limit of the gas).

The gas filling an ion chamber is often air* or a noble gas. It can also be a liquid such as isooctane or liquid argon which has a much higher stopping power for the incoming radiation than a gas. Pioneer 1, 2 and 5 all carried ion chambers for monitoring high-energy charged particles near the Earth. They used an aluminium chamber with a volume of 43 mL filled with argon at a pressure of 1.3 MPa.

4.2.3 Proportional Counters

Like the ion chamber, the basic proportional counter is a Geiger counter run at a lower voltage. It is also often of the central wire anode and outer cylindrical cathode physical design. For the proportional counter though, the voltage is sufficient to lead to one Townsend electron avalanche for each ion pair produced by the incoming particle (i.e. the proportional region in Figure 4.3). The output pulse strength is therefore proportional to the number of ion pairs and if the incoming charged particle is stopped within the detector, so that all its energy goes into ion pair production, then the pulse strength is proportional to the total kinetic energy of the incoming particle. Unlike the Geiger counter therefore, the proportional counter can be used for charged particle spectroscopy – that is, to measure the energy of the charged particle – and if the particle's mass is known, then to determine its velocity as well. Alternatively, if the velocity of the charged particle is already known or restricted to a small range (see Section 5.2) then the proportional counter can identify the nature of the charged particle (electron, proton, alpha particle, etc.).

Like the Geiger counter, the proportional counter is filled with a noble gas (~90% – usually argon) and a quench gas (~10% – usually ethyl alcohol or methane). Liquids, such as liquid xenon, can also be used to fill the chambers. If the proportional counter is sealed, then the quench gas will gradually be consumed. However, proportional counters are often operated with a continuous flow of the gases through the chamber (continuous flow or gas flow proportional counters). The quench gas can then easily be replenished. The disadvantage of the latter type of proportional counter for use on board space-craft is the need for a continuing supply of the gas, but they are so used.

The use of a resistive anode to enable the positions of the avalanches along the anode wire to be determined has already been mentioned. A two-dimensional (2D) measurement of avalanche position can be obtained by making the cathode a flat plate and winding the anode wire in a zig-zag below the plate. More commonly today, the multi-wire proportional counter is used. This also uses a flat plate cathode, but multiple parallel wire anodes. A further improvement is to make the cathode out of thin parallel strips that are orthogonal to the anode wires (microstrip proportional counter). The x and y positions of the Townsend avalanches are then found directly.

The third dimension (the height of the incoming charged particle's track above the anode) is determinable using a drift chamber. Despite the higher operating voltage of a proportional counter, for most of its track an electron from an ion pair is effectively

---

* You may be more familiar with ion chambers than you think. Domestic smoke alarms are ion chambers producing ions in air from an americium-241 alpha-particle source.

within an ion chamber – it drifts through the gas without producing secondary ionisations until very close to the anode wire (Figure 4.2). If the electric field is made as uniform as possible over this outer region (often additional field-shaping electrodes are used to achieve this), then the electron will drift at a constant and known velocity. Typical drift velocities are around 10,000–100,000 $m \cdot s^{-1}$. The time taken for the electron to reach the anode wire can thus be converted into the distance from the wire to the point at which the incoming charged particle produced that particular ion pair. In the MicroMegas (Micro-Mesh gaseous structure*) detector, an additional electrode is used to shorten the drift time so that precise measurements (10 ns) of the time of the interaction can be obtained. The additional electrode is in the form of a fine charged mesh placed some 100 μm above the anode wires and charged to a voltage between that of the anode and cathode. Finally in the time projection chamber (TPC), a magnetic field runs parallel to the electric field enabling the particle's charge and momentum to be determined from the curvature of its track. TPCs, however, are generally physically large devices more suited to the laboratory than to mounting on spacecraft.

Thus, NASA's Advanced Composition Explorer (ACE) spacecraft carries the Solar Energetic Particle Ionic Charge Analyser (SEPICA) instrument.[†] This had proportional counters and solid state detectors (SSDs) (Section 4.2.6) as its primary detectors and used a caesium iodide scintillator as the shield (Section 4.2.4). The instrument's six collimators directed the particles onto the multi-wire isobutene-filled proportional counters and after passage through the counters the energy remaining in the particles was collected by the SSDs. The kinetic energy, the level of ionisation and the atomic species of the charged particles were all measured by the instrument for particles with energies above 200 keV per nucleon.

4.2.4 Scintillation Detectors

The use of zinc sulphide to detect ionising radiation dates back to the earliest days of the study of radioactivity. In 1866, its phosphorescence was discovered by Théodore Sidot and its scintillation (the emission of flashes of light) was noticed by Sir William Crookes in 1903. He found that the light emitted from a zinc sulphide screen close to a small speck of radium was not uniform but comprised numerous individual flashes of light. We now know that each flash is the result of the impact of an alpha particle from the radium with the screen. The alpha particle during its passage through the zinc sulphide excites some of the electrons within the crystal structure from the valence to the conduction band.[‡] The flash of light is emitted as the electron falls back to the valence band. The efficiency of the process and the wavelength of the emissions may be adjusted by doping the scintillator with small quantities of another element – usually silver in the case of zinc sulphide.

---

* MicroMegas and some related detectors are sometimes given the group name of Micropattern Gaseous Detectors (MPGD).

† The SEPICA instrument failed in 2005.

‡ See Section 4.2.6.1, reference A.1.1 or books on solid state physics (Appendix A) for discussions on the electronic band structure of solids.

Many materials are now known that scintillate when bombarded by high-energy charged particles. Amongst those commonly finding application in space-based experiments are sodium iodide doped with thallium and caesium iodide doped with sodium or thallium. For large area detectors, plastic sheet scintillators made from polystyrene, polyvinyl naphthalene or polymethylmethacrylate (etc.) may be used. Other scintillator materials include the following:

Alternative inorganic crystals (e.g. bismuth/germanium/oxygen [BGO], barium fluoride, lanthanum chloride)

Glasses (e.g. lithium or boron silicates doped with cerium)

Organic crystals (e.g. anthracene, stilbene)

Organic liquids, usually aromatic hydrocarbons (e.g. *p*-terphenyl dissolved in toluene)

Gases (e.g. nitrogen, argon)

Sir William Crookes observed and counted his scintillations by eye, but this is not practicable with space-based instruments and inadequate in most other applications where the count number may exceed thousands of flashes per second. The emissions from the scintillator are therefore normally detected by a second detector and this is usually a simple visible light (or UV or infrared [IR]) photon detector. With some scintillators, especially those using organic materials, UV photons are emitted during the initial interactions. The scintillator itself would rapidly absorb these photons and so it is mixed with a suitable fluorescent material that absorbs the UV photons first and then re-emits their energy as longer wavelength photons. The fluorescent material in this context is termed a wavelength shifter. It is often a benzene-related compound and its emissions then occur around a wavelength of 400 nm (blue light). If need be, a second wavelength shifter can be used to match the emitted photons to the most sensitive wavelengths of the light detector.

Visible light detectors such as charge-coupled devices (CCDs), avalanche photodiodes (APDs) and photomultipliers are considered in more detail in Section 1.4. Here, we just note their aspects relevant to charged particle detection.

CCDs are array detectors sensitive from the UV to the near infrared. The total light emission from a single interaction of an incoming charged particle with the scintillation crystal has therefore to be found by adding together all the signals from device's individual pixels. Since 100 eV is typically required for the production of a single photon during the initial interaction, a 1 MeV incoming charged particle that is stopped within the scintillator will give rise to ~10,000 photons.

Although APDs are available in small arrays, they are often used as single detectors. The 'avalanche' part of 'APD' refers to the generation of Townsend-type avalanches within the device. The APD is a solid crystal of (usually) silicon and has a potential difference applied across its surface of about half its breakdown voltage. Absorption of a photon produces an ion pair within the crystal and the potential difference accelerates the electron through the crystal structure, producing further ion pairs via inelastic collisions with the atoms forming the crystal.

The photomultiplier tube (PMT) is also usually a single detector which operates mainly within the visible region (Section 1.4.3.3). PMTs can be made physically large (up to 0.5 m in diameter) at fairly low cost and so are useful in conjunction with large area scintillators. The photon from the incoming charged particle interaction initially interacts with a cathode charged to some −1000 V and coated with a photoemitter such as caesium antimonide. One or more electrons are produced by the photoemitter and these are guided by electric fields to a second electrode (a dynode) charged to around −900 V. The potential difference of ~100 V between the cathode and dynode means that the original electron(s) gains some 100 eV of energy during its passage.* The dynode's surface is coated with a substance which emits electrons when bombarded by electrons (often, but not always, it is the same substance as the photoemitter on the cathode). The 100 eV electron(s) will produce some 10–20 electrons (each) from the dynode's electron emitter. These will then be guided to a second dynode at around −800 V where the process will repeat and result in several hundred electrons. Further multiplications at several more dynodes leads to 10^6 or more electrons being output from the PMT for each visible light photon that entered it. The PMT thus also amplifies its signal via an avalanche of electrons, although the manner of production of the avalanche is quite different (and more controlled) than that of Townsend avalanches.

Thus in terms of the three sections of a charged particle detector which we identified for the Geiger counter, the initial interaction for a scintillation counter is the excitation of an electron (not ionisation) within an atom resulting in the subsequent emission of one or more photons. That photon then goes on to produce an electron (as an ion pair in silicon for an APD and via photoemission in a PMT) which in turn is then multiplied up by a factor of ~10^4 to ~10^7 by avalanche processes. For both APD- and PMT-based scintillator counters, the final output is a pulse of electric charge which may be further processed by conventional amplifiers, and so on.

For CCD-based scintillator counters, the initial photon produces an ion pair in the device's substrate and the electron from that pair is trapped in its individual pixel within the CCD as a single particle. The final output is obtained after a conventional charge transfer within the CCD (but without avalanche amplification) to the device's output electrode.

Unlike the Geiger and related detectors, the scintillation counter has no dead time. However, like them, it is sensitive to both charged particles and to high-energy photons (x-rays and γ rays). With the scintillation counter though, it is possible to discriminate between the signals produced by a charged particle and a high-energy photon. The latter will be absorbed quickly and so its output pulse is short. The charged particle generally takes much longer to be completed stopped within the detector and the resulting pulse therefore has much longer duration than that for a photon.

Recently, the main space application of scintillation detectors has been as anti-coincidence shields rather than direct particle detectors. Thus the ACE spacecraft carried the SEPICA

---

* For comparison, a visible light photon has an energy of between 1.8 eV (red) and 3.2 eV (violet).

instrument (Section 4.2.3). This had a proportional counter and an SSD as its primary detectors and used a caesium iodide scintillator as the shield. Similarly, the Fermi spacecraft's large area telescope (LAT) uses plastic sheet scintillators feeding PMTs as its anticoincidence shield.

4.2.5 Microchannel Plates

Microchannel plates (MCPs) operate in a manner that is closely related to the way that a PMT works – indeed MCPs are widely used for the detection of UV, x-ray and γ ray photons (Sections 1.4.3.2 and 3.2.4) as well as for high-energy charged particles. There is almost no difference between an MCP used for particle detection and one used for photon detection except that the coatings are likely to be different (a nickel–chromium alloy for particles, caesium iodide for photons) and if used for charged particle detection, the MCP must be shielded from the high-energy photons.

The substrate of an MCP is a plate of glass a fraction of a millimetre to a few millimetres thick that is pierced by tens of thousands to millions of small (5–100 μm) tubular holes (Figure 4.4a). Thin metallic electrodes are deposited on the plate's top and bottom surfaces and the walls of the tubes treated to develop a thin semiconducting layer on them. The area occupied by the holes is typically some 50%–70% of the total area of the plate and the length of the tubes is between 40 and 80 times their diameters. The tubes are tilted with respect to the vertical to the plate's surface by a small (5°–15°) angle, called the bias angle (alternatively for the thicker MCPs, the tubes can be given a slight curvature). The tubes are shown as having circular cross sections in Figure 4.4, but other shapes (square, hexagonal, etc.) are also used. The whole assembly is housed in a vacuum chamber at a pressure of 1 or 200 μPa with the charged particles entering the chamber through a thin window, if necessary.

A voltage difference of around 1 kV is applied between the top electrode (the cathode) and the bottom electrode (the anode). An incoming high-energy charged particle colliding with the wall of a tube is likely to do so near the top, because of the bias angle of the tube. There it leads to the ejection of one or more (typically two) secondary electrons. These electrons are accelerated down the tube by the applied voltage and quickly collide with the walls, leading to further electron emissions. The process continues down the whole length of the tube until up to 10^4 electrons exit from the bottom of the tube for the single incoming charged particle (Figure 4.4b). The electron multiplication factor is called the gain of the MCP and increases as the tube's aspect ratio (the length of the tube, in terms of its diameter), increases. However the increase is eventually limited by ion feedback. The residual gas in the chamber will be ionised by the secondary electrons and the resulting ions attracted towards the cathode. When they collide with the tube walls they will produce secondary electron showers just like that of the original incoming charged particle. These after-pulses are spurious and unwanted signals. Their effect can be kept small however, by limiting the MCP gain to $\times 10^4$ or less.

If a higher gain is needed than $\times 10^4$ then this may be obtained by stacking two or more individual MCPs (Figure 4.4c and d). The electron shower from the first MCP feeds directly into the entrance holes of the second which has its bias angles reversed and these

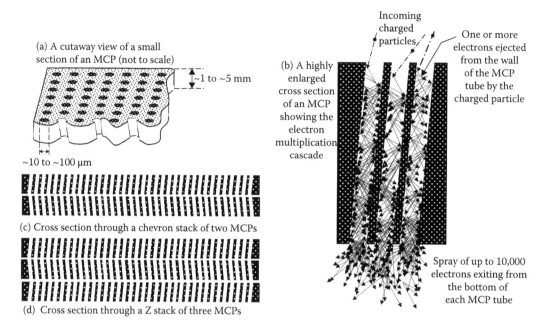

FIGURE 4.4   (a) A magnified section through a microchannel plate (MCP). (b) Electron multiplication within MCP tubes. (c) and (d) MCP chevron and Z stacks, respectively.

electrons in their turn are multiplied. Two MCPs produce a chevron stack (Figure 4.4c), three, a Z stack (Figure 4.4d). However, the ultimate gain even with three MCPs is limited to about ×10⁷ because of the onset of saturation and non-linear response. The latter effects arises because there are two electric currents flowing within an MCP – the first is the flow of the secondary electrons through the tubes. The second is called the strip current and arise because the walls of the tubes have a thin layer of a semiconductor on them so that a small current flows between the anode and cathode. The strip current replaces the electrons lost to the secondary electron showers enabling showers to continue to be produced. However, if the secondary electron current exceeds about 10% of the strip current then the gain of the MCP ceases to be constant. With a typical resistance across the MCP of 1 GΩ and a voltage of 1 kV, the magnitude of the strip current is around 1 μA.

The spatial resolution of an MCP is clearly given by the centre-to-centre spacing of its tubes and so lies in the region of ~10 to ~200 μm. This is not of great importance when the MCP is used for the detection of charged particles, but does matter for MCPs used as the detectors for imaging UV and x-ray telescopes.

The response time to a charged particle of an MCP is in the region of 100 ps to 10 ns. However, an electron avalanche in a single MCP tube lowers the number of electrons available within its walls, especially towards its bottom end, where the avalanche is at its most intense. These electrons must be replenished by the strip current before a second avalanche can be produced. The individual tube in an MCP thus has a dead time of a few milliseconds. However, provided an incoming charged particle hits the MCP at a point other than directly within this recently discharged tube, there will be other fully charged tubes that are able to detect it. For an MCP with, say, a million tubes, the effective dead time is thus in the region of ~10^{-8} s (i.e. such an MCP illuminated by a uniform beam of charged particles can effectively count up to ~10^8 particles per second).

The manufacture of MCP plates is quite a complex process and uses techniques developed for the production of fibre optic cables (see also Section 3.3.2.2). It starts with a thick ring of lead glass whose centre has been filled with a second type of glass that is etchable (dissolvable) in a suitable chemical. This billet of glass is heated in a furnace and converted into a long thin strand by drawing down in the same way that fibre optic cable glass strands are produced. The resulting fibre is cut into a few thousand short lengths and these are fused together to form a new multi-fibre billet of glass with a hexagonal outer shape. The hexagonal multi-fibre glass billet is drawn down a second time reducing the sizes of the cores of the fibres to the size desired for the tubes of the MCP. This second and now hexagonal multi-fibre is again cut into short lengths and the fibres stacked carefully side by side and fused together to form a new glass multi-fibre billet. This billet is cut into slices (or wafers) which have the thickness required for the final MCP and the cut surfaces polished. If the MCP is to have its tubes with a bias angle, then the wafers must be cut from the billet on a slant equal to the bias angle. The wafers are chemically processed to etch out the cores, producing the MCP's tubes. The semiconducting layer on the tubes is then obtained by heating the wafer in an atmosphere of hydrogen so that a small amount of the lead glass is chemically reduced to lead. Finally the electrodes are vacuum deposited onto the two flat surfaces of the MCP wafer.

Many charged particle detectors on board spacecraft use MCPs as their detectors. The four spacecraft making up the Magnetospheric Multiscale Mission (MMS – launched in 2015 to study magnetic reconnection within the Earth's magnetosphere) (Figure 4.5), for example, each carry four dual ion spectrometers (DIS). Each DIS has two top hat electrostatic analysers (see Section 5.2) with fields of view of 10° × 180° and is sensitive to ions in the energy range 10–30 eV. The dual ion spectrometers are spaced at 90° intervals around the perimeter of each spacecraft giving complete coverage of the sky. The top hat electrostatic analysers each use an MCP in the shape of a thick annulus that covers 60% of a circle. The MCP anode is divided into 16 independent sectors so that some positional information is obtained on the charged particle interactions. The MCP is of the chevron configuration (Figure 4.4c) with each of its components 1 mm thick, with 12.5 μm tubes at 15 μm centres. The tubes have aspect ratios of 80:1, nickel–chromium alloy electrodes and bias angles of 8°. The MCP gain varies from about 8×10^4 at a voltage of 2.2 kV to 1.3×10^7 at a voltage of 2.6 kV.

The MESSENGER spacecraft carried the energetic particle spectrometer (EPS) as one of its instruments. This measured the time of flight (ToF – Section 5.2.6.4) of particles. It used an MCP to detect the secondary electrons produced as the primary particles passed through thin polyamide films coated with aluminium (the 'start' film) and lead (the 'end' film) and separated by 60 mm in order to determine the flight time.

The two Van Allen probes are identical spacecraft launched in 2012 to study the Earth's trapped particle belts. Their mission duration was intended to be two years but they are still operational at the time of writing. They have highly elliptical, near equatorial, orbits that enable them to cover a large fraction of the belts with perigees near 600 km and apogees

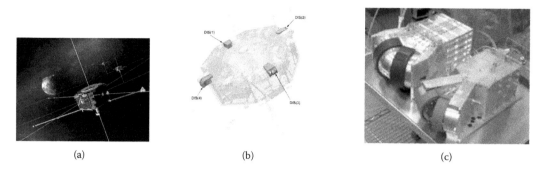

(a)   (b)   (c)

FIGURE 4.5   The dual ion sensor on board one of the spacecraft of the Magnetospheric Multiscale Mission (MMS). (a) An artist's impression of the four MMS spacecraft in space. (b) Semi-transparent sketch of the MMS spacecraft showing the positioning of the four dual ion sensors (DIS). (c) One of the dual ion sensors of the MMS (top). Its companion instrument, the dual electron sensor (DES), is mounted next to it (bottom). (Courtesy of NASA/Goddard Space Flight Center.)

near 37,000 km. The use of two probes is to enable the conditions and changes in the belts to be compared at the same instant over large distances.

The Radiation Belt Storm Probes Ion Composition Experiment (RBSPICE) instruments are designed to study the storm-time ring current and its variability with solar and geomagnetic activity by measuring the composition and pressure of the current. They are sensitive to electrons in the 25 keV to 1 MeV range, to protons and helium nuclei (alpha particles) over the 20 keV to 10 MeV range and to oxygen ions over the 40 keV to 10 MeV range. They use two detectors – an MCP and an SSD (Section 4.2.6) working together (Figure 4.6). The ion's velocity is determined by means of the ion generating secondary electrons as it passes through two thin foils. The secondary electrons are attracted to the MCP by two anodes beneath it and detected there. The time difference* between the arrivals of the two groups of electrons and the distance between the foils gives the velocity. The SSD then measures the ion's total energy. RPSPICE's acceptance angle is 12° by 160° with the latter sub-divided into six sectors. RBSPICE has recently revealed a 'zebra-stripe' pattern within the distribution of electrons of the inner radiation belt arising from the effects of the Earth's rotation – a very surprising result given the slow speed of the Earth's rotation and the relativistic velocities of the electrons.

The van Allen probes' relativistic proton spectrometer (RPS) instruments also use an MCP and SSDs. The RPS instrument detects protons with energies between 50 MeV and 2 GeV. Its purpose is to monitor changes in the inner Van Allen belt arising from solar activity and the contribution made to the belt by cosmic rays. The lower energy protons are picked up by a group of eight SSDs, with only those penetrating through all the detectors being counted. A Čerenkov detector (Section 4.2.9) operates to cover the higher energy particles using a magnesium fluoride radiator and a chevron MCP as the detector. Its threshold energy is about 430 MeV. The MCP uses a bialkali coating (comprising antimony, potassium and caesium) on a window just in front of its top surface to convert

---

* Detectors operating in this fashion are often called time of flight (ToF) instruments – Section 5.2.6.4.

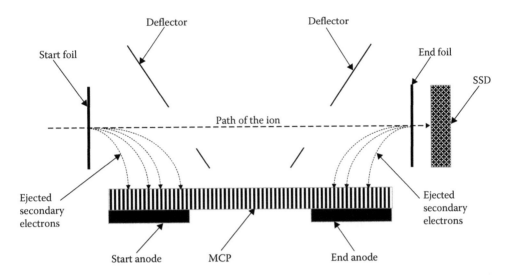

FIGURE 4.6   The particle detector in RBSPICE.

Čerenkov photons in the wavelength range 185–660 nm into photoelectrons* and has a gain of around ×10⁵ at an operating voltage of ~2 kV.

4.2.5.1 Channel Electron Multiplier

A detector that is very closely related to MCPs in the principle of its operation is the channel electron multiplier (CEM†). In effect, the CEM is just one of the tubes of the MCP expanded up to a scale of a few tens of millimetres and with a conical shape. The cone is constructed from a suitable insulator, such as glass, and coated on the inside with a semiconductor. The wide end of the cone has a negative voltage of several kilovolts applied to it, while the narrower end is held at a small positive voltage. An incoming charged particle will produce secondary electrons after a collision with the inner surface of the cone. These electrons will be accelerated towards the narrow end of the cone by the potential difference between the cone's ends, usually colliding with the inner surface of the cone and producing more electrons on the way. The electron shower is collected after exiting the narrow end of the cone by a separate anode. In some designs, the cone is given a curvature of a few tens of degrees (so that its shape resembles that of a cow's horn) in order to increase the chances of the electrons colliding with its inner surface. The gain of a CEM can be up to 10^7.

The van Allen probe's Energetic particle, Composition and Plasma suite (ECT) is a package of three instruments (Sections 4.2.6 and 5.2.6) intended to determine the physical processes and mechanisms involved in changes to the belts and the loss of particles from the belts.

Its HOPE instrument (Helium Oxygen Proton Electron detector) examines particles with energies up to about 45 keV, especially electrons, protons, helium and oxygen ions. It is a ToF detector very similar in design to that of the RBSPICE instrument except that

---

* An MCP used in this fashion is sometimes called an MCP/PMT.
† Also called an electron multiplier (EM) or a continuous channel multiplier (CCM).

two CEMs are used in place of the MCP and its anodes. The CEMs are charged to +4.4 kV at their entrances and to between +6.8 and +9 kV at their exits. The positive charges at their entrances enables the CEMs to act as the start and end anodes in addition to being the detectors. A bias potential difference of −11 kV for ions and +1.5 kV for electrons is applied along the instrument in order to reduce the particles' angular scattering and energy differences.

The start and end foils are made from carbon and are separated by 30 mm. This gives the time of flight for a 1 eV O^+ ion as about 100 ns and this is measured to a precision of 2.5 ns. A particle's velocity is thus measureable over the range 3×10^5 m·s^{-1} to ~10^7 m·s^{-1}. Once the velocity of a particle has been measured, its mass may be found from its energy per unit charge and this is pre-selected by a variable electrostatic energy analyser that precedes the detector (Section 5.2). Each Van Allen probe has five of these units arranged in an array parallel to the spacecraft's spin axis at 36° intervals. The acceptance angle for each unit parallel to the spin axis is about 22.5°. Since the probes are spinning at about 5.5 rpm, the HOPE instrument thus samples about 62% of the whole sky every 11 seconds.

4.2.6 Solid State Detectors

The MCP is marginally definable as an SSD since the interaction of the incoming charged particle and the secondary electron avalanche result from interactions with the (solid) walls of its tubes. Scintillation detectors could, with more justification, also be put into this class. However SSDs, by common consent, are usually regarded as those in which the original interaction and any subsequent electron avalanches (etc.) take place within the bulk solid material itself and that material is usually some type of semiconductor such as silicon or germanium. Usually silicon is chosen because it is around 1% of the price of germanium. Other semiconductors such as gallium arsenide and silicon carbide are used as photon detectors, but are not currently of importance as charged particle detectors. Recently, the Cadmium–Zinc–Telluride Imager (CZTI) instrument (Section 3.2.5.2.3) on board the ISRO's Astrosat has been detecting cosmic rays as well as the intended x-rays.

4.2.6.1 Semiconductors and the Band Structure of Solids

The interaction of a high-energy charged particle with a solid material can be understood in terms of the band structure of the material. In a thin monatomic gas, the individual electrons are attached to individual nuclei and occupy the inner energy levels of the atom (the ground state). Their interaction with the charged particle is either the excitation of an electron to a higher energy level or the ionisation of the atom.

In a solid material, many atoms are in close proximity to each other and their interactions change the energy levels from those observed in an isolated atom. When two well separated atoms of the same element approach each other their energy levels split – into two for an s level, six for a p level, ten for a d level* and so on (Figure 4.7a). When

---

* Details of atomic energy levels and their notation may be found in reference A.1.9 and other books on spectroscopy and atomic structure (Appendix A).

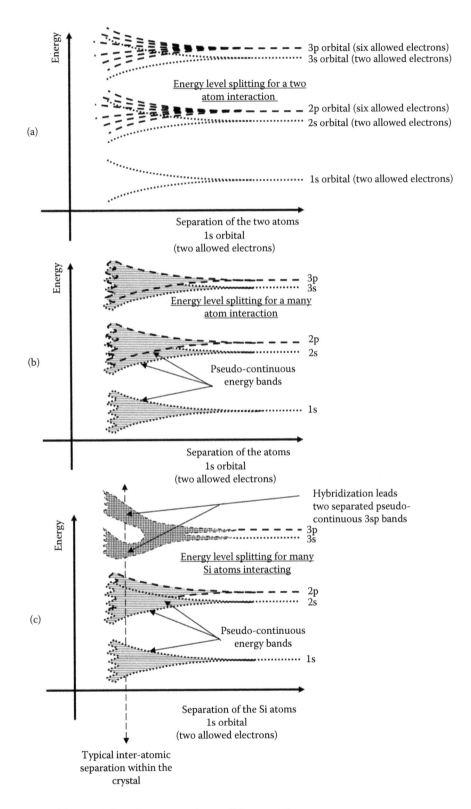

FIGURE 4.7 Electronic band structure in a solid material.

many identical atoms are brought together, each energy level of each atom splits in a similar manner for each of the other atoms that are close enough for interactions to occur. Thus each of the single energy levels possessed by the single atom is now split into a large number of separate levels dispersed over a range of energies (Figure 4.7b). Since each of these levels has a small energy spread due to Heisenberg's uncertainty principle, they overlap to produce an apparently continuous band of allowed electron energies. However, the continuity is only apparent – there is still a limit to the number of electrons that can have energies within the band. If, for example, the original energy level of the atom was an s level and was occupied by two electrons, then the electrons from that level from all the interacting atoms will occupy all the individual levels making up the band and there would be no empty energy levels available to be occupied by any additional electrons.

Within a solid the lowest energy level electrons are still tightly bound to their individual nuclei. The outer energy levels, though, merge and become continuous throughout the solid (Figure 4.8). At low temperatures, the outermost electrons of the isolated atom will occupy the valence band of the solid. If the conduction band is well separated from the valence band and the valence band is fully occupied with electrons, then the solid will be an electrical insulator. The material will be an insulator because, if a voltage is applied across it, the electrons in the valence band will try to move towards the positive electrode. However, movement through the solid means that the electrons will gain energy. Since the valence band is already full, the energy levels that might have represented electrons moving through the solid are already filled by other electrons. Thus, even though the valence band runs continuously through the solid, the electrons are unable to move.

If the gap between the valence and conduction bands is reduced to a small value, however, then thermal energy may enable a few valence electrons to jump up to the conduction band. A small electrical current will then flow when a voltage is applied to the material, carried by electrons in both the conduction and valence bands, the latter being possible because there are now a few free energy levels in the valence band arising from those electrons that have been excited to the conduction band. The material will then be a semiconductor. Finally, if the conduction band overlaps the valence band and/or the valence band is not fully occupied with electrons, then the solid will be an electrical conductor. The electrons in the valence band (and any that have been excited into the conduction band) will have many free energy levels to occupy as they move through the solid in response to the applied voltage.

In intrinsic semiconductors, the energy gap between the top of the valence band and the bottom of the conduction band is quite small (e.g. 1.1 eV for silicon, 0.66 eV for germanium). Thus, thermal excitation of electrons from the valence to the conduction band occurs relatively easily. The absorption of a photon can also excite electrons and a 1.1 eV photon has a wavelength of 1.1 μm while 0.66 eV corresponds to 1.9 μm – so that these are the long wavelength limits for silicon and germanium when used as photon detectors.

The effective band gap, however, can be reduced by doping. This is the addition of a small quantity of another element into the intrinsic semiconductor. Both silicon and germanium

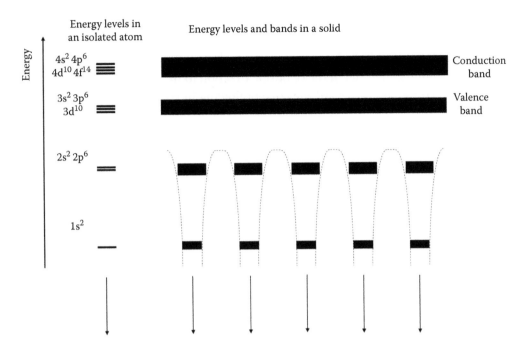

FIGURE 4.8 Valence and conduction bands.

are group 14 elements* and dopants for them are usually chosen from group 13 or group 15 elements. Group 13 elements have a valence of 3 while group 15 have a valence of 5. They are called acceptor and donor elements, respectively. The action of a dopant element is to introduce new, allowed, energy levels into the band gap. The levels are vacant and just above the top of the valence band for acceptor dopants. For donor elements the new levels contain electrons and are just below the bottom of the conduction band. Typically the new levels will be only 0.01–0.1 eV away from their respective bands. Semiconductors that have been doped are known as extrinsic semi-conductors.

Taking the action of the acceptor dopant first, for silicon or germanium, the elements used are normally boron, aluminium or gallium. The new levels will be just above the valence band and electrons from the latter can easily be thermally excited into the new levels (for 0.01 eV, the absorption of a photon of wavelength 120 μm or shorter will also have the same effect). The excitation of valence electrons to dopant energy levels will leave vacant energy levels within the valence band. When a voltage is applied across the extrinsic semiconductor the electrons in the valence band will be able to move and so a current will flow (the dopant energy levels are tied to their donor atoms and so electrons in these levels will remain stationary). In many cases, the moving electrons actually 'hop' through the material – a nearby electron will jump over to an atom with a vacant energy level leaving behind a new vacant energy level, another electron will then hop into that level and so on. Since in a sea of (negative) electrons the absence of an electron will leave a region that appears positive by contrast, it is customary to talk about the *absence* of an electron

---

* Current International Union of Pure and Applied Chemistry (IUPAC) notation – often the older group IV notation will still be encountered for silicon group elements (and groups III and V for the boron and nitrogen group elements).

as being the *presence* of a positive hole and then to treat this as though it were a physical particle.

Viewed then, on a larger scale, it seems as though there is a positive hole moving continuously through the material in the opposite direction to the movement of the electrons. It is thus often convenient to regard the current flowing in the valance band in this situation exactly as though it were a flow of positive charges moving towards the cathode. Since the current is carried by positive charges on this view of the process, the material is called a p-type semiconductor.

The normal donor elements for silicon and germanium are phosphorous and arsenic. Their presence introduces dopant energy levels just below the bottom of the conduction band and these levels contain electrons. The electrons in the donor levels can easily be thermally excited into the conduction band where, under the influence of an applied voltage, they will move through the material and produce an electric current. Since the current is carried by the negative electrons, the material is called an n-type semiconductor.

Now, from the discussion so far, both silicon and germanium should be good electrical conductors since they have four valence electrons each and the 3s and 3p energy levels* (silicon) and 4s and 4p energy levels (germanium) can hold up to eight electrons. However, they have crystal structures similar to that of diamond with each atom being at the centre of a tetrahedron of four other atoms. Each of the four valence electrons links the central atom to one of its neighbours forming a hybrid level, labelled 3sp or 4sp. The electron structure of silicon is thus $1s^22s^22p^63sp^4$ and not $1s^22s^22p^63s^23p^2$ as might be expected. The effect of this hybridisation on silicon's valence energy level is to split it into two halves separated by an energy gap of some 1.1 eV (Figure 4.7c). The four electrons occupy the lower band and the upper one is empty. The lower band is now the valence band and is filled; the upper band is the conduction band and is empty. Similarly, the ground state of germanium is $1s^22s^22p^63s^23p^63d^{10}4sp^4$ and its forbidden energy gap is 0.66 eV. Thus neither silicon, nor germanium, are electrical conductors, but the small values of their forbidden energy gaps class them as semiconductors.

4.2.6.2 Solid State Detectors

For the detection of high-energy charged particles, doped silicon is formed into p–n diodes (see also Section 1.4.1.4). When p-doped and n-doped silicon are placed in contact with each other the electrons in the conduction band of the latter have higher energies than the holes in the valence band of the former. The electrons drift across the junction to occupy the holes and so induce a voltage across it – the n-type silicon becoming positive and the p-type negative. The drift continues until the voltage across the junction rises sufficiently to form a barrier to the flow of the electrons. The diode is then reverse biased – that is an external voltage of up to 100 V is applied to make the n-type silicon even more positive and the p-type even more negative.

---

* The 3d (silicon) and 4d and 4f levels (germanium) have higher energies than the 4s (silicon) and 5s (germanium) levels so that electrons go into these latter levels before occupying the nominally lower levels.

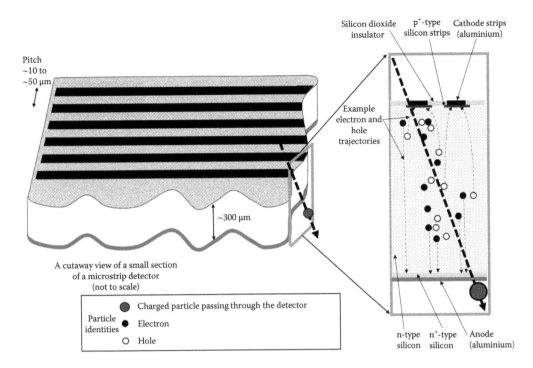

FIGURE 4.9   Schematic cross section through a silicon-based charged particle detector.

When a charged particle passes through the n-type silicon of a p–n diode it excites some electrons from the valence band to the conduction band. Thus both electrons and holes are produced and are usually termed electron–hole pairs. Typically, however, only around 3–4 eV are needed for an electron–hole pair to be produced compared with the 30 eV needed for an ion pair in a gas and the 100 eV for the initial photon in a scintillation detector. Also the stopping power of the solid silicon is much greater than that of a gas so that for a silicon wafer with a thickness of some 300 μm, around 3×10^4 electron–hole pairs are generated by each charged particle interaction. The reverse bias acts to ensure that these electron–hole pairs do not recombine and attracts the electrons to the anode on the n-type silicon and the holes to the cathode on the p-type silicon. The 3×10^4 electron–hole pairs that are generated are not further amplified and the pulse has therefore to be detected using high-quality, low-noise electronics. The typical response time is a few nanoseconds.

The physical structure of a p–n diode when used as a charged particle detector is shown in Figure 4.9. The bulk of the detector is a wafer of n-doped silicon around 300 μm thick. One face has narrow (~5 to ~10 μm) strips or small squares of heavily* p-doped regions which are insulated from each other and each of which is connected to the cathode and to separate read-out electronics. The other face has a narrow region that is heavily n-doped and is connected to the anode. Dividing the cathode into narrow strips produces a microstrip detector and provides positional information for the interaction in one coordinate; using

---

* Heavily doped semiconductors are often given the symbols p+ or n+. The heavy doping results in high conductivities for the semiconductor.

small squares produces a pixel detector that gives 2D positional data but with smaller numbers of electron–hole pairs in each square compared with the strips. The positional accuracy for the pixel detector is around ±5 μm. Two-dimensional position measurement can be obtained with microstrip detectors by using two of them with the strips oriented at 90° to each other – any particle with an energy larger than around 0.25 MeV will pass through both detectors. Alternatively, the n^+ layer can be divided into strips orthogonal to the cathode strips producing a double-sided microstrip detector.

Examples of the usage of silicon-based particle detectors include Payload for Antimatter Exploration and Light-nuclei Astrophysics (PAMELA) and Alpha Magnetic Spectrometer (AMS-02) and RBSPICE and RPS on the van Allen probes (Section 4.2.5).

The Resurs DK No. 1 commercial Earth-observation spacecraft carried, amongst other instruments, PAMELA, the first spacecraft-borne instrument built to investigate cosmic rays and also to measure high-energy charged particles from the Sun, in the Earth's magnetosphere and from Jupiter. The particle spectrometer uses a double-sided microstrip silicon detector giving positional accuracies of 51 μm from cathode strips and 66.5 μm from anode strips. It is sensitive to electrons up to an energy of 400 GeV, to positrons in the energy range 50 MeV to 270 GeV, to protons up to 700 GeV and to nuclei up to carbon with energies up to 200 GeV per nucleon.

AMS-02 is a module attached to the International Space Station* (ISS) in 2011 to study high-energy particles and is also still operating at the time of writing (Figure 4.10). The spectrometer uses, amongst other types of detector, 2264 double-sided microstrip detectors in nine layers that give a total detector area of 6.2 m². It can measure particles' magnetic rigidities† up to an energy of 2 TeV and determines the position of the interactions to about 10 μm.

The van Allen probes also have the relativistic electron proton telescope (REPT), which is the third instrument in the ECT suite and is designed to detect electrons up to ~20 MeV and protons and heavier ions to over 100 MeV. It is simply a stack of nine SSDs at the closed end of a long and thick cylinder. The open end of the cylinder defines the instrument's acceptance angle of 32° and the rest of the cylinder acts as a shield for the detector stack.

The four FIREBIRD CubeSats (Focused Investigations of Relativistic Burst Intensity, Range and Dynamics) that investigated relativistic electron microbursts from the van Allen belts near Earth each carried pairs of 1.5 mm thick silicon SSDs. They were mounted in light-tight enclosures and also used aluminium foil to screen out the lower energy particles. They observed at six electron energies over the range from 250 keV to 1 MeV.

A related detector used, for example, from the orbiters on the Apollo 16 and 17 lunar missions, is the surface barrier detector. This is usually a silicon wafer and so is often called a silicon surface barrier (SSB). The silicon has thin metal electrodes on its two flat surfaces – gold for n-type silicon and aluminium for p-type silicon. The electrodes are

---

* AMS-02 was mounted on the ISS in May 2011 and is still operating at the time of writing.
† The magnetic rigidity (R) of a charged particle is a measure of the curvature of its track when it is moving in a magnetic field. It is given by the product of the radius of curvature of the path (ρ) and the magnetic field strength (B): $R = B \times \rho$. It is also given by the ratio of the particle's momentum (p) to its charge (q): $R = p/q$. The higher the rigidity, the less the path curves.

FIGURE 4.10 AMS-02 mounted on the ISS. (Courtesy of NASA.)

charged to reverse bias the silicon and produce a depletion region in which the charge carriers are driven from below the top electrode. The depletion region must be thick enough to stop the incoming charge particles – about 30 μm for 5 MeV alpha particles in silicon. The charged particles have to pass through the top electrode (i.e. the barrier) and then form the electron–hole pairs within the depletion layer. These are then collected by the electrodes and produce a pulse which provides the detector's output. On Apollo 16 and 17, radon-222 was detected in this way by SSDs collecting the 5.965 MeV alpha particles that are one of radon's decay products.

Germanium could be used in place of silicon in a similar fashion to produce charged particle detectors. However, its much higher cost and the need to cool it to 80 K or so means that silicon-based detectors (which are normally operated at room temperature) are almost always the preferred choice. Germanium is, however, used extensively for x-ray and γ ray detection.

4.2.7 Transition Radiation Detectors

Transition radiation is produced when a relativistic charged particle passes through the boundary between two homogeneous substances with differing dielectric constants. Its existence was predicted in 1946 by Ilya Frank and Vitaly Ginzburg and optical transition radiation was detected experimentally in 1959 by Phillip Goldsmith and John Jelley. Of more significance for charged particle detection was the prediction in 1957 by Gregory Garibian that x-ray transition radiation would be emitted by ultra-relativistic charged particles. He showed that the emission would be forward peaked with most of the emission occurring at an angle of ~$\sin^{-1}(1/\gamma)$* to the particle's path, that the total energy emission would be proportional to γ, that the x-rays would have energies within an order of magnitude of 10 keV and that the individual photon energy increases as the

---

* γ is the Lorentz factor – $\gamma = \left(1 - (v^2/c^2)\right)^{-1/2} = E/(mc^2)$ where v is the particle's velocity, c the velocity of light in a vacuum, E the particle's energy and m the particle's rest mass.

energy of the charged particle increases. The number of x-ray photons emitted however is small – at a single interface the chance of an emission occurring is similar to α (the fine structure constant = ~1/137 = 0.8%). Nonetheless the phenomenon enables the construction of instruments that can detect and discriminate between electrons or positrons and hadrons (especially protons) in the energy range 1 GeV to 1 TeV where other detectors are inefficient. The ability to identify particle types arises from the dependence of the energy emitted as x-rays upon γ. A 100 GeV positron, for example, has $\gamma \approx 2 \times 10^5$ but a proton of the same energy has $\gamma \approx 100$, so that the positron will emit around 2000 times the x-ray energy of the proton (in fact almost no transition x-ray emission occurs for $\gamma < \sim 10^3$).

The small chance of emission of an x-ray photon at a single interface crossing means that practical transition radiation detectors (TRDs) must use many interfaces. For spacecraft-borne TRDs, one of the 'materials' involved in forming the interface is usually a vacuum. Many other substances can be used for the other half of the interface. Generally though, the lower the atomic number of the material the better, since the x-rays are strongly absorbed by higher atomic number elements. Amongst the commonly used materials therefore are aluminium, beryllium, carbon, lithium, mylar, polyethylene and polypropylene. The materials forming the interfaces are called the radiator of the TRD. The physical structure of the radiator may be a stack of tens to hundreds of thin foils, stacks of thin mats of small fibres or thicker layers of foam.

Once the transition x-rays have been produced they must be detected in their turn. This is normally done using relatively conventional x-ray detectors. Ion chambers, multi-wire proportional counters and silicon microstrip detectors as discussed above in connection with the direct detection of charged particles _ can also be used with only minor adaptation for TRDs.

A variant on the proportional counter is also frequently encountered. This is often called a 'straw' since it consists of a long thin cylinder that is typically a few millimetres in diameter and a few hundred millimetres in length. The walls of the tube are formed from an aluminised thin plastic film and it is filled with suitable detector and quench gases. The latter is usually a mix of 80% xenon and 20% carbon dioxide. A wire anode runs down the centre of the straw and is charged to a kilovolt or so with respect to the straw's walls (which form the cathode). The x-rays are detected, like charged particles, by the Townsend avalanche resulting from their production of ion pairs in the gas. In a TRD, many such straws will be used arranged side by side in a layer underneath a layer of radiator. Several such radiator–detector pairs may then be stacked to form the complete instrument.

The PAMELA instrument (already mentioned) carries a TRD in addition to its silicon microstrip detectors. Its TRD uses 10 carbon fibre radiators and 1024 straw x-ray detectors. The straws are 4 mm in diameter and 280 mm long and have a 1.4 kV operating voltage. They use an 80:20 mix of xenon and carbon dioxide. Sixteen straws are arranged side by side to form a layer and then two layers stacked in a close packed format to form a module. The top five radiators are larger than the bottom ones so the larger radiators need four straw detector modules each, while the smaller ones use three modules. Two radiators are used above the first layer of straw modules to increase the signal strength. The PAMELA TRD detects and discriminates between particles up to 1 TeV in energy.

4.2.8 Calorimeter

See also Section 3.2.5.3.

A calorimeter is a detector designed to stop a charged particle that is moving through it, absorb that particle's kinetic energy and determine its value. Calorimeters are subdivided into electromagnetic (ECAL) and hadronic (HCAL) types. The former measures the energy of electrons and positrons, the latter the energy of hadrons, especially protons and anti-protons (and also neutrons). ECALs can also be classed as homogeneous or sampling, but HCALs are only of the sampling type.

A high-energy electron or positron interacts with the calorimeter material by emitting bremsstrahlung radiation as it is slowed down. If the bremsstrahlung photons are γ rays of high enough energy, then they may in turn interact to produce electron–positron pairs which in their turn may produce additional bremsstrahlung radiation. With very high-energy particles the result of the interaction is a shower, called the electromagnetic cascade, of electrons, positrons, x-rays and γ rays within the calorimeter material. Providing all the cascade is contained within the calorimeter and it is of the homogeneous type, the original particle's energy may be determined.

A high-energy hadron also produces a shower within the calorimeter but does so by interacting with the nuclei via the strong nuclear force. The shower primarily contains pions, nucleons, x-rays and γ rays.

A homogeneous calorimeter uses, as its interacting material, a substance that both interacts with the particle and the subsequent cascade and absorbs the energies of the particles and photons – such as sodium or caesium iodide, $Bi_4Ge_3O_{12}$ (BGO) or lead glass. Sampling calorimeters have layers of alternating active detectors, such as scintillators or proportional counters and absorbers, such as tungsten, lead or uranium. Only the energy lost within the active layers is then measured (sampled).

The PAMELA instrument carries a sampling ECAL for studying anti-protons and positrons. It uses 22 layers of tungsten as its absorbers and each tungsten layer has a mosaic of silicon microstrip detectors above and below it. The two detector layers have their strips orthogonal to each other so that positional measurements can be made. The microstrip detectors are each 80 mm × 80 mm in size and are arranged in a 3 × 3 mosaic. Each detector has 32 strips so positional measurements can be made to an accuracy of about 2.5 mm. The calorimeter measures electron and positron energies from a few tens of GeV to 1 TeV.

4.2.9 Čerenkov* Detectors

Čerenkov radiation arises when a charged particle travels through a medium other than a vacuum faster than the phase velocity of light in that medium. The radiation is emitted in a cone centred on the particle's track. Some detectors intercept the whole or a significant

---

* After Pavel Čerenkov 1904–1990.

part of the cone and so see the radiation as a ring. Such instruments are therefore known as Ring Imaging Čerenkov detectors (RICH).

The threshold energy (at which the particle's velocity equals the phase velocity of light in the medium) varies with the mass of the particle and is the lowest velocity at which any Čerenkov radiation can occur. With water as the medium (refractive index, μ = 1.33), the threshold energies* for some of the commonly observed charged particles are the following:

Electrons and positrons	260 keV
Protons and anti-protons	475 MeV
Deuterons	960 MeV
Alpha particles	2 GeV

In air at STP, these values would be a factor of ~1200 times larger (and at a height in the atmosphere of ~20 km, where many charged particles and their resulting secondary particle showers start to emit their Čerenkov radiation, the factor would be ~20,000).

The maximum cone half-angle for the radiation varies from about 1° or less for air (at STP) to ~40° for water and varies from 0° up to these maxima as the particle's velocity increases from the Čerenkov emission threshold to close to the speed of light in a vacuum. In a medium with a refractive index that is constant with frequency, the radiation has a continuous spectrum with the number of emitted photons in a given frequency interval independent of the frequency. The emitted *energy* therefore tends to peak at the shorter wavelengths (hence the blue colour of Čerenkov radiation in the water moderator of some nuclear reactors). However, few substances have such constant refractive indices, so the emission usually peaks near to where its refractive index has its highest values. For protons traveling through air, the peak emission is in the visible part of the spectrum.

Čerenkov detectors can operate by using the threshold energy to distinguish between particles with differing masses, especially if the velocity of the charged particle is already known or restricted to a small range (Section 5.2). The use of two different radiators with significantly different threshold energies (i.e. refractive indices) will also help, since a less massive lighter particle is likely to emit in both materials, while a more massive particle with the same energy (lower velocity) may only radiate in the medium with the lower threshold energy. At velocities close to that of light, the Čerenkov emission intensity varies little with the particle's energy, but varies as the square of the charge (Z) of the particle. This can then be used to distinguish between heavier ions. Čerenkov-based detectors may also be used in a calorimeter mode, determining the particle's energy from the total energy of the Čerenkov emissions.

Particle detectors based upon observing Čerenkov emissions can be enormous in size – the Pierre Auger cosmic ray observatory in Argentina, for example, has 1600 twelve-ton tanks of water distributed over a 3000 km² area of land. Detectors for use in space, however, can bear quite a close resemblance to some designs of scintillation detector and are sometimes referred to as such, although the emission mechanism for

---

* The relativistic formula for kinetic energy must be used here and is $E_K = mc^2 \dfrac{1}{\sqrt{1-(v^2/c^2)}} - 1$ where E_K is the kinetic energy, m the particle's rest mass and v the particle's velocity.

the photons is quite different in the two cases. Terrestrial Čerenkov detectors mostly utilise air (i.e. the atmosphere) or water as the radiating medium and employ photomultipliers to observe the photon emissions. They are often detecting the millions of charged particles in an extensive air shower (EAS) which has resulted from a single very high-energy cosmic ray colliding with an atomic nucleus high in the Earth's atmosphere. Other radiator materials, especially for laboratory-based detectors, can include quartz, glass and perfluorobutane (C_4F_{10}).

Čerenkov detectors for spacecraft (or balloons) use sodium fluoride, aerogel, glass or acrylic plastic radiators and PMTs as their photon detectors. The balloon-borne Super Trans Iron Galactic Element Recorder (SuperTIGER) (Figure 4.11) instrument was flown for 55 days over Antarctica in 2012/13 at an altitude of 39 km in a very successful mission. Its Čerenkov instrument comprised four modules, each with two 2.8 m^2 radiators and forty-two 125 mm diameter PMTs as their photon detectors. The radiators were acrylic plastic with a refractive index of 1.5 and an aerogel with a refractive index of 1.04 and were housed separately in 200 mm deep light reflective boxes. It was designed to study medium atomic number ions in the range from neon to molybdenum.The Martian Radiation Environment experiment (MARIE) (Section 7.2.5) on the Mars Odyssey orbiter used high-refractive index Schott™ glass for its Čerenkov detector. The use of Čerenkov detectors within the van Allen probes' RPS instruments has been discussed in Section 4.2.5.

A space-based Čerenkov detector is currently under construction for mounting on the ISS around 2018. Japanese Experiment Module – Extreme Universe Space Observatory (JEM-EUSO) is of quite different design from SuperTIGER and similar detectors. The radiator for the detector is a large volume of the Earth's atmosphere that is observed by a wide angle camera as the photon detector. High-energy charged particles (>10^{19} eV), upon colliding with a nucleus high in the Earth's atmosphere, initiate an EAS containing millions of nucleons, pions, muons, electrons, x- and γ rays. The shower particles will continue moving along the original track of the incoming particle in a slowly spreading cone, most of them with velocities high enough to emit Čerenkov radiation. The Čerenkov radiation will be emitted close to the line of the original particle's track until it hits the ground. It will then show up as a small (a few kilometres) circular or elliptical patch of light. The Čerenkov emission will not appear as a ring because the many millions of particles contributing to it will have slightly divergent paths. The particles will also emit UV photons via fluorescence as they excite atmospheric atoms and molecules in their passage. These photons will be emitted in random directions and will enable JEM-EUSO to follow the EAS directly in real time. A prototype instrument called EUSO-balloon was successfully flown (by balloon) from Canada in 2014 at an altitude of 40 km to test the concept.

The JEM-EUSO camera will operate in the near UV with three 2.65 m × 1.9 m Fresnel lenses feeding an array of 4932 PMTs. The PMTs are of the multiple anode design with 64 pixels each and can detect individual photons. The field of view of the camera is 60°, so that the angular resolution per pixel is ~0.1° – corresponding to a ground resolution of about 750 m. A single image will cover a circular area of the atmosphere and ground about 400 km across. The instrument's time resolution is 2.5 μs and it will operate continuously. An EAS will be observed from its fluorescent emission first and appear as a fast moving

FIGURE 4.11   The launch of SuperTIGER from near McMurdo Station, Antarctica, on 8 December 2012. (Courtesy of NASA.)

spot travelling downwards in the atmosphere and lasting for about 100 μs. The Čerenkov emission may be observable during the atmospheric passage when occasional photons are scattered towards the instrument, but the main emission will come from scattering from the ground (land or sea) or from clouds. The energy of the incoming charged particle may be estimated from the total photon emission with an accuracy of about ±30%.

CHAPTER 5

Ion Optics and Charged Particle Instrumentation

5.1 INTRODUCTION

The detectors of charged particles discussed in the previous section are equivalent to the eye or charge-coupled devices (CCD), etc., in optical astronomy and, just as in optical astronomy, we need telescopes and other instruments to feed light to the detectors, so, therefore, in studying charged particles, we need equivalent instruments to feed the particles into their detectors. Charged particle instruments are based upon using electric and magnetic fields* to steer the particles in the desired direction(s). Thus the packet/beam/stream of particles may be concentrated into a smaller area/volume (focussed), its direction of motion may be altered (refracted/reflected), its component particles may be sorted according to their velocities/energies, masses or charge-to-mass ratio (spectroscopy) and certain specified components of a mixed packet of particles may be selected out or isolated (filtered). The design, construction and usage of items of equipment which can manipulate particles in these ways forms the topic properly called charged particle optics, but which more frequently nowadays, although less accurately, is called ion optics.

In the subsections below, we examine some of the ion optics-based instrumentation commonly used on board spacecraft and how they operate. NB: The low-power ion thrusters used on spacecraft such as Deep Space 1 also involve many of the concepts and processes discussed below, but applications in this area are beyond the scope of this book.

5.2 ION OPTICS AND CHARGED PARTICLE INSTRUMENTATION

5.2.1 Electrostatic Analyser

In Chapter 4, reference was made to observing particles whose velocities are constrained to lie within a small and known range of values. The ESA is one type of instrument that performs this function and it is the charged particle equivalent of an optical filter

---

* Some familiarity will be needed for these sections with electrostatic, electromagnetic and magnetic physics. If needed, up-to-date books on these subjects or on general physics may be consulted (see Appendix A).

(see also Section 5.2.5). The simplest design is based upon a pair of parallel conducting plates one of which is positively charged with respect to the other and both are in a vacuum. Apart from edge effects, the electric field between the plates is uniform with the lines of force parallel to each other and orthogonal to the plates (Figure 5.1). The electric field strength (E) is given by

$$E = -\frac{V_1 - V_2}{S} \tag{5.1}$$

where V_1 and V_2 are the voltages on the two plates and S is their separation.

A particle with an electric charge (q) placed between the plates will experience a force (F) of

$$F = qE \tag{5.2}$$

Now, if one of the plates has a hole cut into it, a beam of charged particles can be directed into the space between the plates. The particles will follow parabolic tracks between the plates under the influence of the electric field, with, for similarly charged particles, the curvature of the tracks being tighter for those with low momenta and more open for the particles with higher momenta. If, additionally, the particles all have the same mass then the lower velocity particles will have the tighter tracks and the higher velocity ones the more open tracks (Figure 5.2). When the particles return to the plate containing the entry hole, they will thus

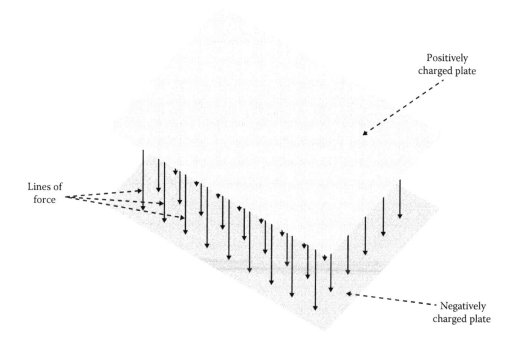

FIGURE 5.1 The lines of electric force between two conducting plates charged to different voltages (edge effects will distort the outer lines for plates of finite sizes).

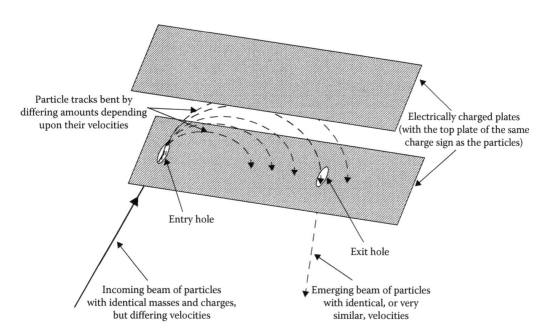

Particle tracks bent by differing amounts depending upon their velocities

Electrically charged plates (with the top plate of the same charge sign as the particles)

Entry hole

Exit hole

Incoming beam of particles with identical masses and charges, but differing velocities

Emerging beam of particles with identical, or very similar, velocities

FIGURE 5.2   A parallel plate electrostatic analyser (ESA).

be strung out into a line with the lower velocity particles closer to the entry hole. A second hole in the plate will then allow those particles with a specific velocity (or a small range of velocities) to emerge. A detector suitably placed to intercept the emerging beam of particles will then be able to determine their flux for the selected velocity and we have a parallel-plate analyser (sometimes called a plane mirror analyser). Particles with a different velocity could then be selected by moving the hole to a different place, but in practice the velocity of the particles in the emerging beam is altered by changing the potential difference between the two parallel plates. By changing the potential difference continuously or stepwise, a velocity spectrum of the original incoming beam may be built up. Alternatively, a linear position-sensitive detector placed along the line where the particles impact the plate can obtain the spectrum in a single observation.

When the incoming beam of particles is non-uniform (i.e. it comprises particles of differing masses and charges) then the ESA sorts them according to their kinetic energy (KE) per unit charge ratios (KE/q). Generally, in instruments used on board spacecraft, this does not cause much confusion. The particles being studied, such as electrons, positrons, protons, anti-protons and ions, have masses so different from each other that only the types of particles of interest will be selected. Thus, although electrons and anti-protons have the same electric charge, their masses differ by a factor of two thousand and if travelling at the same speed their KE/q ratios would differ by the same factor. Where there is some possibility of confusion between different types of particles, then the use of a detector, such as a ToF detector (Section 5.2.6.4), that is sensitive to the particle's velocity may help to resolve the ambiguity (see also Section 5.2.5.1).

If the incoming beam has an angular spread then the ESA may be designed so that the particles with the same energies are brought to a focus at the detector. With the arrangement shown in Figure 5.2, the requirement is met when the incoming beam is at an angle of 45° to the axis of the ESA. For other angles of incidence, focussing may be achieved by separating the entrance and exit apertures from the charged plates (Figure 5.3).

The focussing shown in Figure 5.3 is only along the axis of the ESA; at right angles to this, the exit beam will be broadened by any angular spread of the incoming beam. The particle beam may be focussed in both directions, however, by replacing the flat charged plates with concentric half cylinders, the outer cylinder having at least two and a half times the radius of the inner cylinder and the axis of the cylinders being parallel to the axis of the ESA. This latter design is known as the cylindrical mirror analyser.

An additional problem with these ESAs is that the holes in the charged plate act as weak lenses (see Section 5.2.2) and cause some de-focussing. This may be reduced by placing fine conducting meshes over the holes to make the electric field more uniform.

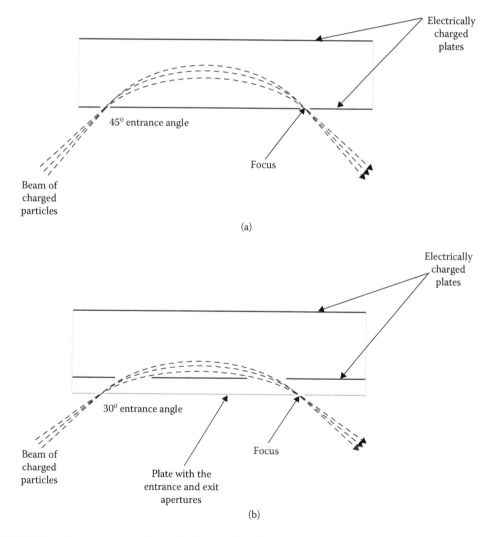

FIGURE 5.3 Cross sections through designs for focussed charged particle parallel-plate ESAs. (a) Beam at 45° to the ESA's axis. (b) Beam at 30° to the ESA's axis.

The parallel plate and cylindrical mirror ESAs both operate by decelerating the particle to zero velocity in the direction orthogonal to the plates and re-accelerating it in the opposite direction to emerge from the second hole with its original energy. They are thus classed as retarding ESAs. A second group of designs change the particle's direction of motion without significantly changing its energy and are called deflection ESAs. The two main types of deflection ESAs have cylindrical or spherical configurations.

The cylindrical deflection ESA, at first sight, looks similar to the cylindrical mirror ESA – being two concentric half cylinders or half sectors. However, a charged particle's path is around the space between the two cylinders, not along the length of the cylinders' axes (Figure 5.4). Focussing occurs at 127.3° (= $\pi/\sqrt{2}$ radians) around the cylinders and only particles with the correct energy emerge from the exit hole to be detected. Like the retarding ESAs, though, the energy selected for the emerging particles can be changed by changing the voltage difference between the cylinders, so allowing the spectrum to be built up.

If needed, the beam may be focussed in two dimensions by using nested hemispheres in place of the sectors of cylinders. The focal point is then 180° around from the entrance aperture and the device is called a spherical deflection analyser.

A cylindrical deflection analyser was on board Mariner 2 which used a Faraday cup detector (Section 5.2.6.2).

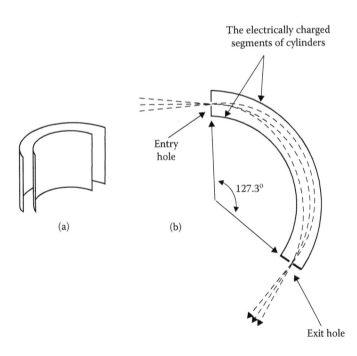

FIGURE 5.4 Particle tracks within a cylindrical deflection ESA. (a) Sketch showing the physical arrangement of the electrically charged segments of cylinders forming a cylindrical deflector ESA. (b) Charged particle tracks within a cylindrical deflector ESA.

A design of spherical deflection analyser that is much used in space applications is called the top hat analyser. Particles enter the analyser through a hole at the top of the outer charged hemisphere (Figure 5.5). A plate just above the entrance hole (the top hat) ensures that only particles coming from a restricted range of angles to the side of the device can enter the hole. However, within this plane, particles can enter over the full 360°. The particles follow curved paths between the hemispheres and exit at their bases. As with the other types of ESAs, only particles with the correct energy (or charge to mass ratio) will emerge – too high an energy and the particle will collide with the outer hemisphere, too low and it will collide with the inner hemisphere. However, the energies of the emerging particles can be selected by changing the voltage between the hemispheres. The particles are picked up by a detector with an annular shape that matches the exit aperture between the hemispheres. By using a position-sensitive detector or by using many small detectors around the annulus (pixels), an instantaneous reading of the particle flux over the whole 360° acceptance angle can be mapped out. Particles entering the top hat analyser on parallel tracks are focussed at their exit point whether they enter the analyser near the centre or edge of the entrance hole.

Mention has already been made of the top hat analysers used in the dual ion spectrometers (DIS) instruments on board the MMS spacecraft (Section 4.2.5, Figure 4.5). These

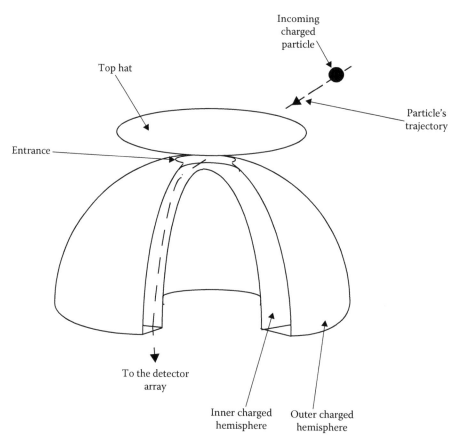

FIGURE 5.5   The top hat analyser.

are constructed from half-hemispheres and so their acceptance area is 10° × 180°. The Hot Plasma Composition Analyser (HPCA) on board the same spacecraft also uses a top hat analyser in combination with a ToF instrument. Its acceptance angle is 360° and this is divided up into 16 pixels. The HPCA is used to study light ions in the energy range ~10 eV to 30 keV. Another example of the use of a top hat analyser on a spacecraft is the Solar Wind Around Pluto (SWAP) instrument on the New Horizons spacecraft (Figure 5.6). This has an acceptance angle of 276° × 10° and detects particles in the energy range from 35 eV to 7.5 keV.

Both Mars Express and Venus Express carried similar electron top hat analysers, as a part of their Analyser of Space Plasmas and Energetic Atoms (ASPERA) packages. The analysers used 16 microchannel plates (MCPs) as their detectors, each covering a 22.5° sector in azimuth. The voltage was swept to enable electrons from 0.8 eV to 30 keV to be selected. The instrument on Venus Express found a strong (10 V) electric field for Venus (Section 7.5.2).

The five THEMIS and ARTEMIS spacecraft each carry twin 180° top hat analysers. They operate from a few eV to 30 keV for electrons and up to 25 keV for ions. The 6° field of view is swept around the whole sky by the 20 rpm spin of the spacecraft and the instruments use MCPs as detectors.

In the future, the Turbulence Heating ObserveR (THOR) proposal is expected to include an ion top hat analyser using 32 channel electron multiplier (CEM) detectors. It is hoped to be able to make some six measurements per second with an angular resolution of about 1.5° and an energy resolution of about 7%.

FIGURE 5.6   An artist's concept of Solar Wind Around Pluto (SWAP) (circled) on the New Horizons spacecraft as it passes Pluto. (Courtesy of JHUAPL/SwRI.)

5.2.2 Electrostatic Lenses

The focussing effects of ESAs upon beams of charged particles have been mentioned above. It is also possible to produce a device that acts more like an optical lens. Such lenses have found little application in instrumentation on board spacecraft, but they do often form part of the ion thruster drive systems used on spacecraft such as Deep Space 1 and Dawn and as micro-thrusters for controlling the orientation and position of many more conventionally powered spacecraft. They also, of course, play vital parts in the operation of electron microscopes and, until the last few years, in televisions and cathode ray tubes. We therefore include a brief discussion of electrostatic lenses for completeness. Electrostatic lenses are relatively weak and are best suited to focussing lower energy (<50 keV) ion beams. When required, magnetic lenses (see below) can be used for beams of particles up to 10 MeV or more.

The majority of electrostatic lenses are based upon two or more tubes made from a conducting material and at differing voltages from each other. The tubes can have almost any cross-sectional shape, though circular or square are the most widely found examples (if focussing is needed only in one plane, then the tubes can be replaced by two parallel flat plates). The length of the tube is generally a few millimetres to a few tens of millimetres, but they can also be so short that they are rings rather than tubes. With a two-cylinder lens the central axes of the tubes are aligned and there is a small gap between the tubes. The electric field across the gap acts to focus the charged particle beam. The position of the lens' focus and magnification may be changed by changing the voltages on the cylinders. If a fixed focus is needed then three cylinders are used in a similar fashion (Figure 5.7)

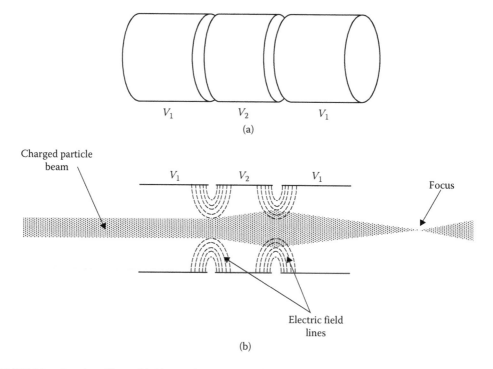

FIGURE 5.7   An einsel lens. (a) Physical structure. (b) Cross section.

and the result is called an einsel* lens. The voltages on the cylinders of an einsel lens are generally the same on the two end cylinders (V_1) and stronger for the central cylinder (V_2). The voltages can be changed to accommodate charged particles with different energies and the focus remains fixed provided that the ratio of V_1/V_2 is kept constant. The voltages on the cylinders can range from a few volts to several kilovolts depending upon the charge, mass and energy of the particles involved.

5.2.3 Electrostatic Prisms

Sometimes, perhaps to direct a beam of charged particles into the entrance aperture of another instrument, it may be necessary to bend the path of a beam without making other changes to it. This can be done by combining two ESAs and the resulting device is called an electrostatic prism by analogy with the way that a right-angle prism can be used to change the direction of a beam of light. A parallel beam of similar charged particles is brought to a focus by an ESA and that focal point made to coincide with the entrance aperture of a second, identical ESA. A parallel beam of particles emerges from the second ESA and by changing the mutual orientations of the two ESAs the beam can be directed in the desired direction.

5.2.3.1 Quadrupole Beam Deflector

A variant on the previous idea is the quadrupole beam deflector (QBD) that is easily switched between diverting a beam of charged particles through ±90° or leaving its path unchanged (Figure 5.8). The particle beam is usually fed via einsel lenses into the device and then further einsel lenses after its emergence from the QBD feed it to the next instrument. Four short quadri-circular conducting rods are arranged in a square inside a grounded conducting enclosure. When none of the rods are charged, the beam passes straight through the device. When two opposite rods are charged to equal but opposite voltages the beam is bent through 90°, the direction of the bend depending upon the charge on the particles and the orientation of the two charged rods. The Ion and Neutral Mass Spectrometer (INMS) carried by the Cassini spacecraft, for example, used a QBD to switch between its open and closed ion sources. During a close passage of Enceladus in July 2006, Cassini flew through one of the cryovolcanic plumes emitted by the satellite. The INMS measurements showed that the plume was largely composed of water vapour.

5.2.4 Magnetic Interactions

5.2.4.1 Introduction

A charged particle moving through a magnetic field experiences a magnetic force given by

$$F = q B v \sin\theta \qquad (5.3)$$

---

* From the German einsellinse meaning 'single lens'.

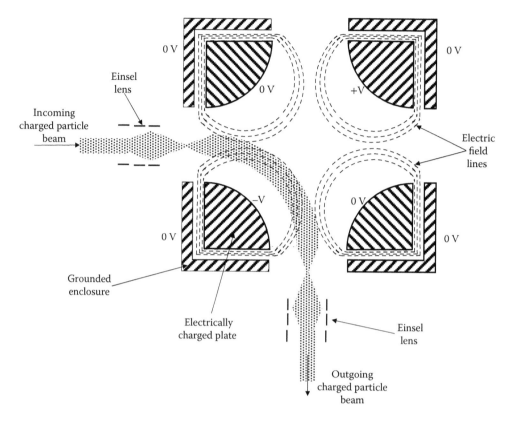

FIGURE 5.8 A cross section through a quadrupole beam deflector operating on a beam of positively charged particles.

where B is the magnetic field strength, v the particle's velocity and θ the angle between the directions of the magnetic field[*] and the velocity.[†]

For a positively charged particle, the direction of the magnetic force is given by the right-hand rule (Figure 5.9). For negatively charged particles, the magnetic force is directed in the opposite direction to that shown in Figure 5.9 – that is into instead of out of the palm. Unlike the electrostatic force, which is constant, the magnetic force is zero when θ is zero – that is when the particle is moving parallel to the lines of the magnetic field. The magnetic force is at its maximum when the velocity is perpendicular to the lines of the magnetic field. For a particle moving perpendicularly to the lines of the magnetic field, the magnetic force is also perpendicular to the velocity. The magnetic force thus cannot change the speed (or energy) of the particle, but it can change the direction of its motion

---

[*] Magnetic field lines are, by convention, regarded as originating at the north magnetic pole and finishing at the south magnetic pole.

[†] Equations 1.3 and 1.4 are usually combined to give the total electric and magnetic force on the particle and the resulting equation called the Lorentz force law:

$$F = q(E + B\,v\sin\theta)$$

or, in vector form

$$F = q(E + B\,x\,\boldsymbol{v})$$

where the vector quantities are shown in **bold** and x is the vector cross product.

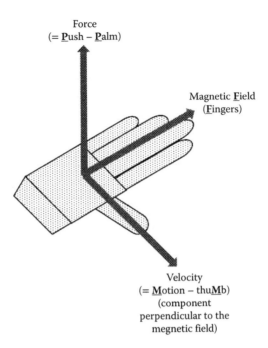

Force
(= **P**ush – **P**alm)

Magnetic **F**ield
(**F**ingers)

Velocity
(= **M**otion – thu**M**b)
(component
perpendicular to the
megnetic field)

FIGURE 5.9   The right-hand rule with some frequently used mnemonics for the interrelationships.

(Figure 5.10) and so the particle will follow a circular path. If there is a component of the velocity along the lines of the magnetic field, then the particle will follow a helical path with a constant circular cross section.

5.2.4.2 Magnetic Lenses

Ion lenses based upon magnetic forces are better suited to focussing higher energy (>50 keV) charged particle beams than are electrostatic lenses (see Section 5.2.2) but have yet to find application within spacecraft-borne instrumentation; a brief summary of them, however, is included here for completeness.

5.2.4.2.1 Solenoid Lens   A thin annular or ring-shaped electromagnet (solenoid) acts as a basic magnetic lens. The magnetic field lines through a solenoid run along its axis and curve around its outside to reconnect. For an incoming beam of charged particles with velocities parallel to the axis, there is thus a component of the field perpendicular to the axis except on the axis and in the plane of the solenoid. The resulting Lorentz forces (Figure 5.9) cause the particles to spiral in towards the axis.

The Lorentz force on a charged particle passing through the central hole of the solenoid is zero on axis and increases towards the inner edge of the coil. Thus, the paths of charged particles within a beam passing through the solenoid will be bent towards the axis through larger angles the further away they are from the axis. The force also increases with the velocity of the particle (Equation 5.3), but when the particles' q/m ratios and velocities are the same (or within fairly restricted ranges), the beam will be brought to a good focus on the axis (Figure 5.11a). The drawback of the magnetic lens for some purposes is that the

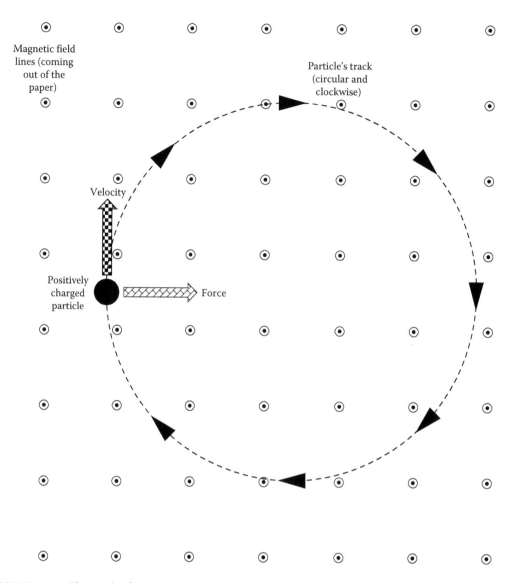

FIGURE 5.10 The track of a positively charged particle moving in a magnetic field that is directed out of the paper.

particles' paths will spiral in towards the axis rather than simply being bent, so that (in optical terms) the 'image' is rotated.

5.2.4.2.2 The Quadrupole Magnetic Lens   The magnetic quadrupole lens has a similar basic physical structure to that of the quadrupole mass spectrometer (Section 5.2.6.3) except that the length of the rods is much shorter and magnetic rather than electro-static forces are involved. Otherwise, two pairs of hyperbolic (or circular) magnetic disks are similarly arranged at the corners of a square. The disks of one pair are south poles, whilst those of the other pair are poles. A beam passing centrally between the disks is focussed in one plane and de-focussed in the orthogonal plane (Figure 5.11b).

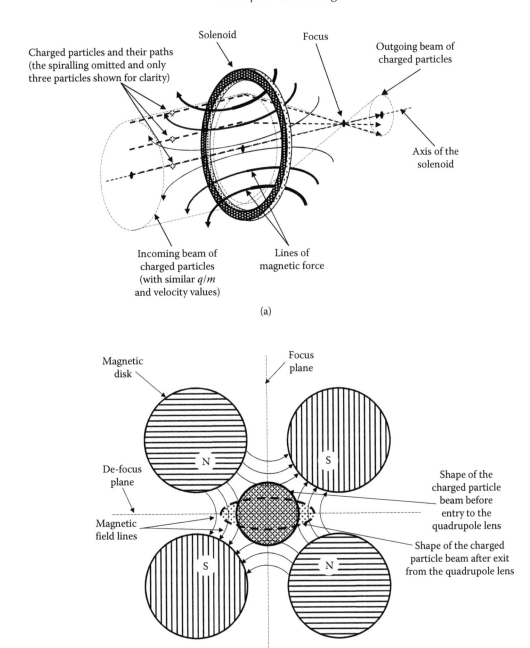

FIGURE 5.11 (a) A solenoid-based magnetic lens. (b) A cross section through the basic quadrupole magnetic lens.

A second quadrupole lens following the first and oriented at 90° to the first, however, will somewhat, but not completely, de-focus in the first plane and focus in the second plane. Residual focussing in both planes therefore remains after the emergence of the beam from the second lens.

5.2.5 Velocity Selectors and Spectrometers

5.2.5.1 Velocity Selector

A requirement that the particles have similar velocities may be fulfilled by using a velocity selector (also known as a Wien filter). The velocity selector combines the effects of electric and magnetic forces so that only particles with the desired velocity pass through it undeflected – particles with other velocities follow curved paths and may be separated out and removed from the beam. For the particles to be undeflected, the electric and magnetic forces on them must be equal and opposite to each other (Figure 5.12). From Equations 5.2 and 5.3, we find that the velocity for which this equality occurs (with θ = 90°) is given by

$$v = \frac{E}{B}\,\mathrm{m\,s^{-1}} \tag{5.4}$$

so that, for example, for $B = 100$ mT and $E = 10^4$ V·m^{-1} we have $v = 10^5$ m·s^{-1}.

5.2.5.2 Velocity Spectrometer

Particle velocity (or energy) spectrometers are essentially identical to mass spectrometers (Section 5.2.6) except that their input is particles with identical masses and charges – such as electrons, protons and alpha particles. The sorting thus occurs according to their velocities (energies). The van Allen probes' Magnetic Electron Ion Spectrometer (MagEIS) package includes four electron-detection units – one for the low energy (20–240 keV) range, two for the medium energy (10 keV to 1.2 MeV) range and one for the high (800 keV to 4.8 MeV) range. There is also a proton and ion detector.

Each of the MagEIS electron detector units is a magnetic field–based particle energy spectrometer using solid state detectors (Section 4.2.6) for detection. The magnetic field of the spectrometer bends the paths of the electrons through 180° in a roughly semicircular track (Section 5.2.6.2). The semicircle's radius increases as the particle's momentum

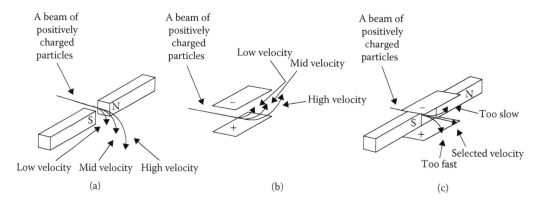

FIGURE 5.12 The electromagnetic velocity selector. (a) The effect of a magnetic field on a beam of charged particles with differing velocities. (b) The effect of an electric field on a beam of charged particles with differing velocities. (c) The effect of magnetic and electric fields on a beam of charged particles with differing velocities.

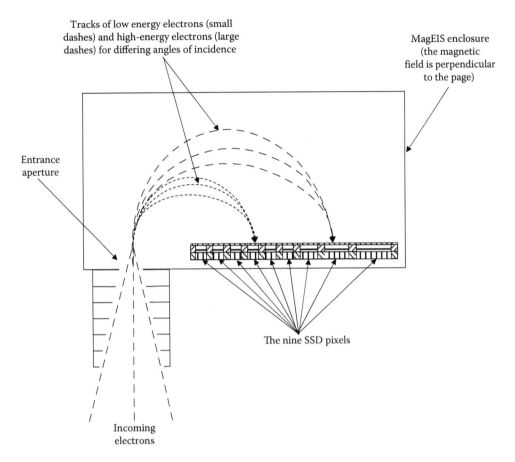

Tracks of low energy electrons (small dashes) and high-energy electrons (large dashes) for differing angles of incidence

MagEIS enclosure (the magnetic field is perpendicular to the page)

Entrance aperture

The nine SSD pixels

Incoming electrons

FIGURE 5.13   Schematic particle paths in a Magnetic Electron Ion Spectrometer (MagEIS) low or medium energy electron unit.

increases. A line of SSDs then detects the electrons (Figure 5.13). A particle's energy is, in effect, measured twice – once by the position of the SSD detecting it and again by the signal generated within the SSD. This enables false signals to be identified and eliminated, so greatly reducing the background noise of the instrument. There are nine SSD pixels in the low and medium energy units made up from two 4.5 mm thick and one 0.5 mm thick silicon layers. The high-energy unit has four SSDs constructed from one 0.3 mm and six 1.5 mm layers of silicon. The pixels for the higher energy electrons are larger than those for the lower energy electrons to allow for the greater scatter in the arrival positions of the electrons on the longer paths through the instrument. The acceptance angle for the units is a 10° × 20° area of the sky; however, electrons with the same momentum are focussed onto the same pixel irrespective of their arrival direction (Section 5.2).

5.2.5.3 Ion Drift Meters

The ion drift meter is basically a pin-hole camera for particles. It has a small entrance aperture and uses a position-sensitive particle detector. The two are separated by the drift space. If the relative velocity between the spacecraft and the plasma is such that the particles travel parallel to the length of the drift chamber, then they will only be observed at the centre of the detector array. If there is some transverse motion between the spacecraft and the plasma, then the particles will be observed displaced away from the centre of the detector array. Knowing the spacecraft's velocity through space, this displacement is easily converted into the plasma's velocity through space.

The Ionospheric Connection Explorer (ICON) carries two such drift meters that can operate over the ±2 km·s^{-1} range to an accuracy of about ±3 m·s^{-1} with a 45° field of view.

5.2.6 Mass Spectrometers

5.2.6.1 Introduction

Mass spectrometers are mostly employed in space applications to analyse planetary atmospheres, surface compositions, comets' emissions and the interplanetary medium, although another application is to monitor the air quality inside manned spacecraft in order to check for the build-up of noxious gases. Additional discussions of their applications within particular missions will be found in Section 7.2.3 along with those of with the gas chromatograph (Section 7.2.3.2) – which in actual instruments is often paired with a mass spectrometer.

The input to a mass spectrometer is a plasma (a beam of ions and charged subatomic particles). If a neutral gas is to be analysed then it has to be converted into a plasma, usually by ionising its constituent atoms with an electron beam. A solid sample has to first be converted into a gas, usually by heating with a laser beam, and then into a plasma.

The charged particles then have to be separated into a mass spectrum and the most commonly used approaches are through magnetic fields (Section 5.2.6.2), electric fields (Sections 5.2.6.3.1, 5.2.6.3.2 and 5.2.6.3.3) and times of flight (Section 5.2.6.4).

The mass spectrometer actually sorts particles by their mass-to-charge ratio. Thus, for example, a singly ionised helium atom may be mistaken for a doubly ionised beryllium-8 atom. However, since the latter has a half-life of just 7×10^{-17} s, it will only rarely be found more than a few tens of nanometres away from its point of origin. The possibility of the confusion being serious is therefore very remote. Other possible misidentifications, such as doubly ionised sulphur-32 for singly ionised oxygen-16, can also usually be resolved by taking into account the expected composition of the source of the sample.

Since the separation processes usually depend upon the particle's charge and velocity, as well as on their masses, it may be necessary to use ESA (Section 5.2.1), Wien filters (Section 5.2.5) or other ion optics prior to the mass spectrometer itself.

Finally, the mass spectrum needs to be detected either by scanning over a point detector or by using a linear detector. MCPs and other charged particle detectors used for this purpose are discussed in Section 4.2.

5.2.6.2 Magnetic Sector Mass Spectrometer

In Figure 5.10, the time taken for the particle discussed above to go once around its 'orbit' does not depend upon its velocity and is a constant for a given field strength and charge/

mass ratio. The 'orbital' frequency is called the Larmor, gyro or cyclotron frequency (v_L) and is given by

$$v_L = \frac{qB}{2\pi m} \, Hz \tag{5.5}$$

where B is the magnetic field strength and q is the particle's electric charge.

The radius of the particle's 'orbit', the Larmor radius (R_L), is given by

$$R_L = \frac{mv\sin\theta}{qB} \, m \tag{5.6}$$

where m is the particle's mass, v its velocity and θ is the angle of the velocity to the magnetic field direction.

For electrons, the Larmor frequency and radius are thus given by

$$v_L = 2.7993 \times 10^{10} B \approx 2.8 \times 10^{10} B \quad Hz \tag{5.7}$$

$$R_L = 5.6856 \times 10^{-12} \frac{v\sin\theta}{B} \approx 5.7 \times 10^{-12} \frac{v\sin\theta}{B} \quad m \tag{5.8}$$

For protons and singly charged ions with atomic mass A, the Larmor frequency and radius are given by

$$v_L = 1.5246 \times 10^7 \frac{B}{A} \approx 1.5 \times 10^7 \frac{B}{A} \quad Hz \tag{5.9}$$

$$R_L = 1.0439 \times 10^{-8} \frac{Av\sin\theta}{B} \approx 1.0 \times 10^{-8} \frac{Av\sin\theta}{B} \quad m \tag{5.10}$$

For particles with an energy of 1 keV, the velocities are

1.9×10^7 m·s^{-1} (electrons)

4.4×10^5 m·s^{-1} (protons)

2.2×10^5 m·s^{-1} (helium ions)

1.2×10^5 m·s^{-1} (nitrogen ions)

1.1×10^5 m·s^{-1} (oxygen ions)

Assuming that the particles are moving perpendicularly to the magnetic field lines ($\theta = 90°$) – which is usually arranged to be the case within most instruments – that the ions are singly charged and that the magnetic field has a strength of 100 mT, then the corresponding Larmor radii are about

1.1 mm (electrons)

43.8 mm (protons)

87.5 mm (helium ions)

163.8 mm (nitrogen ions)

175.1 mm (oxygen ions)

For particles with identical velocities of, say, 10^5 m·s^{-1}, the Larmor radii are about

5.7 μm (electrons)

10 mm (protons)

40 mm (helium ions)

140 mm (nitrogen ions)

160 mm (oxygen ions)

The latter listing gives the essence of the principle of operation of a magnetic sector mass spectrometer; a beam of charged particles with similar velocities but differing charge-to-mass ratios will have different paths through a magnetic field and so may be separated and identified.

The magnetic sector mass spectrometer thus comprises a velocity selector (Section 5.2.5) followed by a magnetic field covering a quarter of a complete circle (the 'sector') and then a linear detector (Figure 5.14).

The magnetic sector mass spectrometer has many variants and geometries,* but the basic principles are similar to those outlined above – some type of velocity selector placed before or after a magnetic particle sorter and followed by a detector. Any suitable charged particle detector can be used for the latter part of the instrument (see Chapter 4), but an array of Faraday cups (or collectors) is often chosen. These are simply small conductive cavities which absorb the ions impacting onto them and so become positively charged. The charge, when discharged through an appropriate circuit (either after a certain time interval to collect the ions, or continuously), provides the measure of the number of intercepted ions.

The Pioneer Venus large atmosphere probe (Figure 5.15), for example, carried a magnetic sector mass spectrometer in order to assess the number densities and their variations with altitude of gases like atomic oxygen and carbon dioxide from a height of 62 km in the atmosphere down to Venus' surface. Neutral atmospheric particles were first gathered by the instrument and then ionised by bombardment with electrons. After focussing, the ion beam passed through an ESA that acted as a velocity selector and then into the magnetic analyser. The ions were detected by an array of CEMs that had a spiral shape to increase

---

* One recent source lists six named designs but also adds that most of the current actual devices do not really fit into any of them.

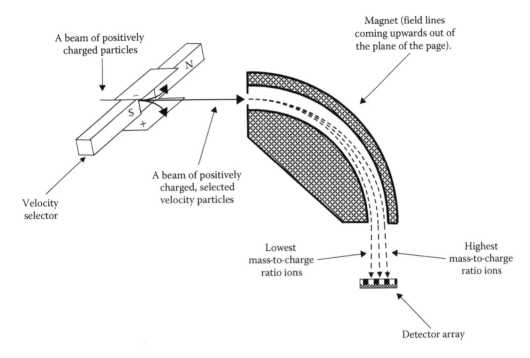

FIGURE 5.14 The magnetic sector mass spectrometer.

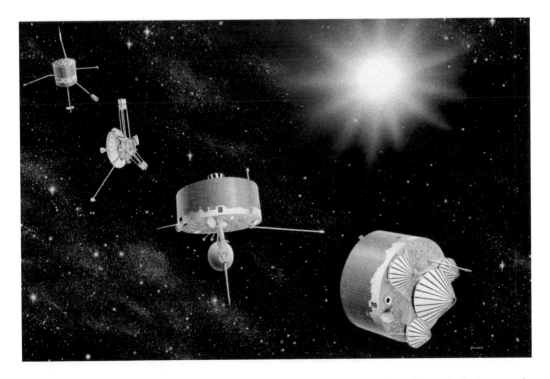

FIGURE 5.15 The Pioneer family of spacecraft. The Pioneer Venus multiprobe, with the large probe at its centre, is to the bottom right of the image. (Courtesy of NASA.)

the interactive area available. Its mass operating range was from 1 to 208 Da and it scanned over this range every 64 seconds. A second magnetic sector neutral mass spectrometer was on the Pioneer Venus multiprobe bus.

The Martian lander, Phoenix, had a miniature magnetic sector mass spectrometer as a component of its thermal evolved gas analyser (TEGA) instrument. Its samples were enriched by the removal of any nitrogen or carbon dioxide before their analysis. The sample was placed into one of eight single use ovens by the robotic arm (Section 7.2.6) where it could be slowly heated up to 1000°C and the vapourised materials fed to the mass spectrometer.

Rosetta's ROSINA (Rosetta Spectrometer for Ion and Neutral Analysis) instrument suite included two mass spectrometers (see also Section 5.2.6.4). The magnetic mass spectrometer covered from 1 to 150 Da with a mass resolution of about 0.02–0.03 Da (i.e. it was able to differentiate, say, between ^{12}CH and ^{13}C). It could examine both neutral gases and ions and amongst its other discoveries was the first detection of molecular nitrogen in a comet.

5.2.6.3 Electric Fields

5.2.6.3.1 Quadrupole Mass Spectrometer   More recent spacecraft than Pioneer Venus have often carried mass spectrometers (or analysers) operating on quite different principles from the magnetic sector spectrometers. They are known as quadrupole mass spectrometers and they utilise both dc and ac voltages to isolate those particles with a particular q/m ratio. The device is based around four conducting rods in a square array (Figure 5.16a

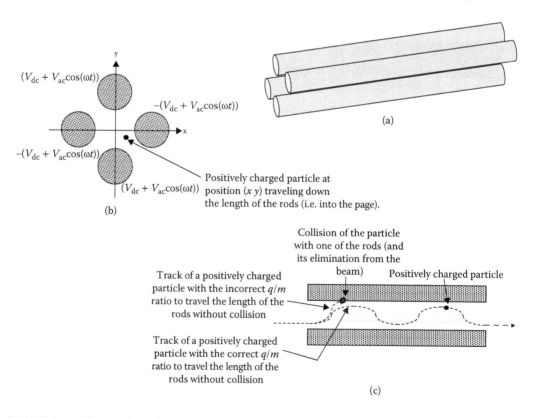

FIGURE 5.16   The quadrupole mass spectrometer. (a) The three-dimensional (3D) arrangement of the rods. (b) Cross section through the rods. (c) Particle tracks between the rods.

and b). The cross-sectional shape of the rods should be hyperbolic but often a circular cross section is adequate. Diagonally opposite pairs of rods are connected together and the two pairs of rods given constant (dc) charges of $(+V_{dc})$ and $(-V_{dc})$. To this is added radio frequency (rf) alternating charges of $(+V_{ac}\cos(\omega t))$ and $(-V_{ac}\cos(\omega t))$, respectively. The positively charged particle shown in Figure 5.16b will thus experience instantaneous accelerations of

$$\frac{d^2x}{dt^2} = -\frac{q}{m}\frac{\left(V_{dc} + V_{ac}\cos\omega t\right)}{r_o^2}x \quad m\,s^{-2} \tag{5.11}$$

and

$$\frac{d^2y}{dt^2} = \frac{q}{m}\frac{\left(V_{dc} + V_{ac}\cos\omega t\right)}{r_o^2}y \quad m\,s^{-2} \tag{5.12}$$

where $2r_o$ is the separation of the inner surfaces of either pair of rods.

However, the accelerations will be circling around as the ac charges change, rotating at a frequency of $2\pi\omega$. Thus the charged particle will 'orbit' around in between the rods. The electric charges do not affect the velocity of the particle along the length of the rods and this remains at its incoming value. The actual path of the particle is thus a spiral down the space between the rods. The radius of the 'orbit' of the particle depends upon its q/m ratio and it may be changed by changing the values of V_{dc} and V_{ac}. Only particles with the 'right' q/m ratio for a particular pair of voltages will be able to avoid colliding with the rods or being flung out sideways between them. The particles with the 'right' q/m value will proceed down the space between the rods and emerge to be detected at the exit end. By changing the voltages, the selected q/m value may be scanned to build up a picture of the particles that are originally present in the beam.

A quadrupole mass spectrometer was carried on the Pioneer Venus Orbiter to analyse Venus' upper atmosphere – the first time this design of instrument had been used in space. The mission also carried four more mass spectrometers, one more on the orbiter, two on the Multi-Probe Bus and one on the large atmosphere probe (Section 5.2.6.2).

The Jupiter atmosphere probe, Galileo, carried a quadrupole mass spectrometer with an electron impact ion source and a CEM detector. Its mass ranges were 1–52 Da and 84–131 Da and it found that on Jupiter carbon is enriched with respect to hydrogen by nearly a factor of three. The Moon Impact Probe (MIP) launched from the Chandrayaan-1 lunar orbiter used a quadrupole mass spectrometer that could operate down to a pressure of 10^{-11} Pa with a mass resolution of 0.5 Da to analyse the very thin lunar atmosphere.

A recent example of the use of a quadrupole mass spectrometer is the within the Neutral Gas and Ion Mass Spectrometer (NGIMS) on board the Mars Atmosphere and Volatile Evolution (MAVEN) spacecraft. The NGIMS (Figure 5.17) is designed to study the upper atmosphere of Mars, in particular the composition and its variations at heights above the surface from 125 to 500 km. The quadrupole rods are hyperbolic in shape and the

FIGURE 5.17   MAVEN's NGIMS. (Courtesy of NASA/Goddard Space Flight Center.)

ac operating frequency is 3 MHz for ions in the range 2–19.5 Da and 1.4 MHz for the
19.5–150 Da range. Four ion lenses are used to transfer the particle beam to the two CEM
detectors. The gas for analysis is either stored in a chamber with only a small aperture to
allow entry of the gas (the closed ion source – for non-reactive particles) and ionised by
a hot filament before analysis or enters directly and is ionised by an electron beam (the
open ion source for species such as nitrogen and oxygen that could be destroyed within the
closed ion source). Switching between the two sources is accomplished using a quadrupole
beam deflector (see above).

Instruments very similar in design to MAVEN's NGIMS are used for many other
missions including the Cassini orbiter, LADEE and the ill-fated COMet Nucleus TOUR
(CONTOUR).*

5.2.6.3.2 Dynamic Ion Traps   Ions may be trapped inside a ring electrode which has separate electrodes as its end caps. The inner surfaces of all the electrodes have convex hyperbolic shapes. Ions (or neutral atoms ionised by an external electron beam) are fed into the
trap through one of the end electrodes. A radio frequency (typically in the MHz region)
AC electric field is applied to the ring electrode, while the end caps are grounded. For a
given AC voltage (the threshold voltage), lower mass/charge ratio ions will have the time

---

* LADEE (Lunar Atmosphere and Dust Explorer). CONTOUR – it probably broke up during a rocket burn in August
  2002.

within a single positive or negative cycle of the AC to accelerate to the walls of the trap and be absorbed. Higher mass/charge ratio ions will still be moving towards the trap's walls when the voltage reverses. Their movement will thus be reversed and they will oscillate about the centre of the trap. A low-pressure gas such as helium may be introduced into the chamber in order to reduce the ions' energies by collisions and so concentrate them into a small central region.

After the trap has been filled with the sample for a few seconds it is closed and the AC voltage is increased. The collected ions' motions become unstable and those nearest the threshold mass/charge ratios do so first. These unstable ions then exit the trap through the second end electrode to the detector. As the AC voltage continues to increase, so ions with higher and higher mass/charge ratios are detected – thus building up the mass spectrum.

The comet lander, Philae, carried an ion trap mass spectrometer, Ptolemy. Despite Philae's brief operating time on the comet, it was able to make some preliminary measurements. It was to have been fed by dust samples drilled from the comet 67P/Churumov–Gerasimenko's crust and heated in one of four small ovens on the instrument to 800°C in order to drive off the volatiles from the sample for analysis by the instrument. However, the low power available from the batteries meant that it could only 'sniff' the gases already nearby. The results showed nitrogen compounds to be in low abundance.

5.2.6.3.3 Electrostatic Mass Analyser   The cylindrical deflection ESA (Section 5.2.1), especially if fed with a beam of particles from a velocity selector (Section 5.2.5), will act as a mass spectrometer. The emerging particles for a given selected velocity and charge on the cylinders of the ESA will have identical q/m ratios – and for singly ionised particles this means that they will have the same masses. Other masses may be selected by changing the velocity and/or the charges on the ESA's cylinders.

Sometimes, also, an ESA is used to feed a mass spectrometer in order to improve the focussing of the ion beam. The Solar Wind and Suprathermal Ion Composition Spectrometer (SWICS) on board NASA's Wind* spacecraft, for example, used an electrostatic deflector and a ToF analyser (below) to study the solar wind ions.

5.2.6.4 Time-of-Flight Mass Spectrometer

The principle behind the time-of-flight (ToF) mass spectrometer is simple – in an assemblage of various particles, all with the same kinetic energy, the more massive ones will have lower speeds than the less massive ones. The ToF device uses this property to sort the particles according (strictly) to their q/m ratios – but since the vast majority of the particles are singly ionised, in practice, the sorting is directly by their masses.

The basic structure of the ToF mass spectrometer is shown in Figure 5.18 (although other configurations are possible). The inside of the instrument is at or close to a vacuum. A small aperture (leak) at one end allows the outside gas to percolate in slowly. The particles accumulate in the first (ionisation) chamber and are ionised by a beam of electrons or by a laser, so becoming positively charged. The ionisation chamber is

---

* Not an acronym.

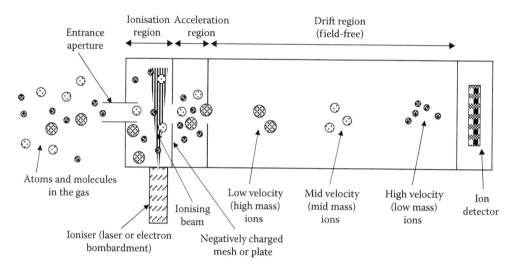

FIGURE 5.18   Time-of-flight mass spectrometer.

held at a slight negative voltage so that the ions accumulate within it. When sufficient ions have been gathered, the back plate of the chamber is switched to a slight positive charge, forcing the ions through a mesh (or plate with a hole in it) into the second (acceleration) chamber. The back plate of the acceleration chamber is then given a strong pulse of positive charge accelerating the ions and sending them through a second mesh (or plate). The particles all gain about the same amount of KE during their acceleration (assuming they are singly ionised). Once into the third (drift) chamber they are in a field-free zone and drift down its length. The faster moving, low mass, ions arrive first at the detector, which is often an MCP, and are detected separately from the slower moving, more massive, ions.

In order to aid the discrimination amongst the higher mass particles they are sometimes reflected back within the instrument to give a longer drift path. A device called a reflectron is used to do this. The reflectron is a stack of ring electrodes at high positive voltages. The ions enter the central aperture of the stack and follow U-shaped paths, emerging with their paths changed by 180°. The reflectron also 'focusses' the particles slightly – if particles of equal masses have gained slightly differing kinetic energies during their acceleration, the less energetic (slower) particles will penetrate a shorter distance into the reflectron than the higher energy particles before being turned around. They will thus travel a shorter path and catch up to some extent on their faster brethren. The pulse formed by the particles with identical masses will hence be narrower (in time) when it arrives at the detector.

The Mercury orbiter MESSENGER carried a ToF mass spectrometer preceded by an ESA velocity selector. The instrument was designed to analyse electrons with energies between 20 and 700 keV and ions with energies between 10 keV and 5 MeV. The ToF start and stop signals were picked up by MCPs and the particles' total energies measured separately using SSD detectors. Pluto Energetic Particle Spectrometer Science Investigation (PEPSSI)

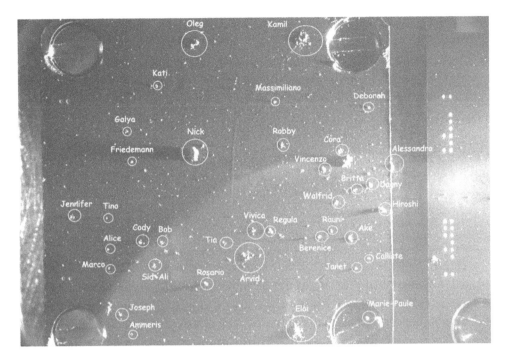

FIGURE 5.19 Dust particles and their identities collected for Cometary Secondary Ion Mass Analyser (COSIMA) analysis. (Courtesy of ESA/Rosetta/MPS for COSIMA team.)

is a ToF spectrometer carried by New Horizons. It was developed from the MESSENGER instrument. Its field of view, defined by a collimator, is 12° × 160° and it is divided into six sectors. 2 × 2 pixel SSDs are used with two of the pixels detecting the ions and the other two the electrons. It operates up to 1 MeV for the protons and ions and up to 500 keV for the electrons.

The Rosetta spacecraft, as already mentioned, carried two mass spectrometers (Section 5.2.6.2). The second was a reflecting ToF mass spectrometer to analyse the composition of dust grains. In order to do this, the solid material of the dust grain must be vapourised before it can enter the spectrometer. The dust grain was first trapped on a target plate (Figure 5.19) which could be moved to centralise the grain at the spectrometer's entrance aperture (microscope viewing of the grains helped with their selection and positioning). Samples from the grain were then removed, layer by layer, by an ion gun targeted on the grain. Masses from 1 to 1000 Da could be measured.

The van Allen probes' instruments, RPSPICE and HOPE (discussed in Sections 4.2.5 and 4.2.5.1, respectively), are also based upon ToF units. Juno observes electrons over the 30 keV to 1 MeV range and ions up to 10 MeV with Jupiter Energetic Particle Detector Instrument (JEDI), a ToF-type instrument with SSD and MCP detectors.

The Europa Multiple Flyby Mission, currently planned for a launch around 2023, will be carrying two mass spectrometers – Mass Spectrometer for Planetary Exploration (MASPEX) and Surface Dust Analyzer (SUDA) plus a related instrument – Plasma Instrument for Magnetic Sounding (PIMS). MASPEX is intended to examine Europa's plumes and exosphere for volatiles and organic* compounds over the mass range 2–1000 Da and will be of the ToF design (Section 5.2.6.4). SUDA will also have a ToF design and will sample dust grains above Europa's surface in the size range 200 nm to 10 μm in order to map the surface composition. PIMS just comprises three Faraday cups (Section 5.2.5) directly sampling the surrounding plasma for ions and electrons in the 0.1–50 eV, 20 eV to 7 keV range (ions) and 10 eV to 2 keV range (electrons).

The planned THOR mission is expected to include four mass spectrometers based upon top hat ESAs and 60 mm ToF analysers. The instruments will be placed at 90° intervals around the spacecraft and have 45° × 360° fields of view, thus providing an instantaneous view of the whole sky. MCPS will be used for detection and measurements made up to six times per second.

5.2.6.5 Future Usage of Mass Spectrometers

Mass spectrometers are very widely used within laboratories, for monitoring commercial chemical processes, testing for atmospheric pollutants, and so on. There are numerous types in addition to those just considered that find applications in these spheres. Often however, commercial considerations mean that the designs are known by different names but are in fact minor variants of the main types of instrument. Amongst those that may find space applications in the future, we may identify the quadrupole ion trap mass analyser and the ion cyclotron resonance analyser.

---

* See footnote about organic molecules on page 93.

CHAPTER 6

Neutral Particles

6.1 INTRODUCTION

Neutral particles are unaffected by either electric or magnetic fields. As will have been apparent above though, when the neutral particle is an atom or molecule, ionisation provides it with a net charge (usually positive) and it may then be examined using the standard charged particle detectors and analysers or minor variants of them (see for example, ROSINA, Section 5.2.6.2).

Three particles of astrophysical interest, however, are intrinsically neutral – the photon, the neutron and the neutrino. The detection and analysis of photons is covered in Part 1. Neutrinos, while emitted in vast numbers by the Sun, are outside the scope of this book. Their detection and scientific study, however, is covered in reference A.1.1 and in many other sources (Appendix A). Neutrons though, are of interest here and so their methods of detection and analysis are discussed below.

Neutrons originate from solar system objects in three main ways. For a few objects, landers and rovers have carried neutron activation analysis instruments to analyse element abundances in the surface layer of the object. These instruments bombard the surface with neutrons from an artificial source. The neutrons interact with nuclei, inducing radioactivity; neutrons may be amongst the decay products. Neutron activation analysis is concerned, however, with detecting the induced γ ray emission and it is discussed in more detail in Section 7.2.3.5; the neutrons are a by-product and are not considered further here.

The second source of neutrons from solar system objects is natural radioactivity, while the third source is emissions induced by cosmic-ray interactions with nuclei in the surface of the object. Clearly, for both these sources of neutrons, only solar system objects with very thin or no atmospheres can be studied by orbiting spacecraft; with thicker atmospheres the neutrons emitted through natural radioactivity will be absorbed in the atmosphere before reaching the spacecraft's orbit and the cosmic rays will be absorbed before they can interact with the solid surface.

Naturally occurring radioactive elements that are present in planetary, satellite, asteroid and cometary surface layers are mainly potassium, thorium and uranium. Of the most abundant isotopes, potassium-40 decays by electron capture or β decay, while thorium-232 and uranium-238 emit α particles. Thus neutrons directly emitted through natural radioactivity are rare and the main source of the observed neutrons is cosmic-ray interactions.

Neutrons, of course, are unstable particles with a half-life (at non-relativistic velocities) of just under 15 minutes. At 1 eV, a neutron has a speed of about 14 km/s and so can travel about 12,000 km during its half-life. The intensity of such neutrons emitted by a source will thus be close to zero at a distance of some 150,000 km from the source. A 1 MeV neutron would have figures 1000 times larger than these. Thus spacecraft orbiting or landing on solar system objects will have no difficulty in observing neutrons largely at their original intensities. But at greater distances (e.g. trying to observe neutrons emitted from the Moon from an Earth-orbiting spacecraft) nothing much will be found.

Cosmic rays are mostly protons and these have energies ranging upwards from 1 GeV. When such a particle collides with an atomic nucleus, the fragmentation products are likely to include neutrons, protons and γ rays with megaelectron volt energies. The remainder of the nucleus is likely to be left in an excited state and may emit further γ rays and also decay radioactively. Some of the neutrons will find their way to the surface and be emitted out into space. Other neutrons will undergo collisions with nearby nuclei, exciting them to higher energy states from which they will emit γ rays at wavelengths characteristic of the element as they return to their stable states. The energy exchange during collisions is most efficient when the particles are similar in mass. Thus neutrons interact strongly with hydrogen and other light element nuclei. It is therefore possible to get an indication of the presence of hydrogen (usually in the form of water) by measuring the relative numbers of high and low energy neutrons. Where hydrogen (water) is abundant there will be many more low-energy neutrons.

6.2 NEUTRON DETECTION

The two main interactions of neutrons with matter that potentially enable their detection are the following:

1. Absorption by a suitable nucleus followed by the radioactive decay of the new nucleus and the detection of its high-energy decay products

2. Elastic scattering reactions in which the 'target' particle speeds off as a high-energy ion and may be detected

Additionally, neutrons may be detected from their kinetic energy deposited into a calorimeter (Section 4.2.8) and can be focussed using polycapillary tubes (Section 3.3.2.2.4) though this has yet to find a spacecraft-based application.

Many different nuclei are suitable for the interactions involved in detecting neutrons via the first approach. Apart from the details of the reaction, however, the resulting instruments are all broadly similar in their design. Typical reactions that occur promptly following the collision are with nuclei such as helium, lithium and uranium. More delayed reactions follow from collisions with nuclei such as aluminium, iron and gold.

Thus, a neutron colliding with a helium-3 nucleus produces the reaction

$$^3\text{He} + \text{n} \rightarrow {}^3\text{H} + {}^1\text{H} + 0.77 \text{ MeV} \tag{6.1}$$

The two resulting hydrogen nuclei (ionised tritium and a proton) speed off, sharing the 0.77 MeV excess energy of the reaction plus the excess kinetic energy of the original neutron. They are thus easily detectable by most of the charged particle detectors discussed earlier.

The interactions of neutrons with uranium-235 are understood by many people since they form the basis for some atomic bombs. The interaction though, somewhat counter-intuitively, occurs most readily with low-energy neutrons. The uranium-235 nucleus briefly forms a uranium-236 nucleus with the neutron and this then fragments in many different ways. For example:

$$^{236}\text{U} \rightarrow {}^{143}\text{Cs} + {}^{80}\text{Rb} + 3\text{n} + \gamma \tag{6.2}$$

or

$$^{236}\text{U} \rightarrow {}^{141}\text{Ba} + {}^{82}\text{Kr} + 3\text{n} + \gamma \tag{6.3}$$

The high-energy charged particles from these reactions are again detectable by many means.

An example of a delayed reaction is aluminium-27, which reacts with the neutron to form aluminium-28; the latter decays via beta particle emission with a half-life of 2.3 minutes. Similarly, gold-197 becomes gold-198 after capturing a neutron and the latter decays, also by beta emission, with a half-life of 2.7 days.

Elastic scattering reactions can occur with any type of nucleus, but the transfer of energy is most efficient when the 'target' particle has a similar mass to that of the neutron. Hydrogen-rich materials are thus most widely used for the interaction medium within detectors based upon scattering. The neutron, after scattering, is still a neutron and may go on to cause further scattering events. The impacted nucleus, now an ion, flies off at high speed to be detected like any other charged particle.

All that remains, once a neutron has interacted with a nucleus via any of the above processes, is to detect the reaction products. Although most detectors are useable for the purpose, those widely employed include liquid scintillation detectors for the products of fast neutrons and gas proportional detectors for slower neutrons (and, of course, fast neutrons can always be slowed down through the use of a moderator).

Examples of neutron detectors on spacecraft include MESSENGER's Gamma Ray and Neutron Spectrometer (GRNS), the two instruments on board the Mars Odyssey – the neutron spectrometer and the High Energy Neutron Detector (HEND) – and Dawn's Gamma Ray and Neutron Detector (GRaND) – GRaND is a descendent of the detectors on Mars Odyssey and also that on the Lunar Prospector spacecraft (which used a He-3-based detector).

The neutron detector component of MESSENGER's GRNS comprised two lithium-glass scintillators sandwiching a boron laden plastic scintillator and used photomultiplier tubes

(PMTs) to detect the scintillations. It detected neutrons from thermal energies (< 0.1 eV) up to ~7 MeV with the lithium-glass scintillators covering the low-energy range and the boron-plastic scintillator the high-energy range. The instrument has, *inter alia*, shown that the darkness of Mercury's surface* arises from carbon. The planned Psyche mission is expected to carry an updated version of MESSENGER's instrument.

Mars Odyssey's neutron spectrometer is based upon four blocks of boron-loaded plastic scintillator with each block viewed by a separate PMT. Neutrons entering one of the blocks lose energy gradually through elastic scattering. Most of the scattering events produce protons that in turn produce multiple ion pairs. Recombination of the ions then releases photons to be detected by the PMT. If the neutron does not escape from the scintillator it will eventually be captured by a boron-10 nucleus in the reaction

$$^{10}B + n \rightarrow {}^{11}B \rightarrow {}^{4}He + {}^{7}Li + 2.8 \text{ MeV} \qquad (6.4)$$

The released energy goes into the kinetic energy of the helium and lithium nuclei and these produce further scintillations. In about 6% of the reactions there is also a release of a γ ray, but this still leaves some 2.3 MeV for the nuclei and the γ ray also induces scintillations via the Compton effect. Low-energy (thermal – about 0.025 eV) neutrons produce just a single output pulse resulting from the helium and lithium interactions plus a lower, longer signal from the later recoils and the γ ray. Higher energy neutrons produce a second pulse arising from the initial elastic scattering interactions.

Mars Odyssey's HEND comprises a suite of five detectors: three helium-3 proportional counters and two scintillators. The use of polyethylene moderators and cadmium shields enables the three proportional counter detectors to cover neutrons in the energy range from 0.4 eV to 1 MeV. The other two detectors use stilbene and caesium iodide as the scintillator materials and PMTs as the detectors. The stilbene block nests inside a 'cup' of caesium iodide and detects 800 keV to 15 MeV neutrons. The caesium iodide scintillator detects charged particles and γ rays, not neutrons.

The Lunar Reconnaissance carries the Lunar Exploration Neutron Detector (LEND) and ExoMars' Trace Gas Orbiter (TGO), the Fine-Resolution Epithermal Neutron Detector (FREND), both instruments being developed from HEND.

Dawn was launched on a mission to Vesta and Ceres in September 2007 and is still operating (in orbit around Ceres) at the time of writing. Dawn's GRaND neutron detector component is based upon scintillator detectors using lithium-loaded glass and boron-loaded plastic. Two detectors are in the form of an 'open sandwich' with a thin layer of the lithium glass optically attached to a thicker slice of the boron plastic. The lithium layer is on the outside and intercepts the thermal neutrons. Higher energy neutrons penetrate into the boron plastic and are detected there. Both layers are viewed by the same PMT. The two sandwiches are located above and below the γ ray detectors and act as neutron shields for them. Neutrons over the energy range from ~0.025 eV to above 700 keV are detected by the scintillator combination. Two more boron-plastic-only scintillators are located at opposite

---

* Mercury reflects only 6.8% of the radiation that it receives – compared with 30.6% for Earth and 11.0% for the Moon.

sides of the γ ray detectors serving as anti-coincidence shields for the γ ray detectors and also detecting neutrons over a similar energy range to that of the sandwich type detectors when viewed by PMTs.

A CubeSat lunar orbiter, Lunar Polar Hydrogen Mapper (LunaH-Map), is planned to carry a neutron detector based upon the scintillator elpasolite* in the form of two arrays using a total of sixteen $20 \times 25 \times 25$ mm blocks. PMTs will detect the scintillations produced by the neutrons. It is intended to map hydrogen (and hence water) deposits down to 1 m below the Moon's South Polar region.

---

* Cs_2YLiCl_6:Ce, also known as CLYC.

III

Direct Sampling and Other Investigative Methods

CHAPTER 7

Direct Sampling Instruments

7.1 INTRODUCTION

Spacecraft have been soft landing now on solar system objects for over half a century. The first successful such landing came in 1966 when, at the USSR's twelfth attempt, the Luna-9 mission landed on the Moon's surface using an airbag to cushion the shock and sent back 27 images (Figure 7.1). The lander was in the form of a half-metre diameter sphere with the top four quarter segments unfolding like flower petals after landing to stabilise the craft and to enable its instruments to operate. It also carried a radiation detector and determined that the dosage at the lunar surface was 300 μGy* per day. Just obtaining images (Chapter 1) from the surface of an object would not count for the purposes of this chapter. However, the radiation meter ensures that Luna-9 also counts as the first successful sampling of a solar system object.[†]

Since Luna-9 there have been many more soft landing missions – to the Moon, Venus, Mars, Titan, asteroids and comets. Landers are now frequently accompanied by mobile companions (rovers) to extend the area covered. The first[‡] such was Lunokhod-1 (Sections 1.2.2.3, 7.2.2.5 and 7.2.6.3.5) which successfully landed on the Moon in November 1970. This was an eight-wheeled vehicle some 2 m in length and massed over 800 kg. It was powered by batteries that were recharged by solar cells and could move at either 1 or 2 km/h. It operated for 10 months and travelled a total distance of 10.5 km. It was followed on the Moon in 1973 by Lunokhod-2 which operated for 4 months and travelled ~39 km.

Between Lunokhods 1 and 2, two missions attempted the first soft landings and the first rovers on Mars. The Mars-2 and Mars-3 missions were launched in May 1971 and intended to be Mars orbiters, landers and rovers. Mars-2's lander crashed together with its rover, but the orbiters operated successfully, as did Mars-3's lander. The latter though ceased transmitting just 14.5 seconds after landing and contact with it was never re-established. The Pribori Otchenki Prokhodimosti-Mars (PrOP-M)[§] tethered rovers, though, had they been deployed, would have

---

* Typical doses for medical x-ray scans are around 1–10 mGy.
[†] Though, of course, from Sputnik 1 onwards, Geiger and other types of radiation detectors have been direct sampling the interplanetary medium and planetary radiation belts.
[‡] An earlier attempt (February 1969) disintegrated during launch.
[§] Russian for 'device for cross country movement'.

FIGURE 7.1 The first image taken from the lunar surface, obtained by the Luna-9 soft lander in 1966.

been rather different from the Lunokhods. They were square-ish flat box-shaped vehicles about 200 mm to a side and 70 mm high. They were to move on small skis at up to 15 mm·s⁻¹ out to a maximum distance from the lander of 15 m (the length of the tether). They would have carried penetrometers (Section 7.2.6.3.5) and ionising radiation sensors (Section 7.2.5).

The Moon (apart from the Apollo 15, 16 and 17 manned landings which carried the Lunar Roving Vehicles) since Lunokhod-2 has only had the Chang'e-3 rover, Yutu, operating on its surface from December 2013 until March 2015. Mars however, after Mars-3, has had four successful rovers – two of which, Curiosity and Opportunity, are still operating (Opportunity now holds the rover distance record – 43.5 km at the time of writing). Mars Pathfinder's Sojourner and MER's Spirit operated for a year and nearly 10 years respectively.

In the near future, the Chang'e-4 and Chandrayaan-2 lunar rovers are proposed for 2018, while 2020 may see the launch of the Lunar-Grunt rover and the Mars rovers, Mars 2020 and ExoMars 2020. A four-wheel rover, Scarab, designed to be able to move autonomously at around 50 mm·s⁻¹ in the dark using lasers to navigate and to obtain cores up to a metre deep, is currently being developed for a possible future mission to the lunar South Pole.

Further into the future, a massive lunar rover, All-Terrain Hex-Limbed Extra-Terrestrial Explorer (ATHLETE) is under development by Jet Propulsion Laboratory (JPL). It is planned to be able to carry up to 450 kg even under Earth gravity and for several vehicles to be able to link together for even bigger loads. It will be able to move either on six wheels or on six legs at speeds up to 10 km/h and climb slopes of up to 35°. At the other end of the scale, there is a proposal for the development of a tiny hopping lander to be called Pico Autonomous Near-Earth asteroid In situ Characteriser (PANIC). It is suggested to be tetrahedral in shape and about 350 mm per side and with a total mass of 12 kg. Up to four instruments could be mounted within it and it would move around (in microgravity) by hopping.

In the case of the Moon, there have also been the manned landings, for Venus, balloons floating in its atmosphere, for Venus and Jupiter, probes of their atmospheres and for Mars, the Mars 2020 lander will carry a small solar powered drone (Mars Helicopter Scout [MHS]) to check the rover's planned route for dangers (e.g. loose sand) and to investigate targets of interest for the rover.

In the future, serious considerations are being given to the possibility of submarine probes for Europa, Enceladus, Ganymede and Titan, aircraft, space shuttles or space planes (like the X-37*) for Titan's and Venus' atmospheres, swarms of drones to explore Mars and perhaps a permanent manned colony on Mars. All these missions and possible missions carry or are likely to carry direct sampling instruments for many purposes.

In addition to these soft landers, direct(ish) sampling is sometimes undertaken by hard landing missions. The Deep Impact mission to comet Tempel 1 deliberately crashed a 370 kg projectile at 10.2 km·s^{-1} into the comet, producing an explosion equivalent to nearly five tons of TNT. The purpose of the explosion was to blast material from the inside of the comet out into space, where the fly-by half of the mission could observe it and where it could also be observed by the Chandra, Hubble Space Telescope (HST), Spitzer and XMM-Newton spacecraft and by terrestrial observatories.

Chandrayaan-1 launched a small probe from its orbit around the Moon that hit the lunar surface near the South Pole in November 2008. The probe carried instruments including a camera and a mass spectrometer (Section 5.2.6) to analyse the lunar atmosphere during its descent and upon impact the ejecta were analysed by the still orbiting mother craft for the presence of water – which it found for the first time on the Moon. The Lunar Crater Observations and Sensing Satellite (LCROSS) mission produced a double impact within the Moon's Cabeus crater. The upper stage of the Atlas V Centaur launch rocket and the spacecraft remained together on a collision course with the Moon until their final approach. Then they separated and the upper stage went on ahead and impacted the Moon at 2.5 km·s^{-1}, producing the equivalent to an explosion of ~2 tons of TNT. The spacecraft followed behind and flew through the debris plume, sampling and analysing the material with an array of cameras and spectrometers before it, too, hit the Moon some six minutes later.

Two other current asteroid impact missions are Hayabusa 2 and Asteroid Impact and Deflection Assessment (AIDA). Hayabusa 2 is presently on course for asteroid 162173 Ryugu around which it is due to go into orbit in July 2018. A 7 kg penetrator containing 4.5 kg of explosive will be released to fall onto the asteroid. The explosion will occur while the main spacecraft is on the far side of the asteroid and so will be observed by a separate deployable mini spacecraft that is basically just a camera and transmitter. The explosion's main purpose is to expose fresh material to be examined by the instruments on the primary spacecraft. Four landers will be deployed (Section 7.2) and also samples returned to Earth (Section 7.2.6.3.6).

AIDA's primary purpose is to investigate the possibility of deflecting an asteroid from a collision course with Earth, but it will also involve a lander (Section 7.3.1) and other instruments to study the asteroid itself. It is currently planned for a 2020 launch and will use two spacecraft. The first, Asteroid Impact Mission (AIM), is the main spacecraft. It is

---

* A miniature version of the space shuttle. It is a US classified military project, so few details are available.

intended to go into orbit around asteroid 65803 Didymos in 2022. Didymos is some 800 m across and has a small (150 m) satellite called Didymoon. The second spacecraft, Double Asteroid Re-direction Test (DART), will be launched in 2021 and will be the impactor. It will be a 300 kg solid cylinder that will hit Didymoon at around 6.25 km·s^{-1} (equivalent to the explosion of 1.5 tons of TNT). The explosion will be observed from AIM for scientific purposes and then Didymoon's orbit monitored to assess the deflection caused by the impact (expected to be a velocity change of about 0.4 mm·s^{-1}).

Many orbiting spacecraft are also crashed into their target objects at the ends of their missions. Sometimes this is just to get, at the last moment, the highest possible resolution images (e.g. the Ranger series of lunar probes, Rosetta and Cassini). On other occasions, the orbit decays after the fuel has been exhausted (MESSENGER and many Earth-orbiting spacecraft). Sometimes though, the crash is used, like that of Deep Impact, to try and probe the target. Thus, the lunar impact crater produced by the LADEE spacecraft was later observed by the Lunar Reconnaissance Orbiter. Similarly, during the later Apollo missions, the Saturn V upper stages and the lunar ascent modules were crashed into the Moon to provide signals for the seismometers left by the earlier missions in the series.

Some details and discussion of sample and return missions are included in Section 7.2.6.3.6. However, since the samples are analysed in terrestrial laboratories, they are technically outside the brief for this book. More importantly, the myriad of scientific analysis techniques available on Earth would require a book 10 times longer than this one for even a cursory discussion. Generally, therefore, the reader interested in this topic will need to search for further information in chemical, biological, microbiological, geological and related sources.

Finally, many spacecraft carry detectors for charged and neutral subatomic particles, atoms, ions, dust particles and magnetic and electric fields (see Chapters 4, 5 and 6 and Section 7.2.5) so that they can make direct measurements of planetary radiation belts and the interplanetary medium.

7.2 SURFACE

Many sampling and drilling instruments probe a few millimetres to a metre or so below the top surface of a solar system object. Strictly, therefore, these instruments should be covered in Section 7.3. However, for most solar system objects, the surface and its immediately underlying layers are mixed together, on time scales ranging from seconds to thousands of years by the effects of weather, temperature changes, meteorites and micro-meteorites, the solar wind, landslips, tidal stresses, and cosmic rays.* Thus the surface and the layers immediately below it will generally form a more-or-less homogeneous collection of materials and structures. We shall therefore take the 'surface' to be a layer up to about 1 m deep and consider those instruments investigating this region in this section.† Instruments probing the interiors of solar system objects at depths ranging from tens of metres to hundreds of kilometres are covered in Section 7.3.

---

* For the Moon, for example, the top 20 mm of its surface is churned over completely in about 80,000 years by micro-meteorites.

† However, γ ray and neutron spectrometers on orbiting spacecraft – such as Mars Odyssey's GRS (Section 3.2.5.2.1) – can identify elements and measure abundances in the top metre of surfaces as well.

The imaging of surface landscapes uses the types of cameras, or small adaptations of them, that were discussed in Chapter 1. The same is also largely true for the cameras used to take close-up images of rock formations and structures for geological studies, for selecting sampling sites and for studying the effect on the surface after samples have been taken. So, these instruments will not be further discussed here. Some aspects of the cameras used for microscopic studies or for mineral identification, are, however, included below.

In this section therefore, we are concerned with the biological, chemical and physical investigations of the top metre or so thick layer of the solid surfaces of solar system objects.

7.2.1 Microscopy*

Close-up images of the surface of a solar system object or of samples from its surface can be obtained with submillimetre resolution by relatively conventional cameras. Thus on the Mars rover, Curiosity (Figure 1.6), there is the Remote Micro-Imager (RMI). The RMI's primary purpose is to select targets for spectroscopy (Section 7.2.2), but it can also image with a spatial resolution of 80 μm at a distance of 2 m. The RMI is an f/4 Schmidt–Cassegrain telescope (Section 1.2.1.4) with a 110 mm aperture operating over the 240–900 nm wavelength range using a 1024 × 1024 charge-coupled device (CCD) (Section 1.4.1.2) as its detector.

Curiosity (Figure 1.6) carries Mars Hand Lens Imager (MAHLI) (see also Figure 7.8) that can obtain colour images with a 15 μm resolution from a distance of 25 mm. It uses an f/9.8 nine-element lens with an 18.3 mm focal length and automatic focussing. The detector is an interline transfer CCD (Section 1.4.1.2) with 1200 × 1600 active pixels and there are white light and ultraviolet light emitting diodes (UV LEDs) for illuminating the samples. MAHLI is mounted on the rover's robotic arm (Section 7.2.6) so that it can be positioned close to its desired target.

The earlier Mars exploration rovers, Spirit and Opportunity (see Figure 7.4 – on the far side of the head of the robotic arm – NB: Opportunity is still operating at the time of writing), carried identical science payloads which included microscopes. The microscopic imagers (MI) actually had a 'magnification' of ×0.4, but their 1024 × 1024 CCD detectors had spatial resolutions of 30 μm over a 31 mm square field of view. They were mounted on moveable arms and imaged over the 400–700 nm spectral range with a fixed focus at 100 mm. Focussing was accomplished by moving the mounting arms. The MIs used triplet f/15 lenses with focal lengths of 20 mm.

The Phoenix Martian lander carried a ×6 optical microscope with a spatial resolution of 4 μm and a field of view of about 1 mm × 2 mm. The microscope used an f/30 doublet lens with a 16.5 mm focal length and it focused on samples 14 mm from the front lens. Its detector was a 256 × 512 frame transfer CCD and red, green, blue and UV LEDs could be used for illumination. Samples for the microscope were delivered to it by Phoenix's robotic arm.

The ill-fated Beagle-2 lander of the Mars Express mission carried a microscope with a ×3.5 magnification and a spatial resolution of 4 μm. It used a triplet lens with a 15.75 mm

---

* See also MicrOmega – Section 7.2.2.2.

focal length and a 1024 × 1024 pixel frame transfer CCD. Red, green, blue and UV LEDs could be used to illuminate its field of view.

Cometary Secondary Ion Mass Analyser (COSIMA) (Section 5.2.6.4) on the Rosetta spacecraft used an optical microscope with a spatial resolution of 14 μm and LED illumination to select its target dust particles (Figure 5.19). The Philae lander also carried a microscope, the Comet Infrared and Visible Analyser (CIVA M/V). This had a 7 μm spatial resolution and could use LEDs at wavelengths of 525, 640 and 880 nm for illumination. A second microscope on Philae was to be used for infrared spectroscopy (Section 7.2.2.1)

The ExoMars 2020 rover will be carrying the Close-Up Imager (CLUPI). The camera will obtain full colour images with a best spatial resolution of 7 μm using a 1768 × 2652 pixel APS detector (Section 1.2.2.2) with fields of view up to 14° and variable focal lengths.

The PolyCam instrument on Origins, Spectral Interpretation, Resource Identification, Security, Regolith Explorer (OSIRIS-REx) is both a Ritchey–Chrétien telescope (Section 1.2.1.3) and a microscope. It is an f/3.1, 200 mm aperture telescope with an angular resolution of 10″. It can be focussed down to a distance of less than a metre giving it a surface resolution of a few millimetres.

Much higher spatial resolutions can be obtained using atomic force microscopes (AFMs). Invented in 1986, two have so far been flown in space. These microscopes are based upon a very fine needle point mounted on a thin arm (often called the cantilever). The needle point and the cantilever are usually made from silicon or silicon nitride. The microscope can operate in several modes. Two of the commoner methods are the Contact mode wherein the needle tip is dragged across the surface of the sample and the Tapping mode (which reduces the possibility of wear on the needle tip or damage to the sample) wherein the needle point is tapped onto the sample at regular intervals. Piezo-electric crystals are used to move the needle point in three dimensions. The deflection of the cantilever is measured either by reflecting a laser beam off it or by measuring the forces needed to keep the cantilever beam in a constant position. The microscope does not need a vacuum, but can operate within an atmosphere or even in a liquid.

The first spacecraft-borne AFM was on board Rosetta and was called Micro Imaging Dust Analysis System (MIDAS). It was used in the tapping mode and took several hours for each scan. Its aim was to study dust particles from comets. The AFM had an array of 16 cantilevers so that a new one could be used as and when a needle wore out. Its spatial resolution was 4 nm.

Phoenix used an AFM which had a spatial resolution of 100 nm – 40 times better than its optical microscope. It could obtain a scan in about 30 minutes. The AFM had eight cantilever arms and needles and used one at a time. A needle that became worn or dirty could thus be broken off and the next one used.

Nanometre spatial resolution can be achieved with scanning electron microscopes (SEMs). Normally, these require the specimens to be in a vacuum chamber, but the Atmosphere or Environmental SEM (ASEM, ESEM) operates with the specimen in a gas. The ASEM is essentially similar to a normal SEM with an electron beam, focussing electrodes or magnets and an image detector or display screen. However, it also has two small (~100 nm to a few times that) open apertures through which the electron beam passes. The specimen is placed below these apertures in the gaseous medium. The gas, of course, slowly leaks through the bottom aperture but most of it is then quickly removed by a powerful pump. This pump does not produce a vacuum, but does lower the gas pressure significantly. The remaining gas can leak through the second clear aperture, but a second pump removes it and above the second aperture the gas pressure is low enough for the electron beam to be produced and focussed. As the electron beam reaches the regions where there is significant gas pressure it starts to lose electrons through scattering by the gas molecules. However, the instrument can be designed so a sufficient amount of the electron beam remains to image the specimen. For a 10 kV electron beam scanning a specimen in the Earth's atmosphere this would require the specimen to be only a few microns below the bottom aperture. However on Mars, with a gas pressure ~0.6% that of Earth, the distance could be much greater and on the Moon and other effectively airless bodies, the ASEM would hardly need the apertures.

While ASEMs have only so far been used on Earth, a miniature version suitable for a spacecraft has been developed by NASA. The Miniaturized Variable Pressure Scanning Electron Microscope (MVP-SEM) has a length of about 250 mm and occupies less than a 10 L volume.

7.2.2 Spectroscopy

Spectroscopy of solar system objects via remote sensing has been covered in Sections 1.3 (extended optical region [EOR]), 2.5 (radio and microwave) and 3.4 (x-rays and γ rays). In this section therefore, we are concerned with spectroscopic techniques used by landers and rovers on the surfaces of solar system objects.

7.2.2.1 Thermal Emissions

This section overlaps in some respects with the next one. The difference is that thermal detectors are usually operating at longer wavelengths than those considered in Section 7.2.2.2 and their scientific purposes are more concerned with studying temperatures (see also Section 7.2.4) and measuring thermal inertia than in determining mineral identification and element abundances. The instrumental designs also tend to be rather different between this and the next section.

The MASCOT Radiometer (MARA) on the Hayabusa-2 spacecraft's lander is designed to make measurements over the 5–100 μm spectral range through six filters. A single broadband filter covers the whole spectral range, while five narrower-band filters select the spectral regions, 5.5–7 μm, 8–9.5 μm, 8–14 μm, 9.5–11.5 μm and 13.5–15.5 μm. Each filter feeds a separate thermopile detector (Section 2.3.2) containing 72 bismuth–antimony junctions with a 20° field of view. MARA's primary purpose is to measure the surface temperature (expected to range from ~120 K to ~450 K) and the thermal inertia of asteroid 162173 Ryugu's surface layers, but its narrow band measurements will allow for some estimation of the mineralogy of the rocks. The Hayabusa-2 orbiter also carries a thermal mapper with an 8–14 μm filter that is identical to that used by MARA. The purpose is to enable MARA to supply ground truth (Section 7.2.7) measurements to calibrate those made from the orbiter.

Opportunity's Miniature Thermal Emission Spectrometer (Mini-TES) (and the similar instrument of the now defunct Spirit – see Figure 7.4 – at the rear of the head of the mast) is mounted on the rover's mast and observes over the 5–29 μm spectral region with a spectral resolution of about 200. It can be moved to point to altitudes up to 30° above the horizon and 50° below and through 60° in azimuth. It is a Fourier Transform Spectroscope (FTS) based (Section 2.5) instrument with an uncooled triglycerine sulphate pyroelectric (Section 2.5) detector in which the hydrogen is in the form of deuterium. The optics are potassium–bromide lenses. As well as making measurements in its own right, Mini-TES provides ground-truth data (Section 7.2.7) for the Mars Odyssey's Time History of Events and Macroscale Interactions during Substorms (THEMIS) instrument (Section 1.2.1.6).

The TES mounted on the (orbiting) Mars Global Surveyor was also based upon an FTS and operated between 6.25 and 50 μm with a spectral resolution ranging from 20 to 160. Its angular resolution was about half a degree

7.2.2.2 Solids

The spectra of solid materials do not have the sharp absorption or emission lines characteristic of the spectra of gases (Sections 1.3.4 and 2.5). Far less information is therefore contained within their spectra. Nonetheless, the spectra from solid materials do show broad variations with wavelength and sometimes these are sufficient to identify particular minerals.

The colour of a solid as seen by the eye or imaged by a colour camera may sometimes be sufficient to identify a material. However, the identification is frequently uncertain – iron pyrites is so often mistaken for gold that its alternative name is 'Fool's Gold'.

More certain identifications may be made if a moderate resolution (R in the tens to the low hundreds) spectrum covering the visible and near infrared (NIR) is obtained. This is especially the case when the probable minerals present in a rock are already suspected from other investigations. Thus from the lunar specimens returned to Earth by the Apollo manned missions and Luna robotic sample and return missions we know that the main minerals on the Moon are ilmenite, olivine, plagioclase (a feldspar) and pyroxene – and their visible and NIR (VNIR) spectral reflectance curves are sufficiently different from each other for their identification (Figure 7.2).

The lunar rover, Yutu, carried by the Chang'e-3 lander had on board a VNIR instrument comprising a VNIR imaging spectrograph and a short wave infrared spectrograph for mineral identification. The VNIR instrument was based upon two acousto-optical tunable filters (AOTFs) (Section 1.3.5). The imaging spectrograph covered the 450–900 nm spectral regions and had a square field of view of 8.5°. It used a 256 × 256 pixel CMOS detector giving it an angular resolution of 2′ and it had a spectral resolution of ~150. It operated by using the AOTF to produce a monochromatic image which was then scanned in wavelength to build up the spectral cube by changing the AOTF's driving frequency. The short wave infrared spectrograph operated in a similar fashion, but with a single detector. It had a spectral resolution of ~250.

The Opportunity rover has MastCam (and also the now defunct Spirit) that may be seen in Figure 7.4 as the black squares at the top of the mast head. This is a pair of cameras

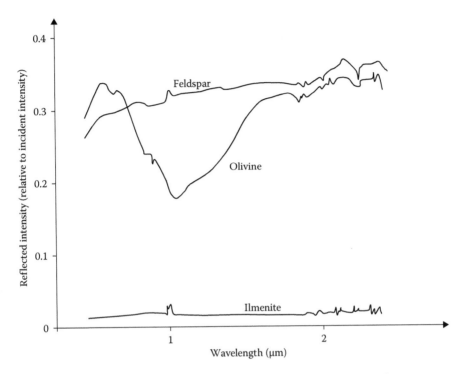

FIGURE 7.2   Schematic reflectance spectra from some common lunar minerals.

mounted on the rover's mast. One camera has a square field of view of 15° provided by an f/8, 34 mm focal-length lens. The other camera also has a square field of view, but one of 5.1°, and this is provided by an f/10, 100 mm focal length lens. The latter has a spatial resolution of 75 mm at a distance of 1 km. The two cameras are separated by 250 mm and can be used for stereo imaging. The detectors are CCDs with 1200 × 1200 pixel illuminated areas. Bayer (Section 1.2.2.2) filters are integrated onto the pixels to enable normal colour images to be obtained. In addition there are 12 narrow band filters spaced over the 440 nm to 1.035 μm waveband. Each camera has four filters in common and four at different wavelengths – thus making available a total of 12 different observation channels. This is enough narrow band filters for rough reflectance curves (Figure 7.2) to be plotted and some minerals identified.

The Mars 2020 rover is planned to carry MastCam-z. This will be very similar to Opportunity's MastCam except that the two cameras will have 3.5:1 zoom lenses giving focal lengths from 28 to 100 mm. They will also use CCD detectors with integrated Bayer filters and with 1200 × 1600 pixels being illuminated. There will be seven narrow band filters in each camera, but only one of these will be in common and the filters will cover the spectral region from 440 nm to 1.10 μm.

Rosetta's Philae lander carried an infrared spectrometer, the CIVA M/I. This would have used a microscope with a 40 μm spatial resolution to obtain images of samples. Spectra were to be obtained by illuminating the sample with monochromatic light from a white light source passed through diffraction grating and then scanning the illuminating radiation over the 1–4 μm spectral bands. Low-resolution reflectance spectra could also be obtained by the descent camera, Rosetta Lander Imaging System (ROLIS). After landing, ROLIS would have focussed on the comet's surface below the lander and used LEDs to illuminate it at 470, 530, 640 and 870 nm wavelengths.

Hayabusa-2's MASCOT lander includes a multispectral microscope that will image areas 5 mm across with a spatial resolution of 20 μm and a spectral resolution of ~250. Called MicrOmega it will, like Yutu's VNIR, use an AOTF. However, the manner in which MicrOmega uses the filter is quite different from that of Yutu. The area to be imaged is illuminated by a white light source whose radiation has been passed through the AOTF. The filter transmits only a narrow waveband of the white radiation and so the camera obtains a monochromatic image at that wavelength. The spectral cube is built up by tuning the AOTF to other wavelengths and obtaining additional images (cf. CIVA M/I). MicrOmega covers the spectral range from 900 nm to 3.5 μm and was developed from an earlier instrument on board Fobos-Grunt.

7.2.2.3 Gases

The spectroscopic study of atmospheres from landers and rovers is covered in Section 7.4.1. Here, we are concerned with the spectroscopic study of gases produced from the solid (and potentially from the liquid) components of the object's surface. This usually involves vaporizing a small portion of the sample using a powerful laser and then studying the resultant gases to determine the composition of the sample.

Curiosity (Figure 1.6) carries the laser-induced breakdown spectrometer (LIBS)[*] – the first such instrument used in space. LIBS uses some tens of laser pulses each lasting 5 ns and with a power of 14 mJ focussed onto a spot a few hundred microns across to evaporate a sample of rock up to 7 m away from the rover (Figure 7.3). The evaporated atoms and ions are in excited states and decay back to their ground levels, producing emission lines. The radiation is collected by a 110 mm Schmidt–Cassegrain telescope (Section 1.2.1.4) and fed by a 6 m long optical fibre to three spectrometers which cover the spectral regions 240–340 nm, 380–470 nm and 470–850 nm with spectral resolutions of about 2700. Their gratings are plane holographic, while concave mirrors are used for the collimators and focussing elements.[†] CCDs are the detectors. The laser can also be used to clear dust off the rock's surface so that its bulk composition may be studied.

LIBS-type instruments developed from that on Curiosity are planned to form part of the payloads of the Mars 2020 (part of the SuperCam package) and possibly the ExoMars 2020 missions in due course.

7.2.2.4 Raman Spectroscopy

Raman spectroscopy enables molecular vibrational lines in the far infrared (FIR) and microwave regions to be observed at UV, visible and NIR wavelengths. It is based upon inelastic scattering wherein the molecules subtract (the Stokes' lines) or add (the anti-Stokes' lines[‡])

---

[*] LIBS, together with RMI (Section 7.2.1.1), form the ChemCam package.
[†] A design known as the Crossed Czerny-Turner.
[‡] Confusingly, the opposite naming convention is also to be encountered.

FIGURE 7.3 A ~50 mm iron meteorite on Mars that has been examined by the LIBS instrument. The small white spots are where the laser has evaporated material from the surface of the meteorite. (Courtesy of NASA/JPL/Caltech.)

energy to the illuminating radiation. The samples (solid, liquid or gaseous) are illuminated by an intense beam of radiation whose wavelength does *not* coincide with any of the molecules' spectrum lines. Today that illumination is invariably from a laser.

The radiation scattered from the sample contains a very strong component at the original wavelength and much weaker components at nearby wavelengths. In detecting the scattered radiation, the original wavelength is eliminated by very narrow band filters. The other radiations (emission lines) are detected and their energy differences from the energy of the original illuminating radiation correspond to the energies of the vibrational (sometimes rotational) transitions of the molecule at longer wavelengths.

The ExoMars 2020 rover will be carrying a Raman Laser Spectrometer (RLS). The RLS should be able to identify organic* compounds directly, to identify minerals produced by biological processes and to study minerals affected by water. Its excitation radiation source will be a laser emitting at 532 nm giving exposures of up to 10 minutes and focussed onto a 50 μm spot. Its intensity of ~25 mW is low enough to ensure that the sample is not thermally damaged. It will be able to observe molecular lines over the 2.5–70 μm spectral region with a spectral resolution of up to ~330. The spectrometer will be based upon a

---

* See footnote about organic molecules on page 93.

holographic transmission grating with a 512 × 2048 pixel CCD as the detector. Samples of the Martian surface will be obtained at depths of up to 2 m, crushed and examined by the RLS in the form of a powder.

The Mars 2020 mission will carry two Raman spectrometers. The first will be based on a neon–copper UV laser operating at 248.6 nm and with a 0.5 pm bandwidth. The UV radiation will additionally cause some minerals to fluoresce, especially those of organic compositions, thus adding to the range of investigations possible. The UV Raman spectrometer is named Scanning Habitable Environments with Raman & Luminescence for Organics & Chemicals (SHERLOC). Its purposes are similar to those of the RLS with the additional task of helping to select samples for a future mission to return to Earth. The illuminated spot, 50 μm in size, will be able to be scanned over a 7 mm × 7 mm area and the UV laser will be mounted on the end of the rover's robotic arm (Section 7.2.6). It will be able to examine natural and abraded surfaces and holes drilled to depths of ~25 mm. The second Raman spectrometer will operate at 532 nm using an neodymium-doped yttrium aluminium garnet (Nd:YAG) laser as a part of SuperCam. It will be able to observe minerals at distances up to 12 m and observe the molecular lines over the same spectral region as RLS with a spectral resolution of ~400. It will use the same telescope and spectrometer as the LIBS instrument (Section 7.2.2.2) on the rover.

If the Europa Multiple Flyby mission goes ahead and if it includes a small Europa lander, then a Raman spectrometer is one of the instruments currently being considered for inclusion on the latter.

7.2.2.5 X-Ray Fluorescence Spectroscopy and X-Ray Diffraction

Fluorescence generally is the phenomenon of the emission of electromagnetic (e-m) radiation from a material following that material's absorption of e-m radiation (usually of a shorter wavelength). It will be familiar to readers from the ubiquitous fluorescent lights wherein UV radiation from a mercury vapour excites the emission of visible light.

When the exciting radiation is in the x- or γ ray region, some of the atoms within materials being bombarded will become ionised. Electrons still attached to the atoms/ions can then cascade down through the energy levels, emitting radiation as they do so. The emission of this secondary radiation, which is also usually in the form of x-rays, is termed fluorescence. The fluorescent x-ray photons will be seen as emission lines with their wavelengths determined by the structure of the atoms' or ions' energy levels. The strongest lines usually originate when a K-shell[*] electron has been ejected, followed by electrons from the L, M, N, … shells dropping down to the K shell. The lines are then named Kα, Kβ, Kγ …, Lα, Lβ, …, Mα, Mβ, … etc., accordingly as the electron is dropping from the next shell out, from two shells out, from three shells out and so on. Just as in spectroscopy at longer wavelengths (Section 1.3.4) each chemical element has its own characteristic spectrum identification which enables its presence in a sample to be deduced and its relative abundance to be measured.

The Sun emits a large amount of x-rays and γ rays and mention of the detection of the fluorescent emissions induced by solar radiation has been made in Sections 3.1.2, 3.2.2,

---

[*] See books and other sources describing the basic structure of atoms for further information.

3.2.5.1 and 3.3.2.2.2. Here, we are concerned with fluorescence induced by x-rays from artificial sources.

Many landers and rovers have carried x-ray fluorescence spectrometers (XRFs). Vikings 1 and 2 carried the first XRFs to Mars in 1976. They used the radioactive isotopes iron-55 and cadmium-109 as their x-ray sources. These both decay by electron capture and have half-lives of 2.6 years and 1.2 years, respectively. The x-rays are actually emitted by the decay products of the radioactive elements – manganese-55 and silver-109, respectively – and have energies of 5.9 and 6.5 keV (Mn) and 22.1 and 25.1 keV (Ag). Their half-lives are long enough to last for the durations of the missions (more than 6 years for the Viking 1 lander).

Soil samples were placed into square tubes by the landers' sampling arms with the x-ray sources illuminating them through the sides of the tubes. The fluorescent x-rays were then detected by four gas-filled proportional counters (Sections 3.2.2 and 4.2.3) also mounted on the sample tubes. The detectors were passively shielded from the x-ray sources and had an energy resolution of 1.2 keV at 5.9 keV. The sample chamber could be emptied via a vibrating trap door at its base in order to receive fresh samples. However if measurements were made using an empty chamber, then the XRFs sampled the elements in the Martian atmosphere instead and so were able to show that argon was present at about the 2% level.

The XRF on board the failed Beagle-2 Mars lander also used iron-55 and cadmium-109 as its x-ray sources. Its detector was a silicon junction diode (Section 1.4.1.4) that needed to be cooled. It was therefore planned to operate the XRF during the Martian night so that the detector could be cooled passively.

An XRF instrument is planned for the Mars 2020 rover, called Planetary Instrument for X-ray Lithochemistry (PIXL); few details of its design are yet available. However, it is expected to use x-ray optics (Section 3.3) to focus its x-rays into a 120 μm spot on the sample and to be able to obtain measurements in a few seconds.

The Moon was first studied using an XRF mounted on the Lunokhod-1 rover. This had two tritium x-ray source and several filters and the detectors were proportional counters. The lunar surface was observed from a height of 300 mm and the instrument could be operated while the rover was moving as well as when it was stationary.

Eight x-ray sources, four each of iron-55 and cadmium-109, were used on the active particle–induced x-ray spectrometer* on the Chang'e-3 lunar rover, Yutu. The unit was mounted on the rover's robotic arm (Section 7.2.6) and used a silicon SSD (Section 4.2).

The Venera 13 and 14 and the Vega 1 and 2 Venus lander spacecraft all carried very similar XRFs to the surface of Venus. These used two iron-55 and one plutonium-238[†] x-ray source. Four proportional detectors were used – three based on krypton and one on xenon. The latter was included to measure elements heavier than iron since it could detect fluorescent x-rays up to an energy of 20 keV. Samples were obtained by drilling into the surface and placing the powder into the soil receivers incorporated into the instruments.

---

* The abbreviation of this, APXS, should not lead to its confusion with the APXS instruments in the next section.
† Emitting x-rays at 13.6, 17.2 and 20.2 keV.

Several XRF spacecraft instruments are made more versatile by using the x-ray source to study samples by diffraction as well as through fluorescence or to have a charged particle source in addition to the x-ray source. Bragg diffraction of x-rays is discussed in Section 3.4.2 and charged particle x-ray spectrometers in Section 7.2.2.6. Here, therefore we look at examples of combined XRF/X-Ray Diffraction (XRD) instruments.

The CheMin instrument on the Mars rover, Curiosity, is an XRF/XRD. It is described in Section 3.4.2 and is still operating at the time of writing.

The XRF/XRD instrument for the ExoMars 2020 rover will use iron-55 as its x-ray source. Three CCD arrays will be the detectors arranged as part of a circle opposite to the x-ray source with the sample at the circle's centre. The CCDs will be used in photon counting mode so that each x-ray photon will have its energy measured and the diffracted photons (at the original energy) may be distinguished from the fluorescence photons (at lower energies).

7.2.2.6 Alpha Particle Spectrometry

Alpha particle (helium-4 nuclei) scattering was studied by Ernest Rutherford, Hans Geiger and Ernest Marsden in their famous gold foil experiments in 1909 which led to the Rutherford planetary model of atoms. In APS, the alpha particles are backscattered and the energy of the backscattered particle (E_s) depends upon the scattering angle (~180°), the particle's initial energy (E_i) and the atomic mass of the scattering atom (A) as shown in Equation 7.1:

$$E_s = \left(\frac{A-4}{A+4} \right)^2 E_i \tag{7.1}$$

In thick samples, alpha particles penetrating far into the sample will lose energy through collisions on their way in and on their way back out. The spectrum of a pure element will then be of a more-or-less constant intensity up to the maximum energy (which is given by Equation 7.1) after which the intensity will drop to zero. With several different elements present, each will have its own spectrum in this form and the total spectrum will have a staircase-like profile with the steps' 'tops' identifying the elements and the 'risers' their relative abundances.

In addition to the backscattered alpha particles, the irradiated sample will emit protons arising from interactions of the alpha particles with the nuclei of elements such as oxygen, sodium and magnesium. Fluorescence x-rays will also be emitted due to the disturbance of all the atoms' electrons by the alpha particle's passage near them. Thus, APS and XRF provide information on the elements present in the sample with APS best suited to light elements and XRF to medium mass elements, while XRD gives crystal structure, enabling minerals to be identified.

The first spacecraft-borne APS was on board Surveyor 5, which soft landed on the Moon in September 1967. This employed six curium-242 alpha particle sources to irradiate the surface. Two detectors measured the energies of the back scattered alpha particles, while four detected the protons emitted from the nuclei. Surveyors 6 and 7 carried similar instruments.

The first Mars rover, Sojourner, carried the alpha particle x-ray spectrometer (APXS) instrument that detected the backscattered alpha particles, the emitted protons and fluorescence x-rays. Curium-244 was used since it emits 5.8 MeV alpha particles while its decay product, plutonium-240, when it is excited or ionised by the alpha particles emits fluorescence x-rays at 14.3 and 18.3 keV. Fluorescence x-rays were also produced within the sample by the alpha particles. The APXS sensor head had nine of the curium radioactive sources. The alpha particles were detected by a silicon SSD and a second SSD acted as an active shield from cosmic rays for the first. The two SSDs then together would have been used to detect the protons, but their intensity was too weak to detect. A silicon junction diode (Section 1.4.1.4) with a spectral resolution of 260 eV near 6 keV detected the fluorescence x-rays and was passive shielded from the radioactive sources by a tungsten collimator. Similar, but improved, instruments were carried by the Mars Exploration Rovers' (MER) Spirit Mars rover and on the still operating Opportunity rover (Figure 7.4 – the APXS is at the back of the sensor head of the robotic arm [Section 7.2.6]). The Curiosity Mars rover (Figure 1.6) also has an APXS, but this lacks the alpha particle detector component. Instead it acts as an XRF (Section 7.2.2.5) using the alpha particles and x-rays from its radioactive sources.

The Philae comet lander had an APXS that was lowered after the final landing to within 40 mm of the surface. Like Sojourner's instrument, this was based upon curium-244 radioactive sources but only had detectors for the backscattered alpha particles and the

FIGURE 7.4 An artist's impression of one of the MER rovers on the surface of Mars. (Courtesy of NASA.)

fluorescence x-rays. Its design was developed from the instruments on Sojourner, Spirit and Opportunity.

An alpha particle detector based upon the direct interaction of the particles in silicon was used to measure radon levels during the Apollo missions and this is discussed in Section 4.2.6.2.

7.2.2.7 Mössbauer Spectroscopy

Although a widely used laboratory technique, so far, Mössbauer spectroscopy has only been used on Mars by MER's Spirit and Opportunity rovers to investigate the abundance of iron. Beagle-2 also carried a Mössbauer spectrometer, but, of course, this was never used.

Particles within atomic nuclei can occupy excited energy levels within the nucleus, just as their electrons can within the surrounding electron cloud. The energies involved when nuclear constituents absorb or emit energy are, however, much greater than that for the electrons, so the absorbed or emitted photons are γ rays. Now photons carry a momentum, p, given by

$$p = \frac{h\nu}{c} = \frac{h}{\lambda} \, \mathrm{kg\,m\,s^{-1}} \tag{7.2}$$

So when a photon is emitted or absorbed, conservation of momentum requires the atom or nucleus to recoil and the recoil velocity then Doppler shifts the photon's wavelength. For radiation of wavelengths longer than soft x-rays, the spectrum lines are sufficiently broad that the recoil is insignificant. However for nuclear transitions, the line width can be as little as 10^{-11} to 10^{-12} of the photon's wavelength (cf. 10^{-4} for the core of the solar H-α line). The recoil for the emission of a 100 keV γ ray would correspond to a wavelength shift of about one part in 10^5 of the photon's wavelength so that it could not be absorbed by the same nuclear transition by another nucleus of the same species. Mössbauer discovered in 1958, however, that by incorporating the atoms into cooled crystals, the recoil was shared by many atoms and the emission or absorption occurred at an almost fixed wavelength.

Thus in order to study the abundance of iron-57, a radioactive source of cobalt 57 is used. Cobalt-57 decays to iron-57 with a half-life of 270 days and emits (amongst other γ rays) at 14.401 keV. This photon can be absorbed by an iron-57 nucleus resulting in a transition from its ground state to its first excited level. When the iron-57 nucleus returns to its ground level the emitted γ ray can be detected. Iron-57 comprises about 2.2% of the stable isotopes of iron, so multiplying its abundance by a factor of 45 gives the total abundance of iron in the sample.

However, the spectrum line is so narrow that if the iron atoms in the crystal are in the form of Fe^{+2} or Fe^{+3} etc., then the changed electrical environment of the nucleus will shift the γ ray's wavelength sufficiently for it not to be absorbed. To overcome this problem, the radioactive source is used while in motion so that the resulting Doppler shift brings the emitted line into coincidence with the wavelength of the line absorbed by the iron in the sample. For iron-57, scanning the source's velocity from -10 mm·s^{-1} to $+10$ mm·s^{-1} suffices for this.

Opportunity (and Spirit before its mission ended) carries a Miniature Mössbauer Spectrometer (MIMOS II) mounted on its robotic arm (Section 7.2.6) based upon cobalt-57 in a rhodium matrix. The source is vibrated at 25 Hz to enable the wavelength shifts to be compensated by the Doppler effect. A second, weaker, source is used for calibration. Four photodiode detectors are used to detect the γ rays emitted by the iron nuclei. Before the intensity of the radioactive source decayed too much, a measurement of the abundance of iron-57 could be obtained in about half a day. A similar instrument was on board the Beagle-2 Mars lander.

7.2.3 Chemical and Biological Analyses

7.2.3.1 Introduction

As carbon-based life forms ourselves, we human beings have an abiding, if parochial, interest in (or sometimes fear of) finding similar life forms that have originated elsewhere in the universe than on Earth. The search for extra-terrestrial life (ET) takes two main forms – firstly, the Search for Extra-Terrestrial Intelligence (SETI), which is hoping to detect signals from life forms on planets (or asteroids, natural satellites, etc.) that belong to stars other than the Sun, and, secondly, Exobiology*, which is trying to find life or evidence of its past existence directly within the solar system.

It is the latter search that is the concern of much of this section. Of course, it may be possible for life forms to exist that are not based upon carbon and there has been much speculation ranging from the wildest nightmares of the SF comic strip writers to the relatively scientific suggestions of silicon substituting for carbon. However, spacecraft missions hoping to detect life have so far not been designed to look for anything except carbon-based life and it is distinctly possible that we could not recognise non-carbon-based life if we found it.[†]

Most of the landers and rovers and many orbiters have carried instruments capable of identifying elements, compounds, and minerals on their target's surfaces, atmospheres and (for Titan) their lakes or seas. Although this section is primarily concerned with direct sampling at the surfaces of solar system objects, discussion of remotely obtained information will be included where significant in order to give the complete picture.

Many instruments are available for chemical analyses and have been covered in other sections – photon spectroscopy (Sections 1.3, 2.5, 3.4 and 7.2.2) and mass spectrometry (Section 5.2.6). Mention of examples of their uses, including from the surfaces of solar system objects, is included in those sections. One widely used instrument, however, has not so far been mentioned: the gas chromatograph, which is briefly described in the next section.

7.2.3.2 Gas Chromatograph

Chromatographs are widely used in the laboratory as a means of separating and identifying the components of mixtures. The technique was first used in 1900 by Mikhail Tsvet. It relies upon the phenomenon of elution which is probably more familiar to the reader than he or

---

* The name of the ExoMars mission derives from its exobiological mission aims.
† The Isaac Azimov short SF story "The Talking Stone" provides an interesting view of this problem.

she realises. Whenever you wash your clothes, for example, the dirt (analyte or sample) is washed from the clothes (stationary phase) by the water (solvent, eluent or mobile phase) and emerges as dirty water (eluate) which you can see (detect) going down the sink.

While washing clothes is not quite a perfect analogy for chromatography it does describe the essentials of the process. The main additional factor to be taken into account is that when the analyte is a mixture of substances, they are carried by the solvent through the stationary phase at different rates. They thus emerge at different times and can be detected or collected separately.

There are many types of chromatographs used in the laboratory with columnar (the stationary phase is contained in a column through which the dissolved sample trickles) and thin film or paper chromatography (the sample is placed on a sheet of paper or a thin film of a suitable stationary phase material held on a flat support and capillary action moves the solvent and the sample across the paper or film) being widely encountered.

A third type of chromatograph uses a gas as the mobile phase and it is this type that is used for spacecraft-borne instruments. The gas chromatograph is similar in principle to that of other chromatographs but uses a gas (helium, nitrogen, etc.) as the mobile phase and a thin layer of liquid or polymer (cyanopropylphenyl dimethyl polysiloxane, carbowax polyethyleneglycol, etc.) on an inert solid support (stainless steel or glass) as the stationary phase. The gases to be analysed are carried through a tube (or column) lined with the stationary phase material by the carrier gas. Their speeds through the tube depend upon their elution rates with the stationary phase and so the differing components of the original mix emerge from the tube at different times.

Gas chromatography can be used to detect and identify atoms and molecules. For example, the Venera 13 and 14 landers analysed the Venusian atmosphere for trace components below a height of 58 km during their descents, finding H_2O, O_2, H_2S, COS, H_2 and Kr. Their gas chromatographs used neon and nitrogen as the carrier gases and columns of polysorbate (emulsifiers widely used by the pharmaceutical industry) as the stationary phases. Ionisation detectors were used in which the samples were ionised and the resulting electric current depended upon the concentration of the component.

The gas exchange experiments (GEX) on the Viking 1 and 2 Martian landers used gas chromatographs as their detectors. GEX took samples of Martian soil and incubated them with various nutrients with and without water. It was expected that if life forms were present then gases such as carbon dioxide, methane, oxygen, hydrogen or nitrogen might be emitted and these would detected by the gas chromatograph. Although oxygen was detected, subsequent analysis suggested that it arose from non-biological chemical reactions.

The Pioneer Venus large atmospheric probe carried a modified version of the Viking gas chromatographs to analyse Venus' atmosphere. Helium was used as the carrier gas and there were two long (15.9 m) and two short (2.4 m) columns all of which were 1.1 mm in diameter.

7.2.3.3 Gas Chromatograph - Mass Spectrometer Instruments

Since gas chromatographs detect atoms and molecules in a non-destructive fashion their eluates can be further studied after their emergence. A mass spectrometer (Section 5.2.6), which identifies the nuclei of elements and their isotopes, is thus commonly used after the gas chromatograph, producing a Gas Chromatograph–Mass Spectrometer (GCMS) package.

GCMSs were carried by both the Viking landers (in addition to the gas chromatograph for GEX – see previous section). Although intended for general purpose chemical analyses, the Viking GCMSs could aid their biological experiments (Section 7.2.3.4) by looking for carbon and its compounds. In fact they showed that there was less carbon present in the Martian samples than had been found in lunar samples returned to Earth for analysis. The GCMSs were fed powdered Martian soil by the landers' robotic arms (Section 7.2.6) and heated it to 500°C to release the volatiles into the gas chromatographs. These used hydrogen as the carrier gas which was later removed by a palladium-based separator. The stationary phase was a column of polymer beads coated with a polyphenyl ether grease. The column was slowly heated to 200°C so that different organic molecules came through at different times. These were then passed onto magnetic sector mass spectrometers (Section 5.2.6.2).

For Cassini's Titan probe, Huygens, the GCMS was based upon a three-column gas chromatograph with each column optimised for a different range of molecules. The columns were tubes 180 or 750 μm across and up to 14 m long, wound into disks and mounted on heaters. Hydrogen was used as the carrier gas. The instruments could operate just as gas chromatographs or the sample gases could be fed into a quadrupole mass spectrometer (Section 5.2.6.3.1) with a mass range of 2–141 Da. Samples were obtained directly from the atmosphere during the probe's descent. A small oven could heat samples to 250°C or 600°C in order to evaporate droplets of liquid (aerosols) in the atmosphere. After landing, the GCMS was also able to analyse Titan's surface composition by heating its inlet valve so that some of the surface material was vapourised. Over 8000 mass spectra were obtained, about 2/3 from the atmosphere and 1/3 from the surface.

Rosetta's comet lander, Philae, carried COSAC (Cometary Sampling and Composition), which was designed for the detection of complex organic molecules. There were eight capillary chromatograph columns with lengths up to 15 m and helium was used as the carrier gas. Ovens could heat samples to 180°C or 600°C. Thermal conductivity detectors were used to analyse the outputs. The latter relies upon the thermal conductivity of the eluate being reduced by the presence of sample molecules in the carrier gas – an electrically heated filament thus has a higher temperature when compared with being in the presence of the pure carrier gas. The gas pressure emerging from the gas chromatograph was too high for the mass spectrometer so a small 'leak' was used to lower the pressure. Since the helium was lost preferentially through the leak, the mass spectrometer received enriched samples. The mass spectrometer was of the reflection time of flight (ToF) design (Section 5.2.6.4).

Organic molecules such as sugars and amino acids can be produced by non-biological processes. Many such molecules however exist in two different three-dimensional structural forms which are mirror images of each other. This is like our left and right hands and the term for the phenomenon, chirality, is derived from the Greek word for hand: χειρ. The two versions of a chiral molecule are known as optical isomers* and are often described as right or left.

---

* Because aqueous solutions of a pure left or right chiral molecule are optically active and rotate the plane of a polarised beam of light in opposite directions.

Samples of chiral organic molecules produced by non-biological processes usually have equal quantities of right and left isomers. The same molecules produced biologically, however, will frequently be formed out of just one of the isomers (homochiral). Thus if a sample can be shown to contain only right or left isomers then it is very strong evidence for the operation of a biological process being involved somewhere.

Optical isomers are *not* separated by gas chromatography. However, they can undergo chemical reactions which convert the two optical isomers into different diastereoisomers* which *can* then be separated by a gas chromatograph. The chemical reaction is known as derivatization.† If homochiral molecules are present in the sample there will be a single detection if the sample has undergone derivatization, but a double detection if both optical isomers were originally present – corresponding to the biological or abiological production of the molecules, respectively.

The design of the Curiosity Mars rover's GCMS was developed from that of the Huygens Titan probe. It uses helium as the carrier gas in its gas chromatograph and feeds a quadrupole mass spectrometer. The gas chromatograph has six 30 m long, 250 μm wide tubes as its columns, wound to form circular disks. Different mass molecules are targeted by using different stationary phases in the tubes and thermal conductivity detectors were used. The mass spectrometer has three different driving frequencies enabling it to detect molecules with masses in the range 2–535 Da (cf. 2–141 Da for Huygens) and uses a CEM (Section 4.2.5.1) as its detector.

Samples are placed by the rover's robotic arm (Section 7.2.6) into cups of which there are 74 mounted on a rotatable carousel. A cup can be moved into one of two ovens and heated to 950°C or 1100°C to drive volatiles into the GCMS. Seven of the cups contain the derivatization agent N-methyl-N-(tert-butyldimethylsilyl)-trifluoroacetamide (MTBSTFA) plus a solvent. Samples placed in these are warmed to 80°C for ~30 minutes to allow the reactions to take place before being heated in the ovens. Although this gives Curiosity the potential to detect biologically produced materials and organic compounds have been found, so far the samples have not been homochiral.

It is planned for the ExoMars 2020 rover to carry a GCMS capable of detecting homochiralty. ExoMars' GCMS will obtain powdered surface samples from the rover's drill and these may be placed into 1 of 20 single-use small ovens. Some of those ovens will be preloaded with the derivatization chemicals. After an oven has heated up the sample to produce the gases, the gas chromatograph will feed them into the mass spectrometer. The gas chromatograph will have four columns with helium as the carrier and the mass spectrometer will be of the ion trap design (Section 5.2.6.3.2).

A second instrument, the laser desorption mass spectrometer (LDMS), will use the ExoMars 2020 rover's mass spectrometer in a different fashion from the GCMS to study the less volatile materials. A sample will be evaporated and ionised by a powerful pulsed

---

* Molecules with the same atomic composition and atomic bonding which differ in their three-dimensional structural forms, but which are not mirror images – like a normal left (say) glove and a left glove in which the fingers have been pushed into the inside of the palm of the glove.

† Suitable chemicals are, for example, tetramethylammonium hydroxide and N-methyl-N-(tert-butyldimethylsilyl)-trifluoroacetamide (MTBSTFA).

laser operating at 266 nm. The ions will then be fed into the mass spectrometer for analysis (cf. LIBS – Section 7.2.2.3). The GCMS and LDMS together make up a package on the rover called Mars Organics Molecule Analyser (MOMA).

In addition to the use of GCMSs to study solar system objects, they are often to be found on manned spaceflights where they monitor the quality of the atmosphere. The ISS, for example, has the Vehicle Cabin Atmosphere Monitor (VCAM) that uses an ion trap mass spectrometer and a 10 m long capillary helium gas chromatograph to monitor the methane, water, nitrogen, oxygen and carbon dioxide levels within the space station once a day.

7.2.3.4 Biological Investigations

Although it may be possible for life to exist on Titan's surface or in its liquid bodies or in the postulated sub-surface oceans of Europa and other natural satellites, so far only Mars has been seriously investigated for the past or current presence of life. The presence of methane in its atmosphere (Section 7.4), which can be produced by some bacteria (methanogens), has sparked some speculation, but abiological sources for the gas cannot be ruled out.

Methane and other hydrocarbons and organic* molecules can be detected by Curiosity's tunable laser spectrometer (Section 7.4.1) and by most of the gas chromatographs (Section 7.2.3.2) and GCMSs (Section 7.3.1.3.3) carried by Martian landers and rovers and even by some orbiters. COSAC on Philae was designed to optimise the detection of complex organic molecules. The Raman spectrometer, SHERLOC (Section 7.2.2.4), planned to be mounted on the robotic arm (Section 7.2.6) of the Mars 2020 rover for close-up work, will also be able to detect organic compounds and its SuperCam Raman spectrometer will be able to detect them at distances up to 12 m from the rover.

However Vikings 1 and 2 in the past, MSL's Curiosity rover and ExoMars 2020 in the future have specifically had or will have experiments designed to detect biologically produced materials. The MOMA package on the ExoMars 2020 rover has been discussed in Section 7.2.3.3 and the Vikings' GEX package in Section 7.2.3.2. Here, we consider the other two Viking biology instruments – the labelled release experiment and the pyrolytic release experiment.

The labelled release experiment used an aqueous nutrient solution (i.e. food) in which the nutrients had been labelled with radioactive carbon-14 atoms. A sample of Martian soil moistened with this solution was then monitored for the release of carbon dioxide containing some of the carbon-14. The expectation being that if found, such carbon dioxide must have been produced by microorganisms in the soil. Radioactive carbon dioxide was indeed produced when the sample was first moistened. However, this was not repeated subsequently when further additions of the solution were made. On Earth, similar experiments where the samples did contain terrestrial microorganisms continued to give off carbon dioxide as more of the solution was added. It is now generally accepted that the labelled release experiments did not detect life and that the results arose from oxidation reactions between the solutions and highly reactive chemicals in the Martian soil produced by solar UV radiation.

---

* See footnote about organic molecules on page 93

The pyrolytic release experiment also used carbon-14 and attempted to see if labelled carbon monoxide and carbon dioxide gases were photosynthesised when a sample of dry Martian soil was illuminated by an arc lamp. Such production of biomass would be ascertained by incubating the sample for five days and then removing the gases. The sample was then baked to 625°C, thus pyrolysing (vaporising) any biological material and monitoring the products for radioactivity. A second sample was heat sterilised first and used as a control against any detections being due to chemical rather than biological activity. The experiment did show that a small amount of carbon was being incorporated into the soil – but so did the control – so the processes involved were again probably chemical and not biological.

Recently, a new type of instrument has been suggested as a possible long-range detector for organic molecules. It is based upon a lidar (Sections 2.4 and 7.2.7) using UV lasers. The lasers would illuminate a dust cloud or perhaps part of the surface at distances up to hundreds of metres, any complex organic molecules would fluoresce (cf. Section 7.1.2.5) and by obtaining spectra of the emitted radiation, the molecules could be identified and perhaps estimates of their abundances made.

7.2.3.5 Neutron-Based Instruments

There are two types of analytical procedures that are based upon the use of high-energy neutrons. In the first, neutron spectrometry, the neutrons themselves are detected (Section 6.2) while in the second, neutron activation analysis, the neutrons produce γ rays and they are detected (Section 3.2) instead.

7.2.3.5.1 Neutron Spectroscopy   Neutron spectrometry is based upon neutrons and protons (hydrogen nuclei) having nearly equal masses, so that during collisions between them the momentum is shared equally. A high-energy neutron entering a hydrogen-rich material will undergo many such collisions and be slowed down rapidly. The momentum exchange between neutrons and heavier nuclei is less efficient and so the neutrons take longer to slow down. Soils containing water, ice or hydrated minerals may thus be identified, distinguished from drier soils and their water content measured.

The procedure is to fire a short pulse of high-energy neutrons into a surface where they will typically penetrate to a depth of some hundreds of millimetres. The neutrons will be scattered many times and some will re-emerge from the surface. The scattered neutrons are detected and their time delays from the initial pulse and energies are measured. When there is little hydrogen present in the soil the neutrons will be moving fast and re-emerge quickly and with high energies. When hydrogen is present then they will be slowed down much more, taking longer to re-emerge with lower energies.

The neutron spectrometer mounted on MSL's Curiosity rover, Dynamic Albedo of Neutrons (DAN), generates pulses of 14 MeV neutrons by accelerating deuterium nuclei and colliding them with a tritium target (a Pulsed Neutron Generator [PNG]). The pulses each contain around 10^7 neutrons and have pulse lengths of about 1 μs. Up to ten pulses a second can be produced. The re-emerging neutrons appear about 1 ms after the pulse with energies of a few tens of electronvolts when the soil is dry; they about three times longer and have energies of a few millielectronvolts when water or ice are present in significant quantities. DAN operates when the rover is stationary and can measure the water (strictly the hydrogen) content to a ±1% accuracy with a 2-minute exposure and about 10 times better than that with a 30-minute exposure. The neutron detectors use helium-3 and the interactions with neutrons are picked up by scintillation detectors. Cosmic ray–produced neutrons are generally of much higher energies than those from DAN's source and so two detectors are used. One filters out the low-energy neutrons before detection, the other detects all the neutrons and so it is the difference between the two detectors' results that is the actual measure of the presence of water. The neutron source is mounted on the front right of the rover (Figure 1.6) whereas the detectors are on the front left.

A neutron spectrometer developed from DAN is planned for the ExoMars 2020 rover. Neutron spectrometers on orbiting spacecraft, such as MESSENGER's Gamma Ray and Neutron Spectrometer (GRNS), Mars Odyssey's High-Energy Neutron Detector (HEND) and Dawn's Gamma Ray and Neutron Detector (GRaND) have been mentioned in Section 6.2.

7.2.3.5.2 Neutron Activation Analysis   Neutron activation analysis (NAA) is essentially the same technique as APS (Section 7.2.2.6), but using neutrons in place of the alpha particles. It is a widely used laboratory technique for the precision analysis of the composition of a large range of materials. In contrast to neutron spectrometry, NAA works best for heavier elements and about three quarters of the naturally occurring elements can be detected.

In NAA, the sample is irradiated by neutrons which are captured by nuclei in the sample. The nuclei will usually then be in excited states and will drop back to their ground states, emitting particles and/or γ rays. These γ rays will have energies specific to the isotope involved – for example, the naturally occurring isotope manganese-55 will be converted to manganese-56 by neutron capture and this emits γ rays at energies of 847 keV, 1.811 MeV and 2.110 MeV. These γ rays may be detected and used for analysis in what is usually called Prompt NAA. Alternatively the nuclei, now back in their ground state, are often radioactive and emit γ rays some time (seconds to years) after irradiation. Detecting these γ rays leads to Delayed Gamma ray NAA (DGNAA).

PNAA is suited to the detection of isotopes where the reaction product is not radioactive, or the half-life is very short or the radioactive γ ray intensities are low. DGNAA is the more widely used technique though and it may be combined with chemical separation of the reaction products in order to speed up the analysis. While NAA is a non-destructive technique so that samples can be analysed by other methods after it has been used, the sample may remain radioactive for some time and so needs careful handling and disposal.

In laboratory NAA, the neutrons are often produced within uranium fission nuclear reactors, but can also originate in radioactive emissions from (e.g.) californium-252, from alpha-particle-induced nuclear reactions (e.g. with beryllium) or from switchable ion sources like the PNG source for DAN. Clearly the last would be the only option for a spacecraft-borne instrument. The neutrons energies can range from a few meV to a few MeV. Germanium-based detectors (Section 3.2.5.2.1) are usually used to pick up the γ rays, but scintillation-type detectors are also to be found (Section 3.2.3).

NAA has yet to be used in space, but there are plans to launch NAA instruments to analyse the lunar surface during the Luna-Glob missions. GeMini-Plus is a γ ray spectrometer based on germanium (Section 3.2.5.2.1) that is small enough with low enough power requirements for landers and rovers that is currently being developed by NASA for future missions. Also an instrument combining NAA and neutron spectroscopy is under development by NASA for use on landers and rovers, called Probing In situ with Neutrons and Gamma rays (PING). A PNG is used to irradiate the surface with 14 MeV neutrons and both the prompt γ rays and the neutrons detected. So this brief mention of NAA has been included here for completeness.

7.2.3.6 Wet Chemistry

The Phoenix Martian polar lander made the first traditional 'wet' chemistry analyses on a planet other than Earth in 2008.[*] It had four identical units, each of which had a 'test tube' (actually a small cubic tank) in which the reactions took place with the sensing electrodes mounted on its walls. Reservoirs of water and chemicals, a stirrer and a funnel to accept the soils samples obtained by the robotic arm (Section 7.2.6) were also part of the package.

The sensors were ion selective electrodes which comprise a membrane that develops a voltage as its target ions diffuse through it from a high concentration aqueous solution on one side to a low concentration one on the other. The membranes can be coated with differing substances to allow differing ions through. The units formed a part of the Microscopy, Electrochemistry and Conductivity Analyser (MECA) package and each analysed about one ml of soil mixed with 25 ml of water. Anions and cations of calcium, magnesium, potassium, sodium, chlorine and sulphates were detected as well as highly reactive perchlorates. The presence of the latter in the form of calcium perchlorate and magnesium perchlorate formed by solar UV action on chlorine-bearing minerals may explain the anomalous results from the Viking labelled release experiments (Section 7.2.3.4).

In the future, capillary electrophoresis may be used to separate and identify organic molecules with a sensitivity some four orders of magnitude better than Curiosity's GCMS (Section 7.2.3.3).

In electrophoresis, the sample is dissolved in a liquid buffer[†] and an electric field is applied across it. Amino acids can ionise when in a suitable buffer solution and so be attracted to the oppositely charged electrode. The molecules' rates of motion through the liquid (ionic mobility) vary from one type of molecule to another and so they will separate out, with the faster ions arriving at the electrodes before the slower ions. In capillary electrophoresis, the separation takes place inside a capillary tube. In addition to the motions of the ions through the liquid, the liquid as a whole flows along the tube under the effect of the electric field.

In operation, the sample is placed into the inlet end of the capillary tube and then that end immersed into a container of the buffer solution while the outlet end is immersed into a second such container. The electric field is applied along the length of the tube and

---

[*] The Viking labelled release experiment pre-dates this (Section 7.2.1.3.4) but was not true wet chemistry. Derivatization (Section 7.2.1.3.3) might also have been included here.

[†] A liquid, usually based upon water, which maintains the correct pH value required for the reactions.

the organic components detected as they emerge from the outlet end of the tube. Several types of detector may be used, but for space applications, UV laser–induced fluorescence of the molecules is proposed (see also Section 7.2.2.5). The technique will be able to distinguish between non-biologically and biologically produced molecules from their chiralities (Section 7.2.2.3).

7.2.4 Physical Properties

Various physical properties of surface materials have been measured by some landers and rovers. These are generally not mainstream experiments and usually involve straightforward physical principles, but a brief review of some of them is given here.

7.2.4.1 Temperature and Thermal Conductivity or Inertia

Thermal radiometers (Section 1.3) can easily measure the surface temperatures of many solar system objects both from Earth and from orbiting spacecraft. The full Moon's average surface temperature was thus determined to be ~70°C as long ago as 1870 by the Earl of Rosse. He measured its infrared emissions with a thermopile (Section 2.3.2) at the focus of his 0.9 m reflector. Modern thermal maps of the lunar South Pole obtained by the Lunar Reconnaissance Orbiter's Diviner instrument (Section 2.3.2) are shown in Figure 7.5.

Thermal radiometers are also used for ground temperature measurements by some landers and rovers such as Hayabusa-2's MARA (Section 7.2.2.2) and the Ground Temperature Sensor (GTS) instrument that forms a part of the Rover Environmental Monitoring System (REMS) on the Curiosity Martian rover. The latter is mounted 1.6 m above the ground on one of the REMS' booms and observes a ground area of about 100 m². It uses three thermopiles to cover the spectral ranges 5.5–14.5 μm, 8–14 μm and 15.5–19 μm. The thermopiles have no optics and their fields of view are limited by simple apertures to be 60° horizontally and 40° vertically. They can operate over the whole 145–300 K temperature range of the

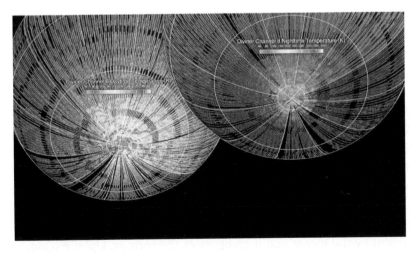

FIGURE 7.5 Day and night temperature maps of the lunar South Polar region obtained by the Lunar Reconnaissance Orbiter's (LRO) infrared radiometer, Diviner. (Courtesy of NASA/JPL.)

Martian surface, measuring the ground temperature to ±5 K. The Martian atmosphere's temperature gradient is very strong so at the height of the GTS above the ground the air temperature could be up to ±40 K different from the ground temperature.

A Thermal Infrared Sensor (TIRS) developed from Opportunity's GTS is being developed for NASA's Mars 2020 rover. It will have three downward pointing thermopiles and two upward pointing to obtain the ground and air temperatures, respectively. The spectral ranges will be upwards of 6.5 μm, 8–14 μm and 16–20 μm (ground sensors) and upwards of 6.5 and 14.5–15.5 μm (air sensors). Simple apertures will again limit the thermopiles' fields of view.

More conventional 'thermometers' can also be used to measure surface temperatures, sometimes in conjunction with thermal radiometers. Philae's Multi-Purpose Sensors for Surface and Subsurface Science (MUPUS) package, for example, was designed to hammer a 350 mm penetrator into comet 67P/Churumov–Gerasimenko's surface in order to measure the hardness of the surface and the temperature profile of its outer layers. Unfortunately, the surface was too hard for the penetrator and the instrument may have broken – however it was able to measure a temperature of 120 K for the comet's surface first. MUPUS also had heat sensors in the harpoons* within Philae's landing legs and Philae carried a separate infrared sensor for thermal mapping (MUPUS-TM). These instruments did operate while the lander's batteries still had power. The harpoons' 16 temperature sensors were fabricated from temperature-dependent electrical-resistance titanium. MUPUS-TM had four thermopiles as its detectors observing an area of the surface of about 1 m² at four wavebands between 5 and 25 μm. The Viking 1 and 2 foot pads also carried thermocouple temperature sensors

Veneras 13 and 14 used resistance thermometers, but based on platinum. These were to measure the atmospheric temperatures during the landers' descents. But the temperature gradient in Venus atmosphere is very slow (~15 K/km), so the ground temperature is likely to be close to the final atmospheric measurements (730 K for Venera 13 and 740 K for Venera 14).

Thermal conductivity was measured by the Thermal Electrical Conductivity Probe for Phoenix (TECP) instrument on the Phoenix Martian lander. This instrument was mounted on the lander's robotic arm (Section 7.2.6) and had four 15 mm long needles which could be pushed into the soil. The temperatures of these probes were then monitored by a chromel-constantan thermocouple. Three of the probes could be heated to make the thermal conductivity measurements.

It is planned for the Interior Seismic Investigations, Geodesy and Heat Transport (InSight) Mars lander to measure heat flow in the surface layers down to perhaps a depth of 5 m. The Heat Flow and Physical Properties Package (HP³) is similar to the 'mole' penetrator, PLUTO (Section 7.2.6), that was on the Beagle-2 Mars lander and is sometimes called a 'self-hammering nail'! Its appearance is that of a small torpedo 350 mm long and about 30 mm in diameter. Inside the torpedo are moveable masses, springs and an electric motor.

To penetrate the surface, one of the masses, the 'hammer', is pulled slowly up the torpedo by a cam driven by the motor and it compresses the springs. Upon release, the hammer

---

* Intended to anchor the lander to the comet's surface.

is driven rapidly by the springs towards the tip of the torpedo. Normally, conservation of momentum would cause the rest of the torpedo to move in the opposite direction to the hammer. This motion, however, is prevented by a second mass, the suppressor, inside the torpedo which moves away from the tip and compresses a brake spring. The suppressor's mass is several times that of the hammer, so that the latter receives most of the kinetic energy involved. When the hammer hits the tip, the torpedo receives a downwards impulse. As the suppressor is brought to a halt by the brake spring, the latter is compressed and drives the suppressor down the torpedo again to where its impact with the tip provides a second downward impulse. Of course, conservation of momentum also means that there are upward impulses acting on the torpedo as well. The upward accelerations however take place over longer time intervals than the downward ones (i.e. the 'G' forces have lower peaks) so that friction between the outer walls of the torpedo and the ground prevents any upward motion and the torpedo has a net downward motion. Repeating the cycle gradually drives the torpedo below Mars surface.

To measure the ground temperature, the HP³ torpedo will be dragging a tether behind it with platinum resistance temperature sensors at 100 mm intervals (the tether will also contain power cables for the motor and signal cables to return the data to the lander). Thermal conductivity will be determined by heating the torpedo and using temperature sensors on the outside of its casing.

The Huygens Titan lander was designed to operate whether it landed on a solid or liquid surface (it landed on a solid surface). For a liquid, though, it could measure its temperature and thermal conductivity. It would have used platinum resistance wires for the temperature measurements and heated them with an electrical current for the thermal conductivity measurements.

7.2.4.2 Electricity and Magnetism

Large-scale (many × 10,000 km) electric fields are rare in the solar system since the interplanetary medium is a plasma and its high electrical conductivity means that any voltage differences that do occur are quickly discharged. The passage of the solar wind magnetic field outwards through the solar system does, though, generate a weak interplanetary electric field. Large-scale magnetic fields however abound and are discussed in Section 7.5. Here, therefore, we are concerned with small-scale (millimetre to 100 km) electric and magnetic fields occurring on and near the surfaces of solar system objects.

Solar system objects both with and without atmospheres may build up static electric charges. Objects with atmospheres may build up the charges within their atmospheres, leading to lightning (Section 2.3.1) and sparks, or at very low pressures, perhaps to glow discharges.* Winds in the atmospheres may also blow surface dust particles around so that they develop static charges during their frictional interactions (triboelectrification).

Objects without atmospheres have solar wind/interplanetary medium particles and UV radiation interacting directly with their surfaces. The solar UV interactions will

---

* The much disputed sightings of light emissions during the much disputed lunar transient phenomena have been attributed by some to glow discharges.

eject electrons from the object's surface, leading to a positive charge. However, this will be strongest at the sub-solar point and reduce to zero at the terminator. The solar wind particles are mainly protons and electrons. The electrons have high velocities and largely random motions, while the protons have lower velocities that are largely radial (to the Sun). Thus at the sub-solar point of the object, the solar wind protons and electrons, being roughly equal in numbers, will have a largely neutral effect when impacting the object's surface, but the solar wind protons' effect, like that of the UV photons, will reduce to zero at the terminator. The solar wind electrons however will bombard the object fairly uniformly from all directions and will produce increasingly net negative static charges on from the sub-solar point to the anti-solar point.

As may be imagined, these various interactions acting in combination on atmosphere-less objects can result in quite complex electric fields. Thus, dust particles on the lunar surface becoming charged by these processes are positive on the illuminated side and negative on the dark side. Repulsion between the particles leads to them levitating during the lunar day and night and sinking to the surface again as the terminator passes over them (when they lose their charges). Some evidence for this process was provided by images from the Surveyor 7 lunar lander. A glow on the sunset horizon was consistent with sunlight scattered over a 1 m high layer of 5–10 µm particles.

While direct measurements have yet to be made there are suggestions for future lunar landers and rovers to carry instruments to study both the general electric field and the static charges. The field would be detected using an electric field mill. This has one or more electrodes immersed in the electric field. A rotating chopper cuts the electric field lines, alternately shielding and exposing the electrode(s) to the field. The resulting electric current provides the measure of the field strength. The Part Time Scientists, a competitor for the Lunar Xprize, intend to include an electric field mill instrument in their rover

The static charges would be collected by an interdigital transducer that generates an electric pulse whenever a charged dust particle lands on its detecting surface. The transducer comprises two comb-shaped metallic electrodes with the teeth of the combs interlocking, but insulated from, each other and thus acting as a capacitor. The combs are supported on a thin flat substrate. The charge from the particle hitting the surface induces a charge in the transducer which is then detected.

The presence of triboelectrified dust particles on Mars is expected and microwave emissions from electrical discharges during Martian dust storms resulting from such charges may have been picked up by terrestrial radio telescopes. However, direct detection from landers or rovers has yet to be attempted. The Schiaparelli Mars lander however did carry an electrical detector called MicroARES (Section 7.4) but this was intended for atmospheric measurements. Similarly, the sounds of micro-electrical discharges (crackles) might have been detected, but were not, by various microphones intended for atmospheric studies (Section 7.4).

The permittivity and electrical conductivity of surface materials have been measured by instruments on a number of landers and rovers. Phoenix's TECP instrument has already been mentioned (Section 7.2.4.1) and this could also be used to measure permittivity and electrical conductivity. Huygens was equipped to measure the permittivity of liquids in case it landed in a lake or sea on Titan. The sensor just comprised electrodes whose mutual capacitance would vary depending on the material in which they were immersed. They could also be used to measure the material's electrical conductivity by passing electric currents between them. Rosetta's comet lander Philae carried five electrodes – one in each of its anchoring legs, one in the penetrator and one attached to the APXS (which was lowered to within 40 mm of the ground). Electric currents between the electrodes measured permittivity and electrical conductivity down to a depth of about 2 m. The asteroid impact mission, AIDA, should be able to measure the permittivity throughout Didymoon using its low-frequency bi-static radars (Section 7.3.1).

Small-scale magnetism (i.e. individual magnetic particles) has been studied by simply incorporating a permanent magnet into the lander or rover's body somewhere and monitoring the particles attracted to it. Thus many of the Surveyor lunar landers had bar magnets attached to one or more of their footpads which could be imaged by their cameras and the particles counted. Surveyor 7, in addition, had two horseshoe magnets attached to its surface scoop. The Mars Pathfinder's lander, Sojourner, also carried magnets.

MER's Spirit and Opportunity rovers had magnets on their robotic arms (Section 7.2.6) and on their front panels and the dust particles that were gathered could be analysed by the rovers' Mössbauer (Section 7.2.2.7) and x-ray spectrometers (Section 7.2.2.5).

Philae used a fluxgate magnetometer (Section 7.5.1.2), called Rosetta lander Magnetometer and Plasma monitor (ROMAP), to investigate comet 67P/Churyumov–Gerasimenko. Hayabusa-2's lander, MASCOT, includes a fluxgate magnetometer with a sensitivity over the ±12 μT range for magnetic field measurements both when it is stationary on the asteroid and during its jumps from one opposition to another.

7.2.4.3 Density and Refractive Index

The densities and refractive indices of the surface materials of solar system objects have not commonly been measured directly, but the Huygens Titan lander did carry instruments to measure both for liquids in case it landed in a lake or sea and for refractive index, only, on any liquid condensates during its descent through the atmosphere.

The density instrument was just a cylindrical float attached by a flexible rod to the wall of the lander. If the float was in a liquid, then strain gauges would measure the bending of the rod and the liquid's density inferred from the measured buoyancy force (knowing the density of the float).

The operating principle for the refractive index instrument was based upon measuring the critical angle for internal reflection (Equation 7.3). A right angle sapphire prism with its third face shaped as a quarter of a circle could have that face illuminated from the inside by a light emitting diode (LED). The angle at which the radiation hit the curved surface changed with position around the surface. The position where total internal reflection ceased gave the critical angle. Refractive indices between 1.25 and 1.45 could be measured to an accuracy of ±0.001, enabling the relative abundances of methane and ethane in a liquid mixture to be determined to ±1% or better:

$$\mu_L = \frac{\mu_S}{Sin(\theta_c)} \tag{7.3}$$

where μ_L is the refractive index of the liquid, μ_S is the refractive index of sapphire (typically ~1.75) and θ_c is the critical angle.

Density can sometimes be estimated from penetrometer (Section 7.2 5.5) measurements or other related investigations. For example, the Surveyor lunar soft landers' robotic arms (Section 7.2.6.1) were used to press onto the Moon's soil and suggested a density around 1500 kg·m⁻³.

Gamma ray densitometers have been carried on board Luna-13, several Veneras and Mars-3. They typically contain a caesium-137 radioactive source of 660 keV γ rays and several Geiger counter detectors. The detectors are shielded from the direct γ ray emissions and so pick up backscattered γ rays from the sample on which the device rests. The scattering of the γ rays arises from Compton scattering by the electrons in the sample's atoms. The more electrons, the more the scattering and the more electrons, the higher the density, so that density can be estimated from the backscattered γ ray intensity. In practice, detectors are placed at different distances from the source and the density inferred from the way that the γ ray intensity varies with distance.

7.2.5 Ionising Radiation Levels

As mentioned in Section 7.1, the first successful soft lander, Luna-9, carried an ionising radiation detector and measured the levels at the lunar surface. This detector was a Geiger counter (Section 4.2.1).

Many orbiting spacecraft use ionising radiation detectors to analyse the solar wind, planetary radiation belts and the interplanetary medium (Chapters 4, 5 and 6) and they are used to monitor the radiation environment during manned missions – the Apollo astronauts, for example, carried an active ion chamber radiation detector (Section 4.2.2) and three passive nuclear emulsion detectors (Section 1.4.3.5.3) to monitor their radiation exposures. But the only other direct surface measurements, so far, have been made by the Mars rover, Curiosity. This has a radiation detector mounted on it. It is a silicon solid state detector stack (Section 4.2.6) combined with caesium iodide scintillators (Section 4.2.4) plus anti-coincidence shields. It can detect ions up to iron, electrons, neutrons and x- and γ rays. It was used to assess the ionising radiation on the rover during its voyage to Mars as well as the radiation environment on Mars' surface since landing.

The Mars Odyssey orbiter carries the MARIE, which measured the radiation environment close to Mars for 20 months from 2002 to 2004. The radiation detector for MARIE was a combination of a stack of solid state detectors (Section 4.2.6) and a Čerenkov detector (Section 4.2.9). There were two ion-implanted SSDs and four lithium-drifted SSDs. The Čerenkov detector used high refractive index Schott™ glass. The package could operate over the particle energy range from 15 to 500 MeV per nucleon. Because Mars' atmosphere is so thin and its magnetic field so weak,* the results from the orbiter should be very similar to those at ground level on Mars. They suggest that the normal surface level rates are about 220 µGy per day, but this can rise to 20 mGy per day during strong solar flares (on most parts of Earth, exposure levels are around 6–7 mGy per year).

7.2.6 Sampling

7.2.6.1 Introduction

Most landers and rovers have had a means of obtaining surface samples of the ground in their immediate vicinities and more recently of being able to go below the surface (see e.g. earlier discussion of InSight's HP³ – Section 7.2.4.1). The first such mechanism however was on Surveyor 3.

Surveyor 3 used a parallelogram extending arm (Figure 7.6). It was not robotic, but was operated from Earth using a TV camera on Surveyor 3 to watch its movements. It was used to conduct strength (hardness) tests by various methods: the end of the sampler arm was simply pressed onto the surface, it was dropped onto the surface and four trenches

FIGURE 7.6 Surveyor 3 showing its parallelogram extending arm (passing just in front of the astronaut) with Apollo 12 in the background. (Courtesy of NASA.)

---

* Approximately 600 Pa and ≤1.5 µT, respectively, for Mars compared with ~100 kPa and 25 to 65 µT for Earth.

were dug into the surface. Most remarkably though, the scoop jaw was used to pick up a nearby rock and exert a pressure of ~700 kPa (100 psi) onto it. The rock survived the experiment (which would have cracked a brick on Earth) showing that it must be a piece of very solid rock.

For landers on objects further away than the Moon, directing their sampling arms under control from Earth is not feasible (because of the time delays in the signals: ≥2.1 minutes for Venus, ≥3 minutes for Mars, etc.) and so they operate automatically/robotically to greater or lesser extents. They may be subdivided into passive and active devices – the passive ones simply scoop up samples as and where they may be available or bring other instruments to the positions required for them to operate while the active variety take actions (abrading, digging, drilling, hammering, penetrating) to obtain their samples.

7.2.6.2 Passive Sampling

This is usually undertaken to provide materials for analysis by other instruments on the lander or rover or to move instruments into positions where they can operate.

The Viking 1 and 2 Mars lander arms just about come into this class. They were telescopic and made from two coiled ribbons forming tubes. They extended and retracted as required like coiled tape measures and were driven by electric motors. They scooped up surface material for the landers' other experiments. However, on some occasions, the depth of the trench was arranged to be deep enough to obtain samples unaffected by solar radiation, which might count as 'digging'.

Mars Polar Lander had an elbow-jointed robotic arm that carried a camera. It was intended for the camera to image scooped-up samples or for the samples to be delivered to its thermal and evolved gas analyser (TEGA) (Section 7.4.1) instrument. The lunar rover Chang'e-3's triple-jointed robotic arm carried only an APXS (Section 7.2.2.6) instrument that could be positioned where desired by a combination of movements of the rover and the arm. The movements were choreographed using images from the rover's hazard avoidance cameras.

Many robotic arms can perform both passively and actively – MER's Opportunity rover's arm, which is rather like a human arm in having joints equivalent to the shoulder, elbow and wrist, for example, in its passive mode, brings instruments such as the Mössbauer spectrometer (MIMOS II – Section 7.2.2.7) or the microscope (MI – Section 7.2.1) that are mounted on the arm into positions where they can operate. Similarly, Mars 2020's arm is planned so that it can move instruments mounted on it, such as its Raman spectrometer, SHERLOC (Section 7.2.2.4), into their working positions.

7.2.6.3 Active Sampling

7.2.6.3.1 Digging   Digging is probably the simplest resort for an active sampler. Any passive sampler if it took a second sample from a previous trench could be counted under this heading. The Phoenix Mars lander, however, had the potential capacity to dig down to a depth of 500 mm using its elbow-type robotic arm. The arm could reach out to 2.35 m and it used a bucket scoop like a miniature version of those on constructors' earth-moving backhoes for its excavations.

7.2.6.3.2 Brushing   Dust covering a desired sample can simply be brushed off – MSL's Curiosity has a motorised wire bristle brush for this purpose in order to prepare areas for analysis by its APXS (Section 7.3.1.2.6) and MAHLI (Section 7.2.1). The abrading tool on Opportunity (and Spirit) also carries brushes (see next section).

Sampling very-low-gravity objects with a drill or a scoop will push the lander away from the surface unless it can be anchored down firmly. A low reaction force brushing mechanism is thus currently under development for obtaining samples from small objects. It will use three rotating brushes that also circle around to brush dust into a container.

7.2.6.3.3 Abrading   Abrading (grinding) the surface may be used to remove dust so that the underlying material can be analysed (cf. previous section), to obtain powdered samples for analysis or to look into the interiors of solid samples. The rock abrasion tool (RAT) mounted on Opportunity's (and also in the past, Spirit's) robotic arm has two high-speed grinding wheels that also slowly rotate in a circle. It can grind holes 45 mm across and up to 5 mm deep in solid rocks.

The Mars lander Phoenix's arm had a rasp in the heel of its scoop that rotated at 5500 rpm and was used to obtain powder samples for its TEGA instrument (Section 5.2.6.2), microscopes (Section 7.2.1) or wet chemistry analyser (Section 7.2.3.6). The grindings from the rasp were channelled directly into the scoop by the rasp's action.

The Beagle-2 Mars lander would have ground flat surfaces for its x-ray and Mössbauer spectrometers (Sections 7.2.2.5 and 7.2.2.7) to analyse and would have produced powder samples for its GCMS (Section 7.2.3) using a short corer/grinder.

7.2.6.3.4 Drilling   Drilling and abrasion differ little except in the aspect ratios of the holes produced (high for drilling, low for abrading). The lunar sample and return mission (Section 7.2.6.3.6) Luna-16, which made the first soft landing on the Moon's dark side, carried a hammer drill on the end of an extendable arm (Figure 7.7). The drill was automatically deployed and operated for seven minutes, drilling a hole some 350 mm deep. The drill rotated at 500 rpm and used a hollow drill bit. The 101-gram soil sample was captured in an elastic tube that was then deposited into a container on the main spacecraft and returned to Earth for analysis. The Luna-20 and 24 missions had similar drills with that on Luna-24 penetrating to a vertical depth of 2 m.

The Venus landers, Veneras 13 and 14, and the lander probes of the Vega-1 and 2 spacecraft carried later versions of the Luna-16 drill (although that on Vega-1 malfunctioned). The pressure of Venus' atmosphere was used to transfer the samples into their containers and these were then moved into the main spacecraft bodies through airlocks. The samples were analysed by XRF instruments (Section 7.2.2.5). The drills also gauged the hardness of the soils from the depths to which they penetrated in their 120 seconds of drilling time.

FIGURE 7.7 Luna-16 – A model in the Moscow Space Museum. The drill unit is on the end of the arm extending to the right. (Courtesy of Vitaly Goodvint.)

Rosetta's comet lander, Philae, carried a drilling system to obtain powder samples for the ovens of its mass spectrometer (Section 5.2.6.3.2). The hollow drill bit had a diameter of 12 mm and was embedded with diamond teeth. The samples would have been obtained by a smaller tube running through the centre of the drill. The drill would have had a potential maximum depth of 500 mm if operated at 500 rpm.

The drill head on the planned ExoMars 2020 rover will incorporate a small infrared imaging spectrometer and lamp, Multispectral Imager for Subsurface Studies (MaMISS), so that the stratigraphy in the walls of the borehole can be inspected directly. The drill will be able to penetrate to a depth of 2 m and obtain 10 mm diameter samples up to 30 mm long. Also for possible launch in 2020, Luna-27 is planned to land on the Moon's South Polar region carrying a drill capable of penetrating to a depth of 2 m.

A hammer drill is mounted on the robotic arm of MSL's Curiosity rover together with a brush. The solid drill bit has a 16 mm diameter and there are two spares. Holes can be

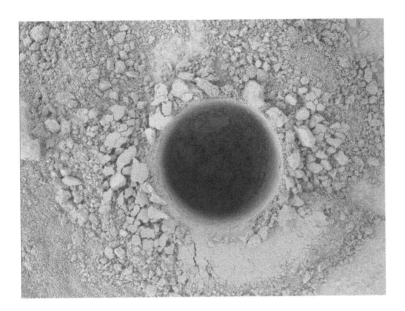

FIGURE 7.8   A 67 mm deep sample hole drilled by the Curiosity rover at a site in the Pahrump hills. The image was obtained using the MAHLI microscope (Section 7.2.1). (Courtesy of NASA/JPL.)

drilled up to 50 mm deep. Powder from the holes can be transported for chemical analysis by other instruments on the rover (Figure 7.8).

Currently under development by NASA for use on possible future planetary rovers is Auto-Gopher. This will use a piezo-electric hammer mechanism and rotation to help it drill through hard rock and to capture powder samples. The drill will be suspended from a tether that carries the power and communications cables and its penetration depth will be limited only by the length of the tether. Samples will be obtained by using the tether to withdraw the drill and return it to the rover, but the drill will then be able to be re-inserted into the hole and continue drilling to greater depths.

Although neither remotely operated nor robotic, brief mention should be made here of the Apollo Lunar Surface Drill (ALSD). This was a battery-powered hammer drill that rotated at 280 rpm and could penetrate to 3 m. The hollow drill bit had a 25 mm outer diameter and a 20 mm inner diameter and used tungsten carbide cutters. It was carried on the Apollo 15, 16 and 17 missions and operated by the astronauts.

7.2.6.3.5 Penetrators   Probably the simplest versions of these tools were to be found on the Surveyor lunar soft landers. Their three legs each had a strain gauge attached and these measured the forces generated during the landings. When combined with the (known) landing velocities and the depths to which the landers' feet penetrated the surface, the bearing strength of the soil could be assessed.

Penetrators to assess the bearing strengths of surfaces and comprising cones pressed into the ground have been carried on the Lunokhod lunar rovers and the Venera and Vega Venus landers. The Venera penetrators were mounted on spring arms and operated rather like a mousetrap in reverse to spring open and hit the surface. The Mars-3 tethered mini rover, PrOP-M (Section 7.1), had a conical penetrometer, but sadly it did not have time to be used.

The nature, especially its hardness, of Titan's surface was gauged by a penetrator mounted to point downwards on the bottom of the Huygens lander. Essentially a spear, it was 100 mm long and was mounted on a piezo-electric sensor that measured the forces acting on it as it pierced the surface when the probe landed.

The comet lander Philae carried a hammered penetrometer, MUPUS, described in Section 7.2.4.1. However under a layer of dust a few tens of millimetres thick, the comet's surface was so hard (like solid ice or pumice) that it failed to go in any further and it may have broken when the hammer's power was increased to its maximum.

The HP^3 penetrator for the planned InSight Mars mission is rather different in its nature from the tools just reviewed: it was discussed in Section 7.2.4.1 and its mode of operation described there. The HP^3 tool is a development of Planetary Undersurface Tool (PLUTO) carried by the Beagle-2 Mars lander. PLUTO would have been deployed by Beagle-2's robotic arm into a vertical position touching the surface and would have used the same internal hammer mechanism as HP^3 to dig down into the soil. A sample would have been obtained by opening a cavity at PLUTO's tip while it was still moving downwards. It would have been capable of reaching a depth of 1.5 m and then the power cable tether to the mothercraft, combined with reversing the hammering mechanism, would have been used to pull it back, with its sample, for analysis. The internal hammer could also have been used to move PLUTO across Mars' surface to obtain samples from (say) the base of a nearby rock.

The Apollo Moon missions are again mentioned here for completeness – during the earlier missions (Apollos 11, 12, 13 [potentially] and 14) sub-surface samples were obtained by coring. The coring tubes were hammered manually into the lunar surface by the astronauts. The tubes were of aluminium with stainless steel tips and 310 mm long. Inside the coring tubes, a smaller (19.5 mm internal diameter) tube received the sample. Tubes could be combined to sample down to about 0.7 m depth.

7.2.6.3.6 Robotic Sample and Return   The return of samples from solar system objects by robotic means started with Luna-16 (Section 7.2.6.3) in September 1970, although five earlier, unsuccessful, attempts had been made in the previous 15 months. Lunas 20 (February 1972) and 24 (August 1976) also performed successful sample-return missions from the Moon, while another three attempts failed. The successful missions brought back 101 g, 30 g and 170 g of lunar material, respectively.* After Apollo 17 (1972), no further lunar sample and return missions have been made although Chang'e-5, with a planned launch in 2017, and Chang'e-6 (2020?) are hoped to return up to 2 kg of material each.

---

* The six successful Apollo missions returned a total of 381 kg. The achievements of the Luna programme should however not be belittled. A state-of-the-art mainframe computer, such as the IBM 360, in the early 1970s occupied a large room, weighed many tons, had less than 10 MB of memory and a speed of under a million instructions per second. A smart phone today (2017) fits in your pocket, has at least 2 GB of memory and operates at over 2 GHz. Furthermore, the IBM 360 in today's values would cost around $50,000,000; the smart phone costs a couple of hundred dollars. The Luna programme was thus a quite remarkable breakthrough given the available information technology. Furthermore, it, rather than the Apollo programme, presaged the way forward for sample and return missions to objects other than the Moon.

Earth-based scientists have, however, more lunar material to examine directly than just that from the Apollo and Luna programmes. Around 50 kg has been found on Earth, mostly in Antarctica or in deserts in Africa or the Arabian peninsula. These lunar samples are in the form of meteorites that have originated from the Moon, being blasted into space during the formation of the Moon's impact craters. The boulders, rocks and dust particles ejected during those impacts do not generally have sufficiently high velocities for them to take up orbits very different from that of the Moon. So they remain in the Earth–Moon vicinity, typically for 10–20 million years, before colliding with Earth (or the Moon). Lunar meteorites may be expected to fall anywhere on Earth, but they are found particularly in Antarctica and terrestrial deserts because they easily show up since their appearance and compositions are noticeably different from those of the surrounding ice or country rocks.

The Fobos-Grunt mission was intended to land on Mars' natural satellite, Phobos, to obtain some 200 g of material and to return it to Earth. Unfortunately, the failure of its rockets meant that it remained in low Earth orbit and eventually burned up on re-entry. Large samples (up to 13 mm) would have been obtained using a robotic arm with a grab on the end. The return capsule would have initially been launched using springs so that it would be clear of the lander (which would remain on Phobos' surface) before igniting its rockets for the return. It is possible that a repeat mission may be attempted in ~2024. The rover on the Mars 2020 mission is intended to obtain samples and stockpile them for later collection by another (so far unplanned) mission. The ESA is contemplating a Mars sample-return mission for the early 2020s, NASA one perhaps for 2024 and Russia perhaps Mars–Grunt in ~2026. In the meantime, like the Moon and asteroids (see below), we already have samples of Martian materials in the form of a few meteorites. Some 140 meteorites have compositions similar to those deduced for Mars by spacecraft (orbiters and landers) and are thought, like the lunar meteorites, to have been ejected from the planet during the formation of large impact craters.

A slightly different sample-return mission may be attempted in 2020/2021. Luna-28 would carry a rover to the Moon which would obtain the samples, while Luna-29 launched some time later would pick them up for return to Earth.

Only one mission since Lunas 16, 20 and 24, however, has so far achieved a comparable success to that programme. The first Hayabusa spacecraft brought back some 1500 dust grains from the surface of asteroid 25143 Itokawa. It was a touch-and-go mission in two senses. First, the spacecraft was not intended to land properly on the asteroid, but to touch it momentarily, obtain its sample and then take off back into orbit. It was also touch-and-go in the more usual sense as well in that the numerous problems (fuel leaks, engine failures, batteries, etc.) encountered during the mission meant that it came close to failing. The sampler system relied upon firing a 5 g tantalum pellet at 300 m·s^{-1} into the asteroid's surface and gathering the ejected debris into a cone-shaped collector. The dust in the collector was then to be transferred to a horn-shaped capsule for return to Earth. On Hayabusa's first sampling attempt an obstacle (or possibly a software problem) prevented the sampler from working normally and instead of 'touching and going' the spacecraft landed for about 30 minutes. Despite the 'proper' sampler not being activated, the touch-down itself at about

FIGURE 7.9   Meteorites from the author's own collection: two sawn, etched and polished slices from iron meteorites and a fragment from a stoney (chondritic) meteorite. The length of the largest iron meteorite slice is 225 mm.

100 mm·s^{-1} disturbed the asteroid's surface sufficiently for some dust to enter the return capsule directly. A second sampling attempt had a similar outcome.

The Hayabusa-2 mission is largely similar to Hayabusa and is currently on its way to asteroid 162173 Ryugu. Two changes from the earlier mission are to equip the sampling horn with teeth so that it may dig into the surface if the pellet does not fire and to add a separate and much larger impactor that will provide material from inside the asteroid for collection (Section 7.1).

OSIRIS-REx is another touch-and-go mission currently on its way towards asteroid 101955 Bennu. It will approach the asteroid's surface at about 200 mm·s^{-1} and use a sampler on the end of a robotic arms to acquire its material in about 5 seconds. The samples will be collected by a surface contact pad, the asteroid's surface having previously been churned up and readied for collection by a nitrogen gas jet. The process will be monitored by one of the spacecraft's cameras for later verification of what has happened. Up to three sampling attempts may be made and 60 g or more of material is hoped to be returned to Earth.

Since most meteorites are thought to originate from the asteroids, terrestrial scientists actually have large quantities (hundreds of tons) of asteroidal materials to work with. Anyone interested in owning their own bit of asteroid can easily and cheaply* purchase small meteorite samples (of the commoner varieties) from several commercial sources (Figure 7.9). Of course, meteorites have undergone processes such as entry into Earth's atmosphere and contamination and weathering on Earth's surface and so may differ in

---

* Less than a ~$1/gram at the time of writing – though *caveat emptor* – it is easy for some types of meteorite or tektite to be imitated, so if you do not have the knowledge or equipment to check them out, only buy from a reputable supplier.

some ways from the material actually on asteroids. One of the results of the Hayabusa mission, though, was to show that the minerals in the returned dust particles, iron sulphide, olivine, plagioclase and pyroxene were identical to those in LL chondrite-type meteorites, thus confirming the reality of the properties of asteroids obtained by inference from meteorites ('ground truthing' – Section 7.2.7).

Three missions have returned samples of dust particles from interplanetary space to Earth – Stardust, Mir and Genesis. They operated in a rather different way from the missions just discussed, simply exposing an inert collector to space and picking up whatever particles happen to hit it. Stardust sampled particles from comet 81P/Wild's coma as well as those in interplanetary and interstellar space in this way in 1999. Its collector used 90 blocks of ultra-low-density aerogel arranged in a circular array and deployed on a boom outside the main spacecraft after launch. Aerogel is largely composed of holes – Stardust's, for example, had a density of ~2 kg·m^{-3}. A dust particle, upon hitting the aerogel, is therefore braked so slowly that it remains intact. After its encounter with comet Wild, the collector was retracted and it returned to Earth in January 2006. Its re-entry capsule hit the Earth's atmosphere at 12.9 km·s^{-1} – the fastest ever for a man-made object, with a peak deceleration force of 34 g. After slowing down, the capsule parachuted safely to the ground. Stardust went on to visit comet Tempel 1, but without its dust collector.

The Mir space station carried an aerogel dust collector for 18 months between 1996 and 1997 to detect both man-made and naturally occurring dust particles. It was one of four packages intended to assess the damage caused to materials in low Earth orbit in order to guide future spacecraft design as much as to study micro-meteorites and other natural dust particles themselves. The Orbiting Debris Collector (ODC) used silica aerogel with a density of 20 kg·m^{-3} arranged in two trays with a total area of 0.63 m^2.

Samples of interplanetary dust can also be collected by using high flying aircraft and balloons. They may also be extracted from deep ice cores in Antarctica and Greenland, from the sediment on the floors of the oceans and from salt deposited millions of years ago now being quarried commercially in deep salt mines.

The particles gathered by Genesis' collector were from the solar wind and were ions of isotopes of atoms such as oxygen, nitrogen and the noble gases, not dust particles. The individual collectors were solid hexagonal or half-hexagonal wafers. They were made from a variety of materials including silicon, sapphire, gold and carbon films. The ions simply crashed into the wafers at about 200 km·s^{-1} and were buried beneath their surfaces. The hexagons were mosaicked into five arrays: one was continually exposed to the solar wind, three could be deployed for specific events such as coronal mass ejections and the last employed electric fields to repel protons and concentrate the lighter ions. One of Genesis' objectives was to provide ground truth (Section 7.2.7) measurements of the solar wind's composition for verifying remote estimates.

The second manned mission of the currently-under-development spacecraft, Orion, could be a very ambitious sample-return project, though it is still rather speculative at the time of writing. It is presently envisaged to be a two-spacecraft mission. First, the Asteroid

Retrieval Mission (ARM) spacecraft would visit a near-Earth asteroid and use robotic arms to secure a boulder up to 6 m in size and up to 30,000 kg in mass. It would lift the boulder off the asteroid and place it into a lunar orbit. The Orion spacecraft with four crew members would then visit the boulder and examine it in situ as well as collecting samples for return to Earth. Launchings of the missions could be around 2025/2026.

7.2.7 Ground Truth

Whenever measurements are made remotely there is always, to a greater or lesser extent, some uncertainty in their interpretation. For example, in the determination of the mineralogy of rocks from their reflectance spectra (Section 7.2.2.2) several minerals could have spectra so similar that they are indistinguishable on this data alone. Methods of deciding which is the correct mineral in such cases are known generally as ground truthing (sometimes called calibration) from the necessity for scientists to often have to go in person to the sites observed by Earth resources satellites to establish exactly what minerals, rocks, soils, crops, plants etc., are being monitored by the spacecraft (or aircraft, balloon, radar, etc.). This does not mean that the remote observations are a waste of time – once the ground truth has been established at one spot it is then assumed that similar observations obtained elsewhere have the same interpretation and so large areas can be mapped or sampled quickly.

For Earth, ground truths can often be established by observing already well-known areas. But special visits may still be needed and this was certainly the case in the early days of Earth resources spacecraft. Even as recently as 2009 and 2010, expeditions had to be made to Adana in Turkey to measure soil salinity so that it could be mapped over a large area using Landsat-7 Enhanced Thematic Mapper plus (ETM+) images (Sections 1.2.1.6 and 1.4.1.4).

Ground truthing for Mars Odyssey Orbiter's THEMIS instrument provided by the Opportunity rover's Mini-TES has been mentioned in Sections 1.2.1.6 and 7.2.2.1, as has that for Hayabusa-2's thermal mapper by MARA (Section 7.2.2.1), the first Hayabusa mission's samples verifying terrestrial studies of meteorites (Section 7.2.6.3.6) and Genesis' confirmation of remote assessments of the solar wind composition (Section 7.2.6.3.6).

Observations of minerals from orbit made by the Mars Reconnaissance Orbiter have recently been ground truthed by the Curiosity Mars rover (Section 7.2.6.3.4). Chang'e-3's observations of basalts in its Mare Imbrium have confirmed many lunar orbiter's observations of the same region.

Serendipitous ground truths were obtained in 1976 when Luna-24 returned samples of lunar surface material from Mare Crisium (Section 7.2.6.3.6). In 2005, SMART-1's D-CIXS (Sections 3.2.5.1 and 3.3.3.2) observed the same area during a large solar flare. The levels of calcium found in the Luna-24 samples nicely confirmed the remotely deduced levels found by SMART-1. To a limited extent, Philae's observations of the nature of the comet's surface confirm the remote observations of the whole of the comet by Rosetta's Comet Nucleus Sounding Experiment by Radiowave Transmission (CONSERT) (Section 2.4.4.6).

7.2.8 Retro-Reflectors

The cut-off corner of a cube, if formed from reflectors, returns any incoming light beam in exactly the opposite direction to that by which it entered (i.e. back towards its source – Figure 7.10). The reflectors found on the rears of most vehicles are made up from arrays of such small 'corner cubes' and if you are a sailor you will also be familiar with them from the radar reflectors carried by most yachts on their masts. Their scientific applications are largely in distance measurement where, when mounted on a distant object, they reflect radiation back to its source. A measurement of the time delay between the emission of a pulse of radiation and its return thus gives a measure of the distance of the object (cf. radar or lidar – Section 2.4).

The Apollo missions to the Moon left arrays of 200 corner cube reflectors on the Moon's surface (Figure 7.11) and these have since been used in reflecting lasers beams from Earth to determine the Moon's distance and its orbit to an accuracy of a few millimetres. A number of Earth orbiting spacecraft have either carried corner cube reflectors or (e.g. the Laser Geometric Environmental Observation Survey [LAGEOS] spacecraft) been just spheres covered in hundreds of such reflectors in order to enable the spacecraft's orbit to be determined very exactly or to measure positions on Earth to very high levels of precision.

Recently, a hemisphere with eight corner cube reflectors was to be deployed by the Schiaparelli Mars lander. It was intended partly to demonstrate the feasibility of providing very precise tracking of rovers on the surface of Mars if they had such retro-reflectors mounted on their bodies (and perhaps also for laser communication between rover and orbiter) and partly for use by Mars orbiters to provide precise measurements of their orbits and hence to determine the gravitational field of Mars.

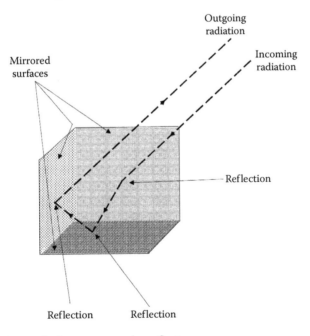

FIGURE 7.10   Radiation paths in a corner cube reflector.

FIGURE 7.11 The Apollo 14 retro-reflector array on the Moon's surface. (Public domain image, but thanks are due to NASA.)

The three Swarm spacecraft each carry four retro-reflectors that enabled their orbits to be measured very exactly using lasers on Earth.

7.3 SUB-SURFACE

Although studies of gravitational fields (Section 8.2.1) provide some information on the interior structure of solar system objects, here, as mentioned at the start of Section 7.2, we are concerned with probing such objects deep below their surface (depths of tens of metres to hundreds of kilometres) by radar and by seismometry. Mars' interior, however, may also be probed in another way in the future. The radio link between the InSight mission and Earth is planned to be used to determine if Mars' rotation has any 'wobble' to it. If so, it could indicate the presence of a liquid interior.

7.3.1 Radar

The principles of radar are discussed in Section 2.4 and ground penetrating radar* is discussed in Section 2.4.4.5. As we see in the latter section, in order to penetrate into the ground the radar needs to operate at low frequencies. Many such radars are bi-static with one spacecraft transmitting and a second spacecraft receiving.

However, Chang'e-3's Yutu rover had a mono-static radar. This could penetrate down to a depth of 400 m and operated at frequencies between 60 and 400 MHz. The planned Mars 2020 mission's rover is also expected to include a mono-static radar, Radar Imager for Mars Subsurface Exploration (RIMFAX) (Section 2.4.4.5), amongst its instrumentation.

---

* Sometimes called sounders.

A possible Venus lander with a surface survival time in the region of 90 days has been investigated by the private company Venusian Solutions. Primarily intended for seismometry (Section 7.3.2) it might also include a mono-static ground penetrating radar operating at 500 MHz using a patch antenna (Section 2.3.2).

Rosetta and Philae would have used a bi-static radar, CONSERT (Section 2.4.4.6), to observe the interior of comet 67P/Churyumov–Gerasimenko. In this case, the transmitter was on board the orbiter (Rosetta) and the receiver on the lander (Philae).

The planned asteroid mission AIDA (Section 7.1) will carry both mono- and bi-static radars. The bi-static radar will operate at a low frequency (50–70 MHz) in order to penetrate right through the asteroid with emitters and receivers on both the orbiter and the lander. The resolution will be around 30 m. The lander, MASCOT-2, is similar to Hayabusa-2's lander, MASCOT (Sections 1.2.2.2 and 7.2) and it will land on Didymoon in 2022. The high-frequency (300 MHz to 2.5 GHz) mono-static radar will be mounted on the orbiter and will probe the surface down to a depth of about 10 m with a resolution of about 1 m.

For completeness, ground penetrating radars on orbiters and manned missions are mentioned briefly here. In December 1972, the orbiting Apollo 17 service module carried the first ground-penetrating mono-static radar on a spacecraft. This used frequencies of 5, 15 and 150 MHz and could penetrate about 2 km below the visible lunar surface.

The Lunar Radar Sounder (LRS) on board the Selene lunar orbiter spacecraft operated at 5 MHz and could penetrate to a depth of 5 km. Its depth resolution was about 75 m. It found several reflecting layers below the lunar surface that were interpreted as ancient surface layers now buried beneath basalt lava flows.

The space shuttle, Endeavour, in 1994 used a ground penetrating radar to find ancient channels of the river Nile, now hidden below the sand. Its operating frequencies were around 1.5, 6 and 10 GHz. Other orbiters with ground penetrating radars have been mentioned in Section 2.4.4.5 and include Mars Express' Mars Advanced Radar for Subsurface and Ionospheric Sounding (MARSIS), Mars Reconnaissance Orbiter's Shallow Radar (SHARAD) and the planned Europa Multiple Flyby mission's Radar for Europa Assessment and Sounding: Ocean to Near-surface (REASON).

7.3.2 Seismometry

7.3.2.1 Introduction

Seismometers are instruments for measuring small shaking motions of the ground. These motions may arise from distant earthquakes, volcanic activity, large explosions, or even just from heavy lorries passing nearby. When the origins of the motions (seismic waves) is a distant quake, the seismic waves follow curved paths through the interior of the object and may be detected (for major earthquakes on Earth) all around the planet. During their journeys the waves may be reflected or refracted at layers within the object's interior.

Seismic waves are of two varieties, labelled P and S. The P waves are essentially sound (Pressure/Primary) waves and so the motions induced are back and forth along the direction of travel of the wave. The S (Shake/Secondary) waves are like the motions in violin strings and so are at right angles to the direction of travel of the wave.* The two wave types travel at different speeds – typically 0–10 km·s⁻¹ for the S waves and 5 to 15 km·s⁻¹ for the P waves. Furthermore P waves can be transmitted through solids, liquids and gases, the S waves, only through solids. The two wave types can, however, partially or wholly inter-convert at boundaries between two layers of differing compositions or densities and can be reflected from the object's surface to pass several times around the object's interior for strong sources.

By studying these interactions, models of the interior may be built up. The main observation is of the time delay between the occurrence of the quake/explosion and the arrival of the seismic waves at the object's surface some distance away. Ideally many observations are needed at many different distances from the source of the seismic waves to build up a comprehensive model of the interior such as that which has been obtained for Earth. This means that many seismometers would be need to be distributed widely over the surface. But even one instrument will at least give an indication of the level of seismic activity in the object. For even a single source, if it is strong enough, a single seismometer may receive many waves that have travelled through different paths through the interior. At its very simplest, if a seismometer receives only P waves, then we may deduce the presence of a liquid layer somewhere between the source and the instrument.

Seismometers generally operate through inertia. A heavy mass suspended by a vertical string can move freely in the horizontal plane. A passing P or S wave will move the string's support in a horizontal plane, but the inertia of the mass will keep it stationary in space. An observer, not realising what had actually happened, would see the mass appear to move suddenly with respect to the support. Attaching a pen to the mass and letting it rest gently on a slowly moving strip of paper mounted on the support will enable the wavelike-motion of the support to be plotted out as a graph. Vertical seismic motions can similarly be measured by suspending the mass on a spring.

In fact, modern seismometers operate like doors rather than pendulums – the mass is attached to a vertical, freely moving hinge and swings from side to side. Three such devices oriented to swing at 120° to each other enable waves from any direction to be detected.

The ground movements detected by seismometers range upwards from less than a nanometre, so in the past, various ways – such as reflecting beams of light off the mass (an optical lever) – were used to amplify the movements before they were recorded on paper or film. Nowadays, electronic sensors are sensitive enough to detect the motions directly and the digitised signals are then recorded by a computer.

Geophones detect the velocity or acceleration of their test masses rather than the change in their positions. The test mass is magnetic and generates a signal directly when it moves inside an electrical coil or (more recently) the mass has been kept stationary and the force needed to do this monitored to give the output signal.

A quite different seismometer design uses a quartz rod. The rod is mounted firmly at one end on a support fixed to the ground and at the other end is separated by a small gap from a pillar also fixed to the ground. A seismic wave alters the distance between the support and

---

* There are also surface waves, Rayleigh waves and Love waves, but these do not penetrate far into the interior.

the pillar so changing the gap. The gap size can then be monitored optically or electronically to provide the signal.

7.3.2.2 Spacecraft Seismometry

The first attempt to land a seismometer on the Moon was with the Ranger 3 spacecraft in 1962. Unfortunately, Ranger 3 missed the Moon by nearly 37,000 km. The Ranger series were impactors, but Ranger 3 had a crushable balsa wood section that was intended to reduce the severity of the impact. Nonetheless the landing would still have been at around 150 km/h with forces up to 3000 g (somewhat euphemistically called a 'rough landing'). Rangers 4 and 5 also carried seismometers but Ranger 5 missed the Moon (though 'only' by 750 km) and while Ranger 4 hit the Moon (on the far side) its solar panels and antenna did not deploy. The seismometers were vertical axis instruments with masses of 3.3 kg and a natural frequency of 1 Hz. Later Rangers did not carry seismometers.

Seismometers were designed and built for the Surveyor and Lunakhod soft landers, but never used. The Surveyor seismometer would have been able to detect movements down to 1 μm, had a mass of 3.8 kg and would have detected over the 0.05–20 Hz range. The Lunakhod seismometer would have been a vertical quartz design operating at 1 Hz.

Thus, the first operational seismometer away from Earth was the one deployed during the Apollo 11 mission in July 1969. This was a passive instrument (i.e. it would only detect natural seismic events – no attempt was made to set off explosions or to crash a part of the spacecraft into the surface – as was the case with later Apollo missions [see also Section 7.1]). The seismometer package comprised a three-axis horizontal instrument that operated over the 0.1–1 Hz range with a sensitivity of 0.3 nm and a vertical single axis instrument operating at 8 Hz with a sensitivity of 0.5 nm. The package was deployed by the astronauts, but unfortunately ceased operating after 3 weeks. It had however detected its first 'moonquake' within a few minutes of starting to operate. Moonquakes can last for tens of minutes since they are not damped out by water as is the case with earthquakes.

Similar seismometers were deployed during the Apollo 14, 15, 16 and 17 missions and Apollos 14, 16 and 17 also deployed geophones. Small explosive charges were set off and the Saturn-5 upper stages and lunar ascent modules crashed into the Moon to provide artificial seismic events for the later Apollo missions (making their seismometers active instruments). The baselines between the four seismic stations were up to about 1000 km, but the lack of a station on the far side has limited the modelling of the lunar interior that could be undertaken using their data. The seismometers continued to operate, recording some 12,500 events, until they were turned off in September 1977. Moonquakes were found to occur down to depths greater than 1000 km (much deeper than earthquakes) and to be linked to the tides produced in the Moon by the Earth's gravity.

One seismometer has so far operated on Mars, although both Viking landers carried them (Viking 1's seismometer failed to deploy). Viking 2's seismometer, however, operated successfully for nearly 2 years. It was a pendulum-based instrument with a best sensitivity of 2 nm operating over the 0.1–10 Hz frequency range. A marsquake some 110 km from the lander was probably detected about 3 months after landing

Veneras 13 and 14 each had passive vertical electromagnet-based 0.88 kg seismometers with a sensitivity of 10 nm. Venera 13 did not detect any seismic activity but Venera 14 registered two small events. Seismometers were also to be carried on the failed Mars-96 and Phobos 1 and 2 missions.

The InSight Mars lander mission had to be postponed recently (now scheduled for a 2018 launch) because its seismometer (Seismic Experiment for Interior Structure [SEIS]) requires a perfect vacuum and its enclosure had developed a persistent leak. SEIS has a sensitivity of about 0.05 nm and an operating frequency from 0.005 to 1 Hz and its three sensors are based upon suspended silicon test masses whose positions are monitored capacitatively. A second seismometer covers the 0.05–40 Hz range.

In the future, the Luna-Glob Moon sample and return mission was originally planned to carry several seismometers, but that requirement has now been abandoned. There are at least two suggestions for possible long duration (~100 days) Venus lander missions that would incorporate seismometers into their instrument packages. The ESA Venus Long Life Surface Package (VL^{2}SP, a post-2030 mission) might be based upon silicon carbide electronics that can operate up to 500°C. It would aim to detect about one seismic event per Earth day with its sensitivity of ~10 nm. The second suggestion is made by a private company (Venusian Solutions – see also Section 7.3.2.1) with few available details except that a lifetime of ~90 days would be aimed at and the seismometer would be embedded in a hole drilled into Venus' surface.

7.4 ATMOSPHERE

7.4.1 Spectroscopy

Only Venus, Mars, Jupiter and Titan have atmospheres and have been visited by landers and/or atmosphere probes. Of these, only Venus and Mars have had their atmospheric gases examined spectroscopically in situ. Almost any of the spectrographs discussed in Section 1.3.3 could be used to examine atmospheric gases, but in practice the constraints of mass and volume for spacecraft mean that specialised instruments have to be designed.

Many of the Venera spacecraft carried instruments for making photometric measurements through filters or low-resolution spectrographs. Veneras 9 and 10 each had two instruments. One measured the flux from 500 nm to 1.2 μm through five filters, each with ~100 nm bandwidths. The second instrument used three narrow band filters centred on CO_2 and H_2O absorption bands and on a part of the continuum within the 800– 900 nm spectral region.

Veneras 11 and 12 carried spectrometers and scanning photometers. The spectrometers were based upon non-parallel interference filters (see Linear Etalon Imaging Spectral Array [LEISA], Section 1.3.5). The filters were ring shaped with half of each ring covering 450–700 nm at a 20 nm spectral resolution and the other half covering from 700 nm to 1.2 μm at a 40 nm spectral resolution. The detectors were single germanium photodiodes and the rings were rotated at 6 rpm to build up the complete spectrum. Light guides fed the radiation into the instruments. Four wide band filters were centred on 490, 710, 1 and 1.2 μm, were mounted inside the interference filters and fed additional photodiodes. Small mirrors scanned the photometers' fields of view vertically. The instruments were mounted externally and protected thermally by insulation. Veneras 13 and 14 carried similar instruments with the addition of a UV channel where a filter with a band pass from 340 to 400 nm fed a gallium phosphide photodiode.

Although the Vega 1 and 2 Venus/Halley missions were fly-bys, the main spacecraft each carried a lander and balloon and these were dropped onto Venus. Both landers carried UV spectrographs to examine the atmosphere during the landers' descents. The ISAV-S spectrometers used a xenon UV light source to illuminate the atmospheric gases which could freely enter the 0.85 m long sample tubes attached to the outside of the pressurized parts of the landers. Mirrors inside the sample tube reflected the UV radiation so that it passed along the tube twice. The remainder of the spectrograph and the light source were contained inside the pressurised part of the lander and a 40 mm thick quartz window transmitted the radiation into the sample tube and then back into the lander. The spectrographs used spherical concave holographic gratings and the detectors were 1 × 512 arrays of photodiodes. The outputs from the diodes were binned so that the UV spectrum was obtained over 13 bands between 220 and 400 nm. Spectra were obtained at 4-second time intervals from a height of 62 km down to the surface, with a height resolution between 40 and 180 m.

The Pioneer Venus large atmospheric probe carried an infrared radiometer with six channels that viewed the solar radiation through a diamond window. The channels observed over the spectral bands 3–50 μm, 4–5 μm, 6–7 μm, 7–8 μm, 8–9 μm and 14.5–15.5 μm. Pyroelectric detectors (Section 2.5) were used since they do not require cooling and two black body sources were used for calibration.

MSL's Curiosity rover carries tunable laser spectrometers for detecting water, carbon dioxide and methane and for studying their isotope ratios. The lasers operate at 2.78 μm (for water and carbon dioxide) and 3.27 μm (for methane) with spectral resolutions in the region of 1.5×10^7. The sample chamber can be pumped out to a vacuum (for calibration) or filled with the Martian atmospheric gases at ambient pressure (70 Pa).The chamber largely comprises a 200 mm long Herriott delay line. This uses two spherical mirrors to reflect radiation back and forth – in the case of MSL's rover, a total of 81 times – giving a travel distance through the gases of nearly 17 m. The diodes are scanned in wavelength by an injection current that partially fills the conduction band, effectively changing the band gap and so the absorbed wavelength and thus producing the spectrum. Methane (CH_4) absorption lines change their wavelengths slightly when a hydrogen atom is replaced by a deuterium atom or when carbon-12 is replaced by carbon-13. The change is small – one absorption band, for example, has eight lines within a spectral region from 3.2701 to 3.2705 μm – but the spectrometer easily resolves these.

The Mars Polar Lander were also to have used a Herriott chamber in its TEGA. This had two laser diodes – one for water detection and the other for carbon dioxide detection. The Herriott chamber gave the spectrometer a 1 m path length in a space of 50 mm.

Methane is a possible biological indicator and on Mars is broken down by solar UV photons in a few tens of years. Its presence now in Mars' atmosphere would thus indicate that it must still be produced. In August 2012, Curiosity's tunable laser spectrometer detected a sudden spike in the methane abundance to a level of about 7 ppb. Previously, Mars Express Orbiter's PFS (Section 2.5) had also detected methane at around 10 ppb. Sources for this methane could include volcanic or other geological activity, meteorites, storage and release in crystals (clathrates) and methane-generating bacteria. If the methane were to be present throughout Mars atmosphere at 10 ppb, then the total mass of the gas would be ~100,000 tonnes. Given the lack of any evidence of life so far in any form on Mars (Section 7.2.3.4), the biological production of such an amount of the gas in a few weeks would seem to be improbable.

Tunable laser spectrometers are likely to be carried on future atmospheric probes to Venus, Saturn, Titan and Uranus, but these projects, except perhaps for Venus, probably lie 15–30 years into the future.

7.4.2 Nephelometry

Nephelometry is concerned with determining the concentrations of solid particles (dust) in gases or liquids and of liquid droplets in gases. The particulate concentration is closely related to the opacity/transparency or optical depth of the liquid or gas. Nephelometers operate by shining a light beam through the liquid or gaseous medium and then detecting the light scattered (forwards, sideways or backwards) from the beam. For this section, it is the suspended particles (solid or liquid) in solar system objects' atmospheres that are of concern.

The first spacecraft-borne nephelometer was on board Venera-9 to measure aerosol levels in Venus' atmosphere during the spacecraft's descent towards the surface. It used a semiconductor infrared (IR) laser and detected the backscattered radiation. The instrument's sensors were mounted outside the spacecraft. Veneras 10, 11, 12, 13 and 14 and the Vega 1 and 2 balloons carried basically similar instruments.

All four of the atmosphere probes of the Pioneer Venus mission carried backscattering nephelometers. Based upon pulsed LEDs emitting at 900 nm and solid state detectors, they analysed particulate matter in Venus atmosphere from a height of 65 km to the surface (and the day probe continued to operate for over an hour whilst on the surface).

The Jupiter atmosphere probe, Galileo's, nephelometer was developed from those of Pioneer Venus. It was deployed outside the main body of the probe after the entry shield was ejected and a conical mirror enabled it to measure both the backscattered and transmitted light and sideways scattered light at five different angles.

The Mars lander, Phoenix, simply shone a laser vertically upwards and picked up the backscattered light to form its nephelometer. The laser was an Nd:YAG emitting at 532 nm and 1.064 μm. It emitted 10 ns pulses, 100 times per second. Amongst other measurements it showed, for the first time, snow falling on Mars. The failed Schiaparelli Mars lander would have assessed dust in Mars' atmosphere by using its high-resolution descent camera (DECA) during its descent to determine the atmosphere's transparency. Although not a true nephelometer, the Mars rover Curiosity has used its cameras to image Ceres and Vesta during the Martian night and so gauge the atmosphere's opacity.

The Mars 2020 rover will monitor the dust content of Mars' atmosphere by using a CCD detector to measure the scattered solar radiation at 880 nm.

7.4.3 Pressure

There are many varieties of instruments for measuring gas pressure and aneroid and mercury barometers are likely to be familiar to all readers. For measuring solar system object's (Venus, Mars, Jupiter and Titan so far) atmospheric pressures, however, diaphragm transducers are more normally used. These operate by having a sealed chamber at a known temperature and gas pressure (or a vacuum), one side of which is flexible (the diaphragm). The diaphragm flexes as the external gas pressure changes and its changing shape is monitored by strain gauges attached to it.*

A guide to atmospheric pressures can also be obtained by the rates of deceleration experienced by landers as they enter the atmosphere (Section 8.2.1).

The surface atmospheric pressure can be estimated using orbiting spacecraft from the 2 µm carbon dioxide absorption band (Section 1.3). This gives the carbon dioxide abundance in a column of the atmosphere and so, knowing the other atmospheric constituents, the total pressure may be gauged. Mars Express and Mars Reconnaissance Orbiter have studied Mars' atmosphere in this way.

Mars' atmospheric pressure was first measured by the Viking lander Viking Meteorology Experiment (VME) packages using stretched diaphragm sensors operating over the 0–1800 Pa range (Mars' surface atmospheric pressure is ~600 Pa, but it varies by up to 30% over the Martian year as the carbon dioxide in the atmosphere freezes out at the polar caps). The sensors were mounted underneath the landers' bodies and connected to the atmosphere by tubes. The Mars Pathfinder lander carried a similar pressure sensor with two ranges: 0–1.6 kPa for use during the descent and 600 Pa to 1.0 kPa for use after landing.

Pressure sensors have since been carried by the Mars landers and rovers, Mars Polar lander, Phoenix, Curiosity and Schiaparelli and are expected to be parts of the payloads of InSight, ExoMars 2020 and Mars 2020.

Venera 4 first measured pressures in Venus upper atmosphere (though Venera 3† could also be said to have provided an estimate by being crushed before reaching Venus' surface – as were Veneras 4, 5 and 6). Venera 4 used a diaphragm sensor and made measurements from a height of 51 km down to 24 km. Veneras 5 and 6 similarly measured Venus' upper atmospheric pressures. Venera 7 survived as far as the surface but did not successfully measure pressure, so Venus' surface pressure was first determined by Venera 8 as being 9.2 MPa. All subsequent Veneras (except the orbiters, Venera 15 and 16) and the two Vega Venus probes have carried pressure sensors.

The Pioneer Venus atmospheric probes carried 12 silicon diaphragm-type pressure sensors with strain sensors to provide the output signals. Each sensor was designed for a different pressure range and the whole array covered from 3 kPa to 10 MPa. The possible future VL²SP mission is also planned to carry a pressure sensor to Venus.

The Galileo Jupiter atmosphere probe carried three pressure sensors for pressures up to 50 kPa, 400 kPa and 2.8 MPa with measurement accuracies of 50 Pa, 400 Pa and 2.8 kPa, respectively. The diaphragms were 15 mm in diameter and fabricated from stainless steel. Changes in magnetic reluctance as the diaphragms were displaced by the gas pressure were used to produce the output signals. The maximum pressure actually measured was 2.4 MPa.

Titan has only been studied by the Huygens lander. This carried eight silicon diaphragm atmospheric pressure sensors connected by a tube to the atmosphere. The bending of the

---

* Actually not that different from an aneroid barometer.
† Its main claim to fame though is that it was the first man-made object to reach another planet.

diaphragm in these instruments was sensed by the change in capacitance between the diaphragm and its back plate. Huygens measured Titan's surface atmospheric pressure as 150 kPa.

Although not quite surface-based, Rosetta's Comet Pressure Sensor (COPS) pressure sensor is probably most appropriately mentioned here. It measured the pressure of the gases surrounding the comet using a Bayard–Alpert gauge.* This operates by ionising the atoms within its chamber and then detecting the number of ions produced. The ions are collected at an earthed electrode and the amplitude of the current provides the measurement of gas pressure.

7.4.4 Temperature

The temperature of the atmosphere generally varies greatly with altitude and also from place to place. It can also differ markedly from the surface temperature (Section 7.2.4.1). Wire (usually platinum) resistance thermometers are widely used by landers and rovers. Thermocouples calibrated by resistance thermometers are also employed (Section 7.2.4.1).

Venus' atmospheric temperature was first measured by the Venera 5 probe which used two resistance thermometers. As with pressure measurements (previous section) all subsequent Venera missions, except 15 and 16, have carried resistance thermometers. The Vega 1 and 2 landers carried resistance thermometers fabricated as thin films not wires.

The Pioneer Venus atmospheric probes also used resistance thermometers. One wire was exposed outside the probe once the heat shield had been ejected; a second was sheltered from the atmospheric flow and could act as a backup if the outer one were to be damaged. The temperature could be measured to ±0.25 K over a 200–800 K range. In use, the sensors were partially shorted out by a film of sulphuric acid condensing on them from the clouds. Fortunately, this evaporated once the probes dropped below the cloud layer.

Thin wire thermocouples were used to measure atmospheric temperatures by the Mars Viking, Pathfinder and Phoenix landers. They were calibrated by a platinum-resistance thermometer and operated over about a range of 100 K. Similarly, MSL's Curiosity rover used thermopiles calibrated by a thin-film platinum resistance gauge.

Schiaparelli carried a platinum resistance thermometer that was a double wire system (exposed and shielded) with 700 mm long, 50 μm diameter wires wound on open frames. Its expected accuracy was ±0.1 K.

Galileo carried temperature sensors developed from those of the Pioneer Venus atmospheric probes. One was directly exposed to the air flow and the other better shielded as a backup – in fact both survived the whole 57-minute life of the probe in Jupiter's atmosphere. The outer sensor was a 100 μm thick and 1.2 m long platinum wire and wound around an open frame of platinum–iridium. The inner sensor was sandwiched between glass films and was a platinum wire 25 μm in diameter and 60 mm long.

The Titan probe, Huygens, had two double wire platinum resistance thermometers with the wires wound onto platinum–rhodium frames.

---

* Also known as a hot filament ionisation gauge.

Atmospheric temperatures can be obtained by orbiters using thermal radiometers (Sections 1.3, 1.4.2 and 7.2.2.1) to observe in the 15 μm carbon dioxide band. The absorption occurs primarily at an optical depth of one and the physical depth (height) in the atmosphere corresponding to this varies with frequency across the band, allowing the temperature profile to be mapped out. Mariner 9, Mars Express, Mars Reconnaissance Orbiter, Viking Orbiter and the Mars Odyssey have all carried instruments to make these measurements by this means. Radio occultations (Section 2.4.4.6) can provide temperature estimates by measuring the refractive index of the atmosphere and this has been done using the Mars Global Surveyor for Mars.

7.4.5 Humidity

Humidity is the water vapour content of an atmosphere. It can be measured as absolute quantity – that is as the number of kilograms of water per cubic meter of the atmosphere – or as a relative quantity – that is as a percentage of the amount of water that the atmosphere at a particular temperature *could* hold *if* saturated in water vapour. If the saturated content is known as a function of temperature, then a relative humidity measurement is easily converted into the absolute value or vice versa. Traditionally, humidity is measured using wet and dry thermometers. Unless the atmosphere is saturated with water vapour, the evaporation of water from the wet thermometer lowers its temperature compared with that of the dry thermometer and the difference provides the measure of humidity.

Clearly wet and dry thermometers are not practicable for spacecraft-borne instrumentation. Several other methods are thus used.

- Capacitative hygrometers in which the dielectric constant of a material changes with humidity. The sensitive material is usually a polymer or metal oxide and it is deposited onto a glass or silicon substrate. When placed between two electrodes the sensitive material's dielectric constant is directly proportional to the relative humidity and so the measured variation in capacitance provides the output signal.

- Resistive hygrometers in which the electrical resistance of a material (usually a salt solution or polymer) changes with humidity and is measured by the variations in an electric current passed through it. The resistance is usually inversely proportional to the relative humidity.

- Thermal hygrometers in which the thermal conductivity of the atmosphere changes with humidity. Two matched thermistors are used and heated to 200°C or so. One is encapsulated in a dry nitrogen atmosphere, the other is exposed to the air. The difference in their temperatures is then proportional to the absolute humidity.

- Coulomb hygrometers in which the electrical conductivity of the atmosphere changes with humidity and is measured by determining the rate of leakage of charge from an electro-statically charged object.

The Vega Venus landers carried resistive and Coulomb hygrometers. In the former, lithium chloride solution acted as an electrolyte and a current was passed through it. As the solution absorbed water vapour from its surroundings, its conductivity increased. Thus as the humidity of the surroundings changes, so does the current and this provided a measure of the humidity.

The Phoenix Mars probe carried a capacitative hygrometer that measured the relative humidity via the changing permittivity of a polymer film. MSL's Curiosity rover also uses a polymer film capacitative hygrometer. The Mars 2020 and ExoMars 2020 missions are expected to include hygrometers amongst the instruments.

Humidity can also be measured from orbiting spacecraft by observing the water vapour absorption bands at 2.7 and 6.3 μm.

7.4.6 Wind

Means of measuring the speeds and directions of winds on Earth are something that, like measuring air pressure, are going to be familiar to all readers – even if it is only from the wind vane on top of the neighbouring church steeple or holding a moistened finger up in the air. To those to whom the words 'rocket science' are a synonym for 'abstruse', 'non-understandable' or 'for experts only', it may therefore come as something of a surprise that some spacecraft-borne wind indicators are no more than wind vanes or damp fingers. Slightly (but not very) more complex instruments, though, do find some applications. Winds have been measured for centuries however, from the relative movement of clouds observed using Earth-based telescopes. Thus, Jupiter's winds have been known to reach speeds of 200–400 km/h from the relative motions of its features as seen from Earth for at least two centuries. Dust storms on Mars can also be observed from Earth and their changes used to estimate wind velocities and direction.

On the Phoenix Mars lander, the wind instrument was a small cylinder suspended on a fibre. It had a mirror underneath it that reflected an image of the position of the cylinder to the lander's camera so that the wind's direction could be observed and some idea of its strength obtained from the degree to which the cylinder was blown away from the vertical. Similarly, the Mars Pathfinder lander had three windsocks attached to its mast which were imaged by its cameras. Mars Pathfinder also had a hot wire anemometer. This was made up from heated platinum–iridium wires 65 μm in diameter. The wire was wound in six sections vertically around a 27 mm diameter cylinder. The overall cooling of the system provided a measurement of the wind's velocity and the differences in the cooling of the sections (like a wet finger), a measure of its direction. Schiaparelli would have measured wind speeds and directions on Mars using a similar device based upon a hot film. It would have been sensitive over the speed range 0.3–30 $m·s^{-1}$ with accuracies of ±1 $m·s^{-1}$ for speed and ±10° for direction

Rotating cup anemometers, such as are found at every Met station on Earth, were carried by Veneras 9 and 10 to Venus. Veneras 13 and 14 meanwhile carried microphones so that the wind speed could be estimated from its sound levels (Section 7.4.7). Winds in Venus' atmosphere were also measured during the Pioneer Venus atmospheric probes descents from the changing sideways velocities of the probes estimated from their Doppler shifts. Winds in Titan's atmosphere were similarly assessed during the Huygens probe's descent.

7.4.7 Acoustics

We have been accustomed to receiving images from other solar system objects since the start of the space age and radio emissions converted into sound have also fascinated many people for almost as long. However, hearing the actual sounds from other solar system objects seems, somehow, to be so much more exotic – perhaps our relative familiarity with the airless and silent Moon leads us sub-consciously to expect all solar system objects to be similarly quiet. Nonetheless microphones have been carried into space by quite a number of missions. Clearly, sounds can only exist where there are atmospheres to carry them and most sounds originate from the effects of those atmospheres. However, nothing much has been heard to date.

Electromagnetic microphones are widely used on Earth and operate using a small coil attached to a diaphragm and immersed in a magnetic field. The movement of the diaphragm under the influence of sound waves causes the coil to move within the magnetic field, so inducing an electric current within the coil which then forms the output signal. Since such microphones are designed to cope with the wildest gyrations of rock musicians they are also ideally suited to enduring the stresses of spacecraft launches and landings.

Veneras 11 and 12 carried electromagnetic microphones to try and detect thunder, although no definitive results were obtained. As mentioned in the previous section, Veneras 13 and 14 carried microphones with the aim of detecting wind noises. The (failed) Mars Polar lander also carried a microphone. The Mars lander, Phoenix, carried a microphone, but this never operated due to an electrical problem.

Cassini's Huygens Titan lander carried instruments both to pick up sound and to measure the speed of sound in its atmosphere. Two small piezo-electric transducers were used to transmit and receive sounds across a small gap within the probe. A third, acting as a sonar, sent sound pulses down towards the surface during the descent.

The Mars 2020 rover may, at last, enable us to hear what the surface of Mars sounds like since it will carry a microphone within its SuperCam package.

7.5 INTERPLANETARY MEDIUM

A spacecraft is obviously automatically in a position to make direct sampling measurements of the interplanetary medium. This has four major components:

Large-scale magnetic fields

Medium-scale electric fields

Dust

Atoms, ions and other subatomic particles

The detection and analysis of atoms, ions and other subatomic particles are discussed in Chapters 4, 5 and 6. The collection of interplanetary dust particles and their return to Earth for analysis has been discussed in Section 7.2.6.3.6, while the collection and robotic examination of dust from planetary and cometary surfaces is covered in Sections 7.2.1 and 7.2.6.

In this section, therefore, it is the magnetic and electric fields plus the dust particles caught and examined robotically by spacecraft in interplanetary space with which we are concerned.

7.5.1 Large-Scale Magnetic Fields

There are strong magnetic fields within the Sun and these extend out into or are dragged outwards by charged particle emissions from solar flares and coronal mass ejections etc. There are also planetary magnetic fields arising mostly from internal movements of material within the planet. The same types of instruments, however, are used for studying all such fields and these are the search coil (or spin coil or induction) magnetometer, the flux-gate magnetometer and the ionised helium magnetometer. To avoid interference from any magnetic fields within the main spacecraft, magnetometers are usually deployed on long (several metres) booms after launch. Often three magnetometer will be used, oriented at 90° to each other, so that the components of a magnetic field may be measured in three dimensions.

An instrument that measures both magnetic and electric fields – the Electron Drift Instrument – is discussed in Section 7.5.2.3.

7.5.1.1 Search Coil Magnetometer

Search coil magnetometers are used when magnetic fields are varying rapidly, or (the more usual situation) when the spacecraft bearing the magnetometers is spinning rapidly within a fairly static field. After the lodestone (and the magnetic compass which grew from it), the search coil magnetometer was one of the earliest magnetometry instruments to be developed. Its operational principle is based upon Faraday's law of induction – if you move an electrical conductor within a magnetic field, so that it crosses the magnetic lines of force, a voltage is induced along the conductor. If you then join up the ends of the conductor, you get a circuit and an electric current will flow. In practice and to increase their efficiency, the conductors are actually formed as simple solenoids, usually with a moderate number of turns and with a magnetically soft metal core.

Many spacecraft have carried search coil magnetometers. Pioneer 1 was the first spacecraft to succeed in taking any magnetometer into space (there had been several previous failed attempts). It was of the search coil design and Pioneer 1 was spin stabilised by a rotation at 110 rpm. Measurements of the Earth's magnetic field throughout the van Allen belts were obtained even though the spacecraft did not go into its intended orbit (it was aimed at the Moon). Pioneer 5, also spin stabilised, used a search coil magnetometer to make the first investigations of magnetic fields between the orbits of Venus and Earth.

The five spacecraft of the THEMIS project all carry three-axis search coil magnetometers as well as flux gate magnetometers (see next section). Three of the spacecraft have highly eccentric orbits taking them through the van Allen belts and two (renamed ARTEMIS – Acceleration, Reconnection, Turbulence and Electrodynamics of the Moon's Interaction with the Sun) orbit the Moon.

Juno, recently arrived at Jupiter, employs a search coil magnetometer within its Waves package. The core of the magnetometer is a 150 mm long mu-metal* rod and the enclosing solenoid is a thin copper wire with 10,000 turns. It is sensitive from 50 Hz to 20 kHz. The spacecraft rotates at 2 rpm with the magnetometer's axis parallel to the spin axis. It also carries flux gate magnetometers and helium magnetometers (see next two sections). The four MMS spacecraft each carry a tri-axial search coil magnetometer (as well as two flux gate magnetometers). The cores are ferromagnetic cylinders 100 mm long and 4 mm in diameter with 10,000 turn windings. They operate from 1 Hz to 6 kHz and are mounted on one of the 5 m booms alongside one of the fluxgate magnetometers.

7.5.1.2 Flux Gate Magnetometer

The operating principle of a flux gate magnetometer is based upon a solenoid with a magnetically soft metal core. When the solenoid is supplied with an alternating current, a dipole-type magnetic field develops that swaps its polarity through 180° in phase with the half-cycles of the AC supply.

If the core is longer than the solenoid, then its projecting half may be enclosed by a second solenoid. The alternating magnetic field will then generate an alternating electric current in this second solenoid (like a transformer). If the apparatus is shielded from external magnetic fields, then the induced current in the second solenoid will be a simple AC current.

If, however, there is an external, constant magnetic field present, then the alternating magnetic field will be stronger during the phase when the external field adds to it than when it is reversed (when the external field subtracts from it). The induced current in the second solenoid will then be stronger in one half of its cycle than in the other – or as we might normally express the situation – there is a DC current with an AC current added to it.

It would be fairly easy to filter out the DC component and use its voltage as a measure of the external magnetic field strength. However, in many instruments, a second soft magnetic core is used lying parallel to the first. The first solenoid is extended and wound around the second core so that the current flows around that core in the opposite sense to the first core. The second solenoid is also extended, but wound around the second core so that the current flows in the same sense to that in its first half. The two induced AC currents now cancel out leaving the DC current as the measure of the external magnetic field strength.

ESA's four Cluster II spacecraft each carry pairs of three-axis flux gate magnetometers operating from 0 to 10 Hz. They have five ranges with a maximum magnetic field strength of 650 µT being measureable. The individual spacecraft operate in Earth orbits that keep them at the four corners of a tetrahedron with inter-spacecraft separations variable from 4

---

* A magnetically soft nickel–iron alloy.

to 10,000 km – so that the interplanetary medium may be studied on many size scales. The four follow-on MMS spacecraft each carry two tri-axial flux gate magnetometers on the ends of 5 m booms. They can measure field strengths between 650 nT and 10.5 μT.

The Saturn orbiter, Cassini, carried both three-axis flux gate and helium (see Section 7.5.1.3) magnetometers. The latter was mounted at the end of an 11 m boom and the flux gate magnetometer at the centre of the boom. Their frequency ranges were 0 to ≥30 Hz and 0 to 10 Hz, respectively. The flux gate magnetometer had four magnetic intensity ranges up to a maximum of 4 μT while the helium magnetometer (a flight spare from the Ulysses spacecraft) had two ranges up to 256 nT.

The Voyager 1 and 2 missions to the outer planets carried flux gate magnetometers – indeed in August 2012, Voyager 1 became the first spacecraft to take a magnetometer into interstellar space and Voyager 2 is expected to follow it at about the time of writing. Voyager 2 also measured the presence of a magnetic field around Uranus that is almost as strong as that of Earth. The Voyagers carried two magnetometers each, one for low-field strengths (≤20 μT) and the second for high-field strengths (≤2 mT).

Explorer 61, also known as MagSat, used two magnetometers to map out the Earth's magnetic field a few hundred kilometres above its surface. The three-axis flux gate magnetometer made the primary measurements and a magnetometer based upon caesium vapour (see next section) was used for calibrating the flux gate magnetometer's data. Rosetta's Philae comet lander used a fluxgate magnetometer, ROMAP, to investigate comet 67P/Churyumov–Gerasimenko showing that it is not magnetised as a whole, but that the surface dust particles did contain some magnetite.

The two van Allen probes carry tri-axial fluxgate magnetometers as well as tri-axial search coil magnetometers. The fluxgate magnetometers have three ranges, 0–256 nT, 0–4.1 μT and 0–65.5 μT, with measurement accuracies of ±5 nT.

Hayabusa-2's lander, MASCOT, includes a fluxgate magnetometer with a sensitivity over the ±12 μT range for magnetic field measurements both when it is stationary on the asteroid and during its jumps from one opposition to another. The Juno spacecraft in its Magnetometer (MAG) package carries two three-axis flux gate magnetometers* deployed on a 12 m boom. These determine the magnetic field's direction and an inboard helium magnetometer (see next section) determines its strength.

The proposed Europa Multiple Flyby mission is expected to include fluxgate magnetometers developed from the MMS designs with helium magnetometers for calibration. The Interior Characterisation of Europa using Magnetometry (ICEMAG) package is intended to probe Europa's postulated sub-surface ocean. The instruments are expected to cover the field strength range ±1.5 μT with a measurement accuracy of 10 pT.

The planned Jupiter Icy Moon Explorer (JUICE) mission to Jupiter is expected to carry two tri-axial fluxgate magnetometers mounted at the end and in the middle of a boom. Their ranges will be up to 8 and 50 μT with resolutions of 1 and 6 pT, respectively. It will

---

* Rather confusingly, it also carries an instrument called the Advanced Stellar Compass but this is not magnetically based and uses stars to determine the spacecraft's orientation.

also carry an optically pumped magnetometer related to the helium design and a tri-axial search coil magnetometer.

7.5.1.3 Helium Magnetometer

Helium has a metastable state into which lower energy electrons can be lifted by the absorption of 1.08 μm photons. The presence of an external magnetic field affects the energy of the metastable state, since the electrons would naturally be in their Larmor orbits (Section 5.2.6.2 – see also reference A.1.9 and other sources in Appendix A). The slight shift of the energy level affects the efficiency of the absorption of the 1.08 μm photons. Thus by monitoring that absorption, the magnetic field could be analysed. In practice, an artificial magnetic field is also applied to the cell containing the helium and then adjusted until the net magnetic field is observed to be zero. The artificial magnetic field is then equal and opposite to the natural field and this provides the output signal for the device. Magnetometers of this type may also be based upon caesium or rubidium vapours, but are essentially similar instruments.

The first spacecraft ever to carry magnetometers into the outer reaches of the solar system, Pioneers 10 and 11, were also the first to leave the solar system (though they have now been overtaken by Voyagers 1 and 2 – and by New Horizons in due course). They carried helium magnetometers. Their magnetometers were mounted on the ends of 6.6 m long booms and they had eight operating sensitivity ranges up to a 140 μT maximum.

International Cometary Explorer (ICE) carried a helium magnetometer with eight sensitivity ranges up to a maximum of 140 mT and with a maximum frequency response of 10 Hz. The Venus fly-by, Mariner 5, also carried a three-axis helium magnetometer although it turns out that at 6 nT, Venus has essentially no magnetic field. The three spacecraft of the Swarm mission carry helium magnetometers to calibrate their flux gate magnetometers. Recently, they have shown the presence of a stream of molten iron in the northern part of the Earth's core moving at some 40 km per year (cf. plate tectonic speeds of ~10 to ~100 mm per year).

7.5.2 Electric Fields

7.5.2.1 Langmuir Wave Instruments

As noted in Section 7.2.4.2, large-scale electric fields are rare in the solar system (and, indeed, in most other places in the universe) since the interplanetary medium is a plasma and its high electrical conductivity means that any voltage differences that do occur are quickly discharged. If magnetic fields, however, are moving through a plasma, then they can set up relatively small-scale electric fields. This is because electrons are 2000 times less massive than protons but both have electric charges of the same magnitudes. Thus, the magnetic forces on the electrons and protons have the same magnitudes (though, of course, opposite directions) but the electrons' lower mass means that they accelerate 2000 times more rapidly than the protons (and the situation is even more extreme for the heavier ions). The electrons in the plasma are thus moved rapidly away from the more-or-less stationary protons and ions by a moving magnetic field.

This imbalance in the positions of the negatively and positively charged particles within the plasma will generate an electric field. The electric forces from this field will try to move the charged particles back to their equilibrium positions. When the magnetic field ceases moving or weakens or when the electrical field that has been generated can overcome the magnetic effects, then the electrons do move back towards their equilibrium positions. But when they arrive there they are moving rapidly and so overshoot, thus setting up an electric field in the opposite direction which sends them back again – and so on. The electrons continue to move in simple harmonic motions about their equilibrium positions in what is called an electron plasma oscillation or a Langmuir wave. When the thermal velocities of the electrons are small compared to velocities induced by the electric field ('cold' electrons), the frequency, ν_L, of the Langmuir wave is determined by the square root of the electron number density in the plasma (Equation 7.4):

$$n_e = \left(4\pi^2 \, m_e \, \varepsilon_0 \, \nu_L^2 \right) / e^2 \quad \text{electrons m}^{-3} \tag{7.4}$$

where n_e is the electron number density.

Thus, measuring the Langmuir frequency will enable the electron density of the plasma to be determined and an estimate made of its total density.

The protons and ions also undergo oscillations but with much lower frequencies than those of the electrons because of their higher masses.

A planetary-scale electric field can develop through the migration of electrons to the top of the atmosphere. So far this has only been detected for Venus where the top hat analyser that was part of the Analyser of Space Plasmas and Energetic Atoms (ASPERA) package (Section 5.2.1) measured a 10 V field. The field is thought to arise because the temperature at a height of 200 km in Venus' atmosphere is around 50°C so that free electrons have velocities in excess of 100 km/s, while the carbon dioxide and other atmospheric molecules and ions are moving at less than 1 km/s. The free electrons can thus easily escape from the atmosphere while the heavier ions have much more difficulty (Venus' escape velocity is 10.4 km/s). This quickly sets up the electric field and this in turn reduces the number of escaping electrons to be the same as the number of escaping ions. However, ions can then be pulled out from the atmosphere by the electric field until they do escape. In this way, Venus may have lost its water – the water molecules being dissociated by sunlight and the hydrogen and oxygen ions then escaping. Venus Express' measurements suggest the current mass loss rate is around 100 tonnes per year – about 4×10^{14} kg since the formation of the solar system

Langmuir waves can be studied directly by measuring the voltage induced in an antenna immersed in the plasma.* Thus, the Voyager 1 and 2 spacecraft carry two 10 m long beryllium–copper tubes arranged at 90° to each other and which act as the antennas for both the Plasma Wave Subsystem (PWS) and the Planetary Radio Astronomy (PRA) instruments. The PWS detects over the frequency range from 10 Hz to 56 kHz (1 to 4×10^7 electrons/m³) and measures the voltages to about ±6%. While passing through Jupiter's bow shock, the Voyagers measured the Langmuir frequencies dropping from ~6 kHz to ~1 kHz corresponding to the electron density dropping from ~4.4×10^5 to ~1.2×10^4 electrons/m³.

---

* The electron oscillations also emit radio waves at much higher frequencies that can be detected normally (Chapter 2).

The Voyager spacecraft are still operating at the time of writing in what is now called the Voyager Interstellar Mission. The termination shock is the point where the speed of the solar wind drops from super- to sub-sonic. Voyager 1 crossed this in 2004 (at 94 AU) and Voyager 2 in 2007 (at 84 AU).* Both Voyagers detected 311 Hz Langmuir waves as they went through the shock – corresponding to 1.2×10^3 electrons/m³. Voyager 1 then crossed the heliopause, where the solar wind's speed reduces to zero, in 2012, at a distance of about 122 AU from the Sun and is now in interstellar space. After crossing the heliopause the number of ions detected by the spacecraft fell by a factor of 10 to 15.

The Pioneer Venus Orbiter's Orbiter Electric Field Detector (OEFD) instrument used a pair of electric dipoles with an effective length of 750 mm and frequency channels at 100 Hz, 730 Hz, 5.4 kHz and 30 kHz. Near Venus bow shock with the solar wind, wave detections were made at the two middle frequency channels and attributed to ions. Langmuir waves were detected at 30 kHz – corresponding to an electron density of $\sim 10^7 \cdot m^{-3}$.

The second instrument in Juno's Waves package (Section 7.5.1.1) is a plasma wave sensor. Like the Voyagers its antenna is two rods. Each rod is 2.8 m in length and the instrument operates from 50 Hz to 40 MHz. At its top operating frequency, it will be able to measure up to 2×10^{13} electrons/m³ – well above the highest plasma densities to be found around Jupiter (within the Io torus).

7.5.2.2 Langmuir Probe Instruments

Both the plasma wave antenna and the Langmuir probe are basically conducting rods stuck out into the plasma. The difference is that the plasma wave induces voltages into the antenna whereas the Langmuir probe has a voltage applied to it to attract the electrons or ions in the plasma towards it. The Langmuir probe is generally much smaller than the antennas discussed in the previous section and can be used to measure plasma density, electron temperature and the plasma potential relative to the spacecraft.

A simple conductor without any applied voltage, when first placed into a plasma, will tend to collect electrons because their velocities are higher than those of the heavier particles ensuring that they arrive at the probe first. This gives the probe a negative voltage which then starts to repel the electrons and attract the ions. When equilibrium has been reached there is no net current and the probe is left at a negative voltage called the floating voltage.

If the probe now has an external negative voltage applied to it (a negative bias – typically around –10 V), then electrons will be repelled and ions attracted. A current due to the attraction of the positive ions accumulating at the probe's tip on the electrons within the conductor will then flow through the conductor towards the probe tip. If the bias is reduced towards zero, then eventually plasma electrons will also start to flow to the probe. The total current arising within the conductor will be the result of the sum of the (negative) contribution of the electrons and the (positive) contribution of the ions and so will reduce towards the floating voltage as the bias reduces towards 0 V. By making the bias increasingly positive, the ion current will reduce to zero and the whole of the current within the

---

* At the start of 2017, Voyager 1 was about 137 AU from the Sun and Voyager 2 about 113 AU.

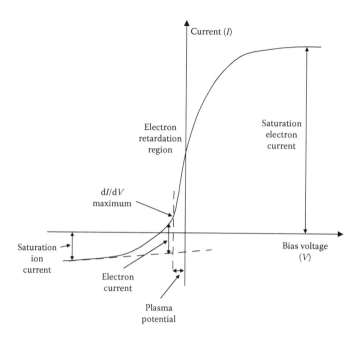

FIGURE 7.12 Schematic Langmuir probe $I–V$ curve.

conductor will be away from the probe tip and arise from the repulsion of the accumulated plasma electrons.

Plotting the variation of current with bias voltage produces the $I–V$ curve (Figure 7.12). The plasma potential is the bias voltage at which the first derivative of the $I–V$ curve is a maximum. The electron and ion densities are obtained from the saturation electron and ion currents, while the electron temperature comes from the exponential change in the electron current in the electron retardation region.

The Langmuir instrument on the Pioneer Venus orbiter, Orbiter Electron Temperature Probe (OETP), used two Langmuir probes in the form of cylinders 76 mm long and 2 mm in diameter. They were mounted on booms to remove them as far as possible from any influences of the main spacecraft on the plasma. One probe was aligned with the spacecraft's spin axis, the other was radial to it. For electron densities above about $10^9 \cdot m^{-3}$ the measurement errors for electron densities and temperatures were better than 10%.

The Spacecraft Potential, Electron and Dust Experiment (SPEDE) on the Small Missions for Advance Research in Technology-1 (SMART-1) lunar orbiter carried two probes on 600 mm booms. The sensors were titanium nitride foils. Electron temperatures up to 100,000 K and electron densities from 10^5 to 10^9 m^{-3} could be measured. The plasma surrounding comet 67P/Churyumov–Gerasimenko was analysed by Rosetta's Langmuir Probe (LAP). This comprised two 50 mm diameter titanium spheres mounted on 2 m booms. Near the comet electron densities of 10^7 to 10^8 m^{-3} and electron temperatures of up to 10^6 K were found.

Mars Atmosphere and Volatile Evolution's (MAVEN) Langmuir probes are 400 mm long cylinders just over 6 mm in diameter mounted on 7 m long booms. They operate over electron densities from 10^8 to 10^{12} m^{-3} and electron temperatures from 500 to 5000 K analysing Mars' ionosphere. The three Swarm spacecraft each carry two Langmuir probes that are charged oppositely so that electrons and ions can be measured simultaneously. Cassini carried a Langmuir probe comprising a 50 mm titanium sphere on the end of a 1 m long rod as a part of its Radio and Plasma Wave Science (RPWS) package. It could measure at 0.25 V intervals between ±32 V.

The planned Jupiter orbiter, JUICE is expected to carry four Langmuir probes operating from 0 to ~1.5 MHz mounted at the ends of four separate 3 m booms. The probes will be 100 mm spheres and probably have a diamond film surface coating to enhance their durability.

7.5.2.3 Electron Drift

Both electric and magnetic (Section 7.5.1) fields can be measured by an entirely different type of approach. The electron drift instrument (EDI) is based upon sending out from the spacecraft a weak (<1 µA) beam of electrons with energies of a few hundred electronvolts. If sent out in the correct direction, then in the presence of a magnetic field the electrons will trace out their Larmor orbit (Section 5.2.6.2) and return to the spacecraft, where they may be detected.

The time taken for the electrons to go around the Larmor orbit once gives the magnetic field strength. In the presence of an electric field as well, the orbits will no longer be quite circular and the position of the return beam will drift slightly. The magnitude of the drift gives the electric field strength.

EDIs are currently used on the four Cluster II and the four MMS spacecraft and were first employed during the Equator-S mission. The Cluster II's instruments were identical to those on Equator-S while MMS's were developed from those earlier designs. All the spacecraft use two heated tungsten filament electron emitters (called guns) and two microchannel plate (MCP) detectors.

7.5.3 Interplanetary Dust

Dust particles in interplanetary space are probably better known as micrometeorites. Travelling at several tens of kilometres per second, when they hit another (larger) object they are themselves completely evaporated and a small pit (i.e. crater) is left on the surface of the other object. When that object is a spacecraft those pits can be found using a microscope and examined – so in one way all spacecraft are detectors of interplanetary dust. Examination of some of the windows from the returning Apollo capsules has revealed some 10 pits and suggests an impact rate of one dust particle per square metre every two days or so. Other estimates suggest that between 250 g and 1.5 kg of particles in the size range 50 µm to 1 mm are hitting Earth every second. At the larger end of this size range, the dust particles may be seen as meteors. The origin of most interplanetary[*] dust is thought to be from the surfaces of asteroids and comets.

---

[*] For example the Lunar Reconnaissance Orbiter's left narrow angle camera was hit by a 0.8 mm micrometeoroid traveling at a relative velocity of 4.4 km/s in 2017, although afterwards it continued to operate normally.

Micrometeorites are, of course, a problem if you are building a spacecraft and instruments and structures may need protection against the damage caused by them. However, if it is the dust particles that are of interest, then no information on their structure will be obtained since this is totally destroyed by the impact. It may, though, be possible to infer the dust particle's mass and velocity and there are many possible ways of doing this. Only by decelerating the particle slowly – as with Stardust's aerogel (Section 7.2.6.3.6) – may the dust particle's structure be found.

Most dust detectors are designed for the purpose, but sometimes dust impacts may be caught by other instruments. Thus, the 10 m long plasma wave instrument antennas on Voyager 1 and 2 (Section 7.5.2.1) also function as dust impact detectors. The particle upon impact is evaporated into a cloud of ionised gas. When this cloud sweeps over an antenna it generates a pulse that is picked up by the instrument. Both spacecraft detect such pulses at a rate of one every 10–15 minutes.

The Vega 1 and 2 Venus/Halley missions carried three types of impact detectors in order to cover the mass range from 10^{-19} to 10^{-9} kg. The first type was essentially a microphone. A thin circular membrane 125 mm in diameter had three piezo-electric detectors attached to it. A dust impact would set up a wave in the membrane and this would be picked up by two of the piezo-electric detectors (the third detector was used for calibrating the main detectors).

The second type of detector covered the smaller masses and collected the ionisation products following the impact. A dust particle would hit the bottom of a chamber containing two electrodes at a voltage difference of about 2 kV. The plasma produced by the impact was separated into its ions and electrons by the voltage, the electrons collected at the anode and their total charge measured to determine the dust particle's mass.

The third type of dust detector (DUCMA – Dust Counter and Mass Analyser), one that has since been used on several other spacecraft, was based upon a permanently polarised* polyvinylidene fluoride (PVDF) membrane. The membrane had two aluminium–nickel electrodes coated onto it but without a bias voltage between them. It was 28 μm thick (thick enough not to be penetrated by particles travelling at up to 12 km/s and with masses up to 10^{-13} kg) with an area of 7500 mm². A high-velocity dust particle impact results in a large and fast (ns) current pulse between the electrodes. The pulse amplitude depends upon the impacting particle's mass and velocity. The pulse is thought to arise from the destruction of the metal layer by the impact and thus the exposure of the permanently polarised molecules of the PVDF. This in turn induces an electric field fringing the impact crater so producing the pulse of current. The detector was able to count up to 100,000 impacts per second and would continue to function even if penetrated by a large particle.

---

* In some molecules, the distribution of the electrons is not symmetrical and one part of the molecule has a net negative charge and the other a net positive charge (a dipole). Normally, the distribution of these dipoles is random and there is no net electric field to the material as a whole. Applying an external electric field, however, will cause the dipoles to line up and produce an electric field opposing that of the external field. This is called induced polarisation. Some materials however can be persuaded to have their dipoles aligned permanently – for example, by allowing them to solidify in the presence of a strong electric field. The material will then have a permanent electric field (like a permanent magnet) and this is known as permanent polarisation.

Stardust used two dust detectors in addition to its aerogel collector (Section 7.2.6.3.6). One was a PVDF-based detector for studying dust particles up to a mass of 10^{-7} kg and impact speeds up to 6 km/s. The instrument used two separate circular membranes. The larger one was 160 mm across and 28 μm thick, the smaller was 50 mm across and 6 μm thick. The second dust detector was for detecting more massive particles. It comprised two quartz piezo-electric sensors mounted on the spacecraft's dust shield. This picked up the vibrations in the shield (sound waves) caused by these larger impacts.

The High Rate Detector (HRD) on Cassini used two separate membranes in a design similar to that of the Stardust detectors. In 2006, when Cassini went through one of the emission plumes from Enceladus, it detected a large increase in the number of dust particles near the satellite – thus supporting the theory that it was the main source of material for Saturn's E ring.

PVDF-based detectors have also been used in the Space Dust (SPADUS) instrument of the Advanced Research and Global Observation Satellite (ARGOS) mission for detections of particles up to 200 μm diameter (10^{-8} kg). The Student Dust Collector (SDC) on New Horizons comprises 12 membranes, each 65 × 140 mm in size and 28 μm thick. It operates over the mass range 10^{-15}–10^{-12} kg. The Aeronomy of Ice in the Mesosphere (AIM) mission uses 14 PVDF detectors that are nearly identical to the SDC.

The Galileo Jupiter orbiter carried an impact ionisation detector that was identical to the one on Ulysses.[*] The instrument operated over the 10^{-19} to 10^{-9} kg mass range and also measured the particles 'speeds, directions and electrical charges. Cassini, in addition to its PVDF dust detectors, also carried an impact ionisation detector with a spherical target whose area is 0.09 m². The lunar orbiter LADEE used an ionisation dust detector developed from those of Galileo, Ulysses and Cassini to study the dust environment of the Moon. In the future, it is planned that the Europa Multiple Flyby mission will incorporate an impact ionisation dust detector in its instrumentation package.

Rosetta's Grain Impact and Dust Accumulator (GIADA) comprised three detectors. The first detected the entry of a dust grain into the system non-destructively. This detector was essentially the same as a nephelometer (Section 7.3.2). It used four laser diodes to create a light curtain some 3 mm thick. A dust particle passing through the curtain scattered some of the light and this was picked up by four photodiodes.

The second detector was of the microphone type. It used a 500 μm thick aluminium diaphragm and five piezo-electric detectors. The delay between the first and second detections of a particle enabled an estimate of its speed to be made while the second detection gave its momentum. The mass of the particle could thus also be determined.

The third detector measured the total accumulation of dust particles. It used five pairs of quartz crystals oscillating at 15 MHz. One crystal of each pair was exposed to the environment whilst the second was shielded. The resonant frequency of the exposed crystal changed as its mass increased with the accumulating dust layer. The second, shielded, crystal did not accumulate any dust and was used to calibrate the measurements of the exposed

---

[*] Ulysses was a solar spacecraft, but it used a gravity assist from Jupiter to change its orbital inclination to the ecliptic to 80°. In so doing it was able to make many observations of Jupiter and its environment, including dust particles.

crystal. They had acceptance angles of about 40°. One of the crystal pairs was aimed at the nucleus while the others pointed sideways.

A possible dust detector for future interplanetary missions could be based upon a mesh of conductors. Dust impacts would break some of the wires forming the mesh, changing its overall resistance. The parameters of the impact would be obtained from the change in resistance, which would, in turn, depend upon the number of wires broken.

CHAPTER 8

Accelerations

8.1 INTRODUCTION

As we all know from watching astronauts floating around inside the International Space Station (ISS), spacecraft moving freely in space provide a weightless environment. This does not mean that there are no accelerations present – a spacecraft in a simple circular orbit is constantly experiencing a force towards the centre point of that orbit and so is being accelerated (in the sense of changing its direction of motion though not its speed). The reason why this is not apparent to the astronauts is that the force of gravity is acting upon them in exactly the same way as it does on the spacecraft – they therefore follow the same orbit and so appear weightless (along with stray pens, sandwiches, notebooks and any other loose objects that are lying around).

If the orbit is not a simple circle – it could be elliptical, parabolic, hyperbolic or there could be more complex shapes (perturbations) when the spacecraft is under the influence of (say) a planet and one or more of its satellites – the appearance of weightlessness still remains because the gravitational field continues to impart the same motion to the space-craft and to all its contents.

However, if the spacecraft's motion should cause it to move through the outer reaches of an object's atmosphere, then a feeling of some weight returning will be noticed – and if the spacecraft re-enters the Earth's atmosphere the weightlessness will be replaced by the feeling of weight which is several times that normally experienced on the surface of Earth. This, of course, is because the spacecraft is now under the influence of two forces – gravity and the drag of its passage through the atmosphere, while the passengers inside still only directly feel gravity. The spacecraft and its passenger are moving on different orbits and so the passengers feel some part of the spacecraft pressing against them (i.e. weight).

Instruments to measure accelerations (Section 8.2) will be affected in the same way as astronauts – in free fall they will measure no forces, but they will detect forces acting on them when atmospheric drag affects the spacecraft, or when the spacecraft is using its rockets etc. Thus, accelerometers can be used to measure and analyse effects and forces acting on the spacecraft other than gravity.

In many investigations, however, it is the gravitational field itself that is to be studied. This may be accomplished, not by using accelerometers, but by measuring the orbit remotely using radar, lidar, Doppler ranging etc., based at a stable observing site not influenced by the gravitational field affecting the spacecraft. Alternatively, instruments on board the spacecraft may be used to measure its position with respect to a remote and stable reference point. Investigations of the nature of the gravitational field of an object in this manner are discussed in Section 8.3.

Other spacecraft-gravity missions have been/are/will be attempting to test out the general theory of relativity (e.g. Gravity Probe-B and Lageos) or to detect gravitational waves* from the universe (e.g. LISA pathfinder and eLISA[†]) these missions are outside the remit of this book and are not further considered here.

8.2 SPACECRAFT-BORNE ACCELEROMETERS

Spacecraft-borne accelerometers are essentially identical to the seismometers discussed in Section 7.3.2. They are also closely related to spring-based weighing machines. We form very poor quality accelerometers ourselves by being able to feel the change in our weight when going up or down in lifts etc. Most readers will also know of accelerometers' actions from the way that the images shown on the screens of their digital cameras or portable computers rotate to remain upright in all positions of those devices.

The basic principle of an accelerometer (also called an inertial measurement instrument) is based upon a mass suspended from a spring. In free fall, the mass will be at its equilibrium position. When accelerated along the line of the axis of the instrument, the mass will be displaced from its equilibrium position and the degree of displacement may be sensed and so provide the output signal. Alternatively, the test mass may be held motionless and the forces required to do this then provide the measure of the acceleration.

It will be seen from this description of the principle of the accelerometer that a stationary accelerometer on the surface of Earth will register an acceleration of 1 g (9.8 m·s[−2]). Thus, accelerometers are also very closely related to the gravimeters used on the Earth's surface during geophysical surveys (except that these usually have their zero points set to the average value of Earth gravitational attraction and measure the deviations from that).

Accelerometers are a part of the instrumentation of many spacecraft – for monitoring the action of rockets, thrusters and gyroscopes, for triggering the release of parachutes in landers and for scientific purposes. All four of the Venus Pioneer atmosphere probes, for example, carried accelerometers. The large probe had four units to provide three-dimensional data (with two aligned to deceleration axis for redundancy) and there was one each on the small probes, also aligned to their deceleration axes. The test masses were permanent magnets held in place within solenoids. The currents through the solenoids required to maintain the masses' positions provided the measure of acceleration. The instruments had four switchable ranges enabling them to operate from 4×10^{-6} to 6000 m·s[−2]

---

* Not to be confused with waves at an interface between two fluids within a gravitational field (such as the sea's surface and the atmosphere) which are also sometimes called gravitational waves.
[†] Laser Interferometer Space Antenna and evolved Laser Interferometer Space Antenna.

with accuracies around ±0.005%. They were used to estimate the probes' velocities and hence heights within Venus' atmosphere during their descents and also provided data on atmospheric density.

The three Swarm spacecraft carry accelerometers based upon quartz cube test masses. The masses are free to move and rotate and their motions are monitored by four electrodes that measure the capacitance changes. The instruments are used to measure atmospheric drag and solar radiation pressure on the spacecraft.

The Viking Mars landers carried six accelerometers each – three to monitor events such as parachute deployment and three for scientific purposes such as measuring the decelerations during descent through Mars' atmosphere. The accelerometers were all identical instruments. They were also the same as those on the Mars Pathfinder and Mars Atmosphere and Volatile Evolution (MAVEN) spacecraft and earlier designs were carried by the Venus Express mission. These accelerometers used test masses mounted on flexible arms and sensed the departure of the masses from the equilibrium position capacitatively. The magnitude of the current in a coil mounted on the test mass and immersed in a strong magnetic field that was required to return the test mass to the equilibrium position provided the output signal. They had three switchable ranges and operated from 0.16 to 400 m·s⁻².

Galileo's Jupiter atmosphere probe used an accelerometer to trigger the release of its parachute. The triggering was delayed by 53 seconds and it later transpired that the accelerometer had been mis-wired. Fortunately, the Galileo parachute did eventually deploy. The Genesis solar wind sample and return spacecraft was not so lucky. It crashed on its return to Earth due to its accelerometers being installed backwards and its drogue parachute not deploying at all.

Additional accelerometers on Galileo were used to probe Jupiter's upper atmospheric structure from the rates of deceleration. They were similar to the Viking accelerometers and their dynamic range was from 3×10^{-5} to 4100 m·s⁻².

8.3 GRAVITY STUDIES

Studies of the nature of the gravitational fields of solar system objects from perturbations of spacecrafts' orbits date back to the first lunar orbiter, Luna-10, in 1966. The deviations from its expected orbit showed that the Moon's gravitational field must be more distorted than previously foreseen.

The causes of these distortions were found in 1968 when the orbits of the five Lunar Orbiter spacecraft were analysed in detail by NASA's Paul Muller and William Sjogren. The large circular maria on the moon – such as Mare Imbrium and Mare Orientale – were all found to have stronger gravitational fields than the lunar average (a positive gravitational anomaly). This was particularly puzzling since these areas are actually lower than the normal lunar surface and so they would be expected to have weaker gravitational fields because of the missing material.

The current explanation for these positive anomalies is that the material underlying the circular maria is denser than average. There is thus a concentration of mass under them (which has led to the phenomenon being labelled as a mascon). The reason for the increased density is that it is thought that the circular maria are the very largest examples of impact craters. The extraordinary violence of an impact capable of producing a crater with a diameter in excess of 1000 km must have melted and compacted the underlying material – resulting in the increased density. Mascons have also been found on Mercury and Mars.

Although almost any spacecraft's orbit can be measured and used to analyse gravity fields, the most accurate data comes from drag-free satellites. The basic principle of a drag-free satellite is to have a test mass that is shielded from all forces, except gravity, by an external enclosure. The test mass is unconnected to the enclosure and follows an orbit determined purely by gravity. The enclosure monitors the position of the test mass inside it and uses thrusters to keep that mass centralised. The movement of the enclosure thus also follows a path purely determined by gravity even though it is experiencing other forces.

The Earth orbiting Gravity Field and Steady-State Ocean Circulation Explorer (GOCE) operated as a drag-free satellite even though its orbit (just 250 km high) took it through the outer parts of the Earth's atmosphere. It used ion thrusters and was of an aerodynamic design to reduce the drag as much as possible. Rather than a single test mass, GOCE used three pairs of accelerometers (Section 8.2) oriented orthogonally to each other. These employed cubic test masses, levitated electrostatically, whose positions were monitored capacitatively. Each pair of accelerometers was separated by 500 mm and the difference between the accelerations that they measured determined the gradient of the Earth's gravity field.

The Earth's gravitational field has also been measured by a different approach by the two Gravity Recovery and Climate Experiment (GRACE) spacecraft. The two spacecraft (nicknamed Tom and Jerry) were in identical polar orbits separated by about 200 km. They used a K-band (18–27 GHz) microwave ranging system to measure their separation to an accuracy of about 10 μm. When the leading spacecraft encountered a positive gravitational anomaly it would speed up, so increasing the separation. It would then slow down as it passed the anomaly. When the second spacecraft encountered the same anomaly it would also speed up, shortening the separation and then also slow down again. Monitoring these speed and separation changes combined with accurate positions from the Global Positioning System (GPS) satellites and electrostatic accelerometers enabled the resulting gravity maps of Earth to be an improvement by a factor of around a thousand over the previous measurements. In the future, the GRACE-FO (GRACE-Follow-On) mission will be very similar to the GRACE mission, except that in addition to the microwave link it will try out a laser-based measuring system with a potential 20-fold increase in accuracy.

The two Gravity Recovery and Interior Laboratory (GRAIL) lunar orbiters mapped the gravitational field of the Moon in a manner similar to GRACE's mapping of Earth. GRAIL A (Ebb) and Grail B (Flow) were separated by between 175 and 225 km and orbited the Moon at a height of 50 km.

The Selenological and Engineering Explorer mission (SELENE – also known as Kaguya) comprised three spacecraft – the mothership, a relay (RSAT – relay satellite – also known as Okina) and a radio transmitter (VRAD – very long baseline radio source – also known as Ouna). RSAT relayed the main orbiter's signals to Earth when the orbiter was behind the Moon. The main gravitational field measurements were made by the same method as GRACE and GRAIL – by range and Doppler measurements between the main spacecraft and VRAD. In addition, the orbits of VRAD and RSAT were tracked very precisely using terrestrial radio interferometers (Section 2.2).

Terrestrial Doppler measurements of the transmissions from the asteroid probe, Dawn, can determine its velocity to about 0.1 mm·s^{-1}. This allows the asteroids' interiors to be probed and has shown that Ceres is differentiated into a dense core and lighter mantle and that the same structure is possible for Vesta. Similarly, the transmissions from the Mars orbiters, Mars Global Surveyor, Mars Odyssey and Mars Reconnaissance Orbiter have shown that Mars' crust is about 50–125 km thick.

An apparent gravitational anomaly was suspected from Doppler measurements of the transmissions from the Pioneer 10 and 11 spacecraft at distances from the Sun in excess of 20 AU. The slowing of their velocities was larger than expected by a very small amount (~9 × 10^{-10} m·s^{-2} – equivalent to a change in the distance from the Earth of about 400 km/year). Numerous theories to explain the anomalies were proposed in the early years of this century, including drastic changes to basic physics such as modifications to the law of gravity or to inertial behaviour. However, it now appears that the effect was not gravitational but due to anisotropic radiation forces on the spacecraft from the thermal emissions arising from the radioisotope energy generators on board each spacecraft.

APPENDIX A

Bibliography

A.1 A SELECTION OF OTHER BOOKS BY C.R. KITCHIN

(A.1.1) *Astrophysical Techniques*, 6th Edition, 2014, Taylor & Francis, ISBN: 978-1-4665-1115-6

(A.1.2) *Telescopes and Techniques – An Introduction to Practical Astronomy*, 3rd Edition, 2013, Springer, ISBN: 978-1-4614-4890-7

(A.1.3) *Exoplanets – Finding. Exploring, and Understanding Alien Worlds*, 2012, Springer, ISBN: 978-1-4614-0643-3

(A.1.4) *Galaxies in Turmoil – The Active and Starburst Galaxies and the Black Holes That Drive Them*, 2007, Springer, ISBN: 978-1-84628-670-4/978-1-4471-6126-4

(A.1.5) *Illustrated Dictionary of Practical Astronomy*, 2002, Springer, ISBN: 1-85233-559-9/978-1-85233-559-5

(A.1.6) *Solar Observing Techniques*, 2002, Springer, ISBN: 978-1-8523-3035-X/978-1-8523-3035-4

(A.1.7) *Photo-Guide to the Constellations – A Self-Teaching Guide to Finding Your Way around the Heavens*, 1998, Springer, ISBN: 3-5047-6203-5/978-3-5047-6203-4

(A.1.8) *Seeing Stars* (with R. Forrest), 1998, Springer, ISBN: 3-540-76030-X/978-3-540-76030-6

(A.1.9) *Optical Astronomical Spectroscopy*, 1995, IOP Publishing, ISBN: 0-7503-0345-X/978-0-7503-0345-3/0-7503-0346-8/978-0-7503-0346-0

(A.1.10) *Journeys to the Ends of the Universe – A Guided Tour of the Beginnings and Endings of Planets, Stars, Galaxies and the Universe*, 1990, Adam Hilger, ISBN: 0-7503-0037-X/978-0-7503-0037-7

(A.1.11) *Stars, Nebulae and the Interstellar Medium*, 1987, Adam Hilger, ISBN: 0-85274-580-X/978-0-85274-580-9/0-85274-581-8/978-0-85274-581-6

A.2 SPACECRAFT/SPACECRAFT MISSIONS/SPACECRAFT INSTRUMENTATION

A.2.1 Spacecraft – Orbiters, Landers and Rovers

(A.2.1.1) *NASA Mars Rovers Manual: 1997–2013 (Sojourner, Spirit, Opportunity and Curiosity) (Owners' Workshop Manual)*, 2013, David Baker, J H Haynes & Co Ltd, ISBN: 0857333704 /978-0857333704

(A.2.1.2) *Soviet Robots in the Solar System: Mission Technologies and Discoveries*, 2011, Wesley T. Huntress Jr. and Mikhail Ya Marov, Springer, ISBN: 1441978976/978-1441978974

(A.2.1.3) *Robotic Exploration of the Solar System*, Paolo Ulivi and David M. Harland, Springer. Part 1: *Golden Age, 1957–1982*, 2007, ISBN: 0387493263/ 978-0387493268

Part 2: *Hiatus and Renewal, 1983–1996*, 2009, ISBN: 0387789049/978-0387789040
Part 3: *Wows and Woes, 1997–2003*, 2012, ISBN: 0387096272/978-0387096278
Part 4: *The Modern Era 2004–2013*, 2014, ISBN: 1461448115/978-1461448112

(A.2.1.4)   *An Introduction to Space Robotics*, 2000, Alex Ellery, Springer, ISBN: 185233164X/978-1852331641

A.2.2 Mercury

(A.2.2.1)   *Planet Mercury: From Pale Pink Dot to Dynamic World*, 2015, David A. Rothery, Springer, ISBN: 3319121162/ISBN-13: 978-3319121161

A.2.3 Venus

(A.2.3.1)   *The Scientific Exploration of Venus*, 2014, Fredric W. Taylor, Cambridge University Press, ISBN: 1107023483/978-1107023482

A.2.4 Earth and Moon

(A.2.4.1)   *Moonshots & Snapshots of Project Apollo: A Rare Photographic History*, 2015, John Bisney and J.L. Pickering, University of New Mexico Press: ISBN: 0826355943/978-0826355942

(A.2.4.2)   *Monitoring Earth Resources from Aircraft and Spacecraft*, 2012, Robert N Colwell, CreateSpace Independent Publishing Platform, ISBN: 1478254939/978-1478254935

(A.2.4.3)   *The Moon in Close-up: A Next Generation Astronomer's Guide*, 2011, John Wilkinson, Springer, ISBN: 3642148042/978-3642148040

A.2.5 Mars

(A.2.5.1)   *Mars: A New View of the Red Planet*, 2016, Giles Sparrow, Quercus, ISBN: 1786483726/978-1786483720

(A.2.5.2)   *Seeing Like a Rover: How Robots, Teams, and Images Craft Knowledge of Mars*, 2015, Janet Vertesi, University of Chicago Press, ISBN: 022615596X/978-0226155968

(A.2.5.3)   *Mars Up Close: Inside the Curiosity Mission*, 2010, Marc Kaufman, National Geographic, ISBN: 142621278X/ISBN-13: 978-1426212789

(A.2.5.4)   *The Scientific Exploration of Mars*, 2009, Fredric W. Taylor, Cambridge University Press, ISBN: 0521829569/978-0521829564

A.2.6 Jupiter, Saturn, Uranus and Neptune

(A.2.6.1)   *The Cassini-Huygens Visit to Saturn: An Historic Mission to the Ringed Planet*, 2015, Michael Meltzer, Springer, ISBN: 3319076078/978-3319076072

(A.2.6.2)   *NASA Voyager 1 & 2 Owners' Workshop Manual 1977 Onwards (Including Pioneer 10 & 11)*, 2015, Christopher Riley, J H Haynes & Co Ltd, ISBN: 0857337750/978-0857337757

(A.2.6.3)   *Living among Giants: Exploring and Settling the Outer Solar System*, 2014, Michael Carroll, Springer, ISBN: 3319106732/978-3319106731

(A.2.6.4)   *Alien Seas: Oceans in Space*, 2013, Editors Michael Carroll and Rosaly Lopes, Springer, ISBN: 1461474728/978-1461474722

(A.2.6.5)   *NASA'S Voyager Missions: Exploring the Outer Solar System and Beyond*, 2nd Edition, 2008, Ben Evans and David M Harland, Springer, ISBN: 1852337451/978-1852337452

(A.2.6.6)   *Cassini at Saturn: Huygens Results*, 2007, David M. Harland, Springer ISBN: 038726129X/978-0387261294

(A.2.6.7)   *Jupiter Odyssey: The Story of NASA's Galileo Mission*, 2000, David M. Harland, Springer, ISBN: 1852333014/978-1852333010

A.2.7 Small Solar System Objects

(A.2.7.1) *Moons of the Solar System: From Giant Ganymede to Dainty Dactyl*, 2016, James A. Hall III, Springer, ISBN: 3319206354/978-3319206356

(A.2.7.2) *Pluto & Charon: The New Horizons Spacecraft at the Farthest Worldly Shores*, 2016, Codex Regius, CreateSpace Independent Publishing Platform, ISBN: 1534960740/ 978-1534960749

(A.2.7.3) *New Horizons to Pluto*, 2015, Mike Goldsmith, CreateSpace Independent Publishing Platform, ISBN: 1515200612/978-1515200611

(A.2.7.4) *Dawn of Small Worlds: Dwarf Planets, Asteroids, Comets*, 2015, Michael Moltenbrey, Springer, ISBN: 3319230026/978-3319230023

(A.2.7.5) *Asteroids: Relics of Ancient Time*, 2015, Michael K. Shepard, Cambridge University Press, ISBN: 110706144X/ISBN-13: 978-1107061446

(A.2.7.6) *Dwarf Planets and Asteroids: Minor Bodies of the Solar System*, 2014, Thomas Wm Hamilton, Strategic Book Publishing & Rights Agency, ISBN: 1628577282/978-1628577280

(A.2.7.7) *Moons of the Solar System*, 2013, Thomas Wm Hamilton, Strategic Book Publishing & Rights Agency, ISBN: 1625161751/978-1625161758

(A.2.7.8) *New Horizons: Reconnaissance of the Pluto-Charon System and the Kuiper Belt*, 2009, Editor Christopher Russell, Springer, ISBN: 0387895175/978-0387895178

(A.2.7.9) *The Hunt for Planet X: New Worlds and the Fate of Pluto*, 2009, Govert Schilling, Copernicus, ISBN: 0387778047/ISBN-13: 978-0387778044

(A.2.7.10) *Pluto and Charon: Ice Worlds on the Ragged Edge of the Solar System*, 2nd Edition, 2005, Alan Stern, Wiley, ASIN: B00SLTX5FE

A.3 PLANETARY STRUCTURES

A.3.1 Interiors

(A.3.1.1) *Interiors of the Planets*, 2009, A. Cook, Cambridge University Press, ISBN: 052110601X/ 978-0521106016

A.3.2 Surfaces

(A.3.2.1) *Volcanism on Io: A Comparison with Earth*, 2014, Ashley Gerard Davies, Cambridge University Press, ISBN: 110766540X/978-1107665408

(A.3.2.2) *Introduction to Planetary Geomorphology*, 2013, Ronald Greeley, Cambridge University Press, ISBN: 0521867118/978-0521867115

(A.3.2.3) *The Geology of Mars: Evidence from Earth-Based Analogs*, 2011, Mary Chapman, Cambridge University Press, ISBN: 0521206596/978-0521206594

(A.3.2.4) *Planetary Surface Processes*, 2011, H. Jay Melosh, Cambridge University Press, ISBN: 0521514185/978-0521861526

A.3.3 Atmospheres

(A.3.3.1) *Solar System Astrophysics: Planetary Atmospheres and the Outer Solar System*, 2nd Edition, 2014, Eugene F. Milone and William J.F. Wilson, Springer, ISBN: 1461490898/ 978-1461490890

(A.3.3.2) *Drifting on Alien Winds: Exploring the Skies and Weather of Other Worlds*, 2011, Michael Carroll, Springer, ISBN: 1441969160 / 978-1441969163

(A.3.3.3) *Planetary Atmospheres*, 2010, F.W. Taylor, Oxford University Press, ISBN: 0199547424/ 978-0199547425

A.3.4 Planets etc. as a Whole

(A.3.4.1) *Mars: An Introduction to Its Interior, Surface and Atmosphere*, 2014, Nadine Barlow, ISBN: 1107644879/978-1107644878

(A.3.4.2) *Titan: Interior, Surface, Atmosphere, and Space Environment*, 2014, Editors Ingo Müller-Wodarg, Caitlin A. Griffith, Emmanuel Lellouch and Thomas E. Cravens, Cambridge University Press, ISBN: 0521199921/978-0521199926

(A.3.4.3) *Fundamental Planetary Science: Physics, Chemistry and Habitability*, 2013, Jack J. Lissauer and Imke de Pater, Cambridge University Press, ISBN: 052161855X/978-0521618557

(A.3.4.4) *Planetary Geology: An Introduction*, 2nd Edition, 2013, Claudio Vita-Finzi and Andrew Dominic Fortes, Dunedin Academic Press, ISBN: 1780460155/978-1780460154

A.4 PHYSICS AND MATHEMATICS

A.4.1 Electricity and Magnetism

(A.4.1.1) *Electricity and Magnetism*, 2016, Elisha Gray, CreateSpace Independent Publishing Platform, ISBN: 1540385329/978-1540385321

(A.4.1.2) *Electricity and Magnetism*, 2015, Munir Nayfeh and Morton Brussel , Dover Publications Inc., ISBN: 0486789713/978-0486789712

(A.4.1.3) *Electricity and Magnetism*, 3rd Edition, 2013, Edward M. Purcell and David J. Morin, Cambridge University Press, ISBN: 1107014026/978-1107014022

A.4.2 Spectroscopy

See also reference A.1.1, Chapter 4 and reference A.1.9.

(A.4.2.1) *Introduction to Spectroscopy*, 5th Edition, 2014, James R. Vyvyan, Donald L. Pavia, Gary M. Lampman and George S. Kriz, Brooks/Cole, ISBN: 128546012X/978-1285460123

(A.4.2.2) *Astronomical Spectroscopy for Amateurs*, 2011, Ken M. Harrison, Springer. 2011, ISBN: 1441972382/978-1441972385

(A.4.2.3) *Reading the Lines in Stellar Spectra*, 2007, Keith Robinson, Springer. 2007, ISBN: 0387367861/978-0387367866

A.4.3 Particle Physics

See also reference A.1.1, Chapter 1.

(A.4.3.1) *The Particle Zoo: The Search for the Fundamental Nature of Reality*, 2016, Gavin Hesketh, Quercus, ISBN: 1784298670/978-1784298678

(A.4.3.2) *Neutrinos in High Energy and Astroparticle Physics*, 2015, José Wagner Furtado Valle and Jorge Romao, Wiley, ISBN: 3527411976/978-3527411979

(A.4.3.3) *Modern Particle Physics*, 2013, Mark Thomson, Cambridge University Press, ISBN: 1107034264/978-1107034266

(A.4.3.4) *Particle Physics, Dark Matter and Dark Energy*, 2011, David Chapple, Abramis, ISBN: 1845494776/978-1845494773

A.4.4 Solid State Physics

(A.4.4.1) *Solid State Physics: An Introduction*, 2nd Edition, 2015, Philip Hofmann, Wiley, ISBN: 3527412824/ISBN-13: 978-3527412822

(A.4.4.2) *Solid State Physics*, 2nd Edition, 2014, H.E. Hall and J.R. Hook, Wiley, ISBN: 8126551372 /978-8126551378

(A.4.4.3) *The Oxford Solid State Basics*, 2013, Steven H. Simon, Oxford University Press, ISBN: 0199680779/978-0199680771

(A.4.4.4) *Introduction to Solid State Physics*, 8th Edition, 2012, Charles Kittel, WIPL, ISBN: 8126535180/978-8126535187

A.4.5 Telescopes

See also reference A.1.1, Chapters 1, 2 and 3 and reference A.1.2.

(A.4.5.1)   *Getting Started in Radio Astronomy: Beginner Projects for the Amateur*, 2013, Steven Arnold, Springer, ISBN: 1461481562/978-1461481560

(A.4.5.2)   *Radio, Submillimeter, and X-Ray Telescopes*, 2013, Editor N.G. Basov, Springer-Verlag, ASIN: B01BQWKOAG

(A.4.5.3)   *Amateur Telescope Making*, 2013, Editor Stephen F. Tonkin, Springer, ISBN: 1852330007 /978-1852330002

(A.4.5.4)   *The Principles of Astronomical Telescope Design*, 2010, Jingquan Cheng, Springer, ISBN: 1441927859/978-1441927859

(A.4.5.5)   *The Radio Sky and How to Observe It*, 2010, Jeff Lashley, Springer, ISBN: 144190882X/ 978-1441908827

(A.4.5.6)   *The Telescope: Its History, Technology, and Future*, 2007, Geoff Andersen, Princeton University Press, ISDBN: 0691129797/978-0691129792

A.4.6 Fourier Transforms

(A.4.6.1)   *A Student's Guide to Fourier Transforms: With Applications in Physics and Engineering*, 3rd Edition, 2011, J.F. James, Cambridge University Press, ISBN: 0521176832/ 978-0521176835

(A.4.6.2)   *Introduction to Fourier Transforms in Physics*, 2007, K.A.I.L. Wijewardena Gamalath, Cambridge University Press, ISBN: 052170054X/978-0521700542

Timeline of Solar System Investigations

SOLAR SYSTEM DISCOVERIES – TIMELINE

The topics included in this timeline are a personal selection of what I see as significant milestones in the development of our understanding of the solar system (the Sun is generally excluded for reasons already discussed). Doubtless some of the choices are idiosyncratic and doubtless other astronomers would add entries which I have omitted. In particular I have usually only listed the first natural satellite(s) of planets and asteroids to be found and have only included those spacecraft and their discoveries that have made significant advances in knowledge (spacecraft names are in italic and their entries generally appear under their year of launch). However anyone who reads (or even just skims through) the timeline should acquire a good synoptic view of both our present ideas about the solar system and of how those ideas were developed and evolved.

TABLE B.1   The Era of Unaided Visual Observations (Prehistory to ~1609 AD)

Year	Discovery, event, fact, idea, possibility or wild speculation
Pre-recorded history	Someone notices that although most objects in the sky keep the same relative positions, a few move around. Five star-like heavenly bodies (or occasionally seven because the dawn and evening apparitions of Mercury and Venus were sometimes thought to be different objects) are identified that move regularly with respect to the thousands of fixed objects in the sky. The name, planets, given to these moving bodies derives from the Greek *asters planetai* (wandering star). We now name these planets Mercury, Venus, Mars, Jupiter and Saturn.
c. 2000 BC	Chinese observations show that Jupiter returns to the same part of the sky about every 12 years (i.e. – what we now call its sidereal orbital period).
c. 1700–1500 BC	Possible original date of the Venus Tablet of Ammisaduqa – a cuneiform tablet recording observations of Venus, now known through copies made around 710 BC.
c. 1100 BC	The angle between the Earth's orbital and equatorial planes (the obliquity of the ecliptic) known with reasonable accuracy in both India and China.
c. 530 BC	Pythagoras (c. 570–c. 495 BC) proposes the idea of a spherical Earth, although he may have adopted the concept from earlier Greek philosophers.
c. 400 BC	Philolaus (c. 470–c. 385 BC) suggests that the Earth, planets, Sun, Moon and an anti- or counter-earth (also known as the Antichthon) all move around a 'central fire'. This is the first suggestion that the Earth is a planet and is moving through space (although not around the Sun).
c. 380 BC	Plato (428 or 427–348 or 347 BC) teaches that all celestial objects move along perfect circles at uniform speeds – an idea that will bedevil astronomy for two millennia. The resulting models for the solar system had to contain numerous deferents, epicycles, eccentrics and equants in order to account for the observed non-uniform movements. Ptolemy's model of the solar system, for example, contained 80 circles in order to give a reasonably accurate prediction of the positions of the planets. Even Copernicus' heliocentric model retained 34 such circles. Not until Kepler introduced the concept of elliptical orbits in 1609 was Plato's dictum finally refuted.
c. 350 BC	Heracleides of Pontus (c. 390–c. 310 BC) suggests that the heaven's apparent rotation is actually due to the counter rotation of the Earth. He may also have proposed a Sun-centred model for the solar system or at least suggested that Mercury and Venus go around the Sun. Similar claims or beliefs are also attributed at around this time to Hicetas of Syracuse (c. 400–c. 335 BC) and Ecphantus (fourth century BC) although it is possible that these were characters in Heracleides' writings, not real people.
c. 350–c. 330 BC	Aristotle (384–322 BC), basing his ideas upon observation of the natural world, advocates the Earth-centred model of the universe. Such is his influence upon subsequent thinking that no alternative is seriously considered by mainstream European workers for nearly two millennia. From the shape of the Earth's shadow on the Moon during a lunar eclipse, he also argues that the Earth is spherical.
c. 270 BC	Aristarchus of Samos (c. 310–230 BC) proposes a Sun-centred model for the solar system, with the planets in their correct order, and explains the lack of observed parallax motion for the stars by suggesting that they are so far away that the size of the Earth's orbit is negligible in comparison. He also makes the first good estimate (of 60 times the radius of the Earth) for the Earth–Moon distance and suggests that the distance to the Sun is 19 times larger than this, although the true value is 400 times larger.
c. 200 BC	Eratosthenes estimates the Earth's circumference at 252,000 stadia. If there are 185 m to the stade (the most likely conversion), this gives the circumference as 46,000–47,000 km (cf. the present day value, through the poles, of 40,008 km). He made the measurement by noting that at the summer solstice the Sun was overhead at Syene (present day Aswan), but at the same time was 1/50 of a circle away from the zenith at Alexandria, some 5000 stadia north of Syene.

(Continued)

TABLE B.1 (*Continued*)   The Era of Unaided Visual Observations (Prehistory to ~1609 AD)

Year	Discovery, event, fact, idea, possibility or wild speculation
c. 135–130 BC	Hipparchus, when compiling his catalogue of ~850 star positions, notices that they differ by about 2° from those measured by Timocharis some 150–160 years earlier. In particular, Timocharis recorded αVir (Spica) as being 8° west of the position of the autumnal equinox (the point on the sky where the Sun passes from the northern hemisphere to the southern) whilst Hipparchus found it to be 6° to the west. The change arose from the motion of the position of the equinox around the sky and this in turn from the wobble in space of the direction of the Earth's rotational axis – a phenomenon that we now call precession. Hipparchus' data suggested that the equinox would move right the way around the sky in about 27,000 years – the modern value is ~25,750 years.
129 BC?	Hipparchus used a solar eclipse that was total at one place and partial at another to estimate the Earth–Moon distance as being around 550–600 times the separation of the two places. It is uncertain which eclipse was involved and from where the observations were made, but a total solar eclipse was visible from the Hellespont (the Dardanelles today) in 129 BC. Another likely place for an observation then would have been Alexandria. The Hellespont is about 1000 km north of Alexandria, suggesting that Hipparchus determined the Earth–Moon distance as about 550,000–600,000 km (cf. actual value 384,400 km).
	Total solar eclipses, however, also occurred in southern Egypt in 136 BC and 127 BC. Alexandria and somewhere south of Syene (modern Aswan) are thus alternative sites for the observations (also with a North–South separation of about 1000 km).
c. 100 BC ↑ **BC (BCE)** **AD (CE)** ↓	Lo Hsia Hung advocates a moving Earth as an explanation for the seasons.
c. 150	Claudius Ptolemy completes his *Great Treatise* – now better known as the *Almagest*. It is a geocentric mathematical model of the solar system which, by using various geometrical constructions such as epicycles, deferents and equants, is able to predict the positions of the five known planets and of the Moon reasonably accurately. Since it continues in widespread use until well into the seventeenth century (i.e. ~1500 years), it must be counted as the most successful theory ever produced by science, despite its erroneous physical basis.
c. 160	The Syrian writer, Lucian of Samosata, writes the first book to speculate about space travel and alien life. Intended as a satire, his *True History* relates how a group of travellers are carried to the Moon on a giant water spout or whirlwind. The Moon, planets, stars and even the Sun are inhabited by fantastic humanoid type creatures, such as the Vulture Dragoons, Garlic Fighters and Flea Archers. The creatures living on the Moon and Sun are engaged in a war when the travellers arrive.
c. 490	Aryabhata (476–550), who was probably born in what is now central India, proposes that the Earth rotates and may have espoused a Sun-centred model for the solar system.
c. 800–c. 1200	In *The Adventures of Bulukiya*, a story from *A Thousand and One Nights*, the hero's quest for immortality lead him to travel across space to many other inhabited worlds, some larger than the Earth.
c. 830	Al Mamun has his astronomers re-measure the value of the obliquity of the ecliptic, obtaining the value 23° 35′ (its actual value at that time would have been about 23° 35′ 29″, the modern value 23° 26′ 21″).
c. 840	Al Farghani (c. 800–c. 870) determines distances to the Sun, Moon and planets. His values, based upon an Earth radius of about 5000 km, are reasonably good for the Moon and for Mars at opposition, but are otherwise much too small.

(*Continued*)

TABLE B.1 (*Continued*)   The Era of Unaided Visual Observations (Prehistory to ~1609 AD)

Year	Discovery, event, fact, idea, possibility or wild speculation
1032	Abū Alī Sīnā (Avicenna – c. 980–1037) observes a dark spot on the Sun which he takes to be Venus in transit, although a naked-eye sunspot is also a possibility. If correct, this is the first recorded observation of a planetary transit.
1044	The book *Wu Jing Zong Yao* (Collection of the most important military techniques) was written by Zeng Gongliang, Ding Du and Yang Weide and published in China. It describes a 'Long Serpent' weapon. This appears to be a rocket-propelled arrow and may have been in use up to 150 years earlier. It is the first recorded use of a rocket (gunpowder had been known and used in China from about 850 AD onwards).
c. 1070	Abū Ishāq Ibrāhīm al-Zarqālī (Arzachel, 1029–1087) notes that Mercury's orbit is oval in shape.
1281	Qutb al-Din al Shirazi (1236–1311) publishes his *Nihayat al-idrak fi dirayat al-aflak* (The Latest Knowledge in the Science of Planetary Orbits) in which the planetary orbits are modelled using uniform circular motions.
c. 1350	Ibn al-Shatir (1304–1375) produces a revised version of the Ptolemaic model of the universe that eliminates the need for eccentrics and equants. Although Earth-centred, al-Shatir's system is mathematically equivalent to Copernicus' Sun-centred solar system model.
1377	In his *Traité du ciel et du monde* (Treatise on heaven and the world), Nicole Oresme queries the prevailing Aristotelian idea of a fixed Earth at the centre of the universe by suggesting that the apparent rotation of the heavens might actually be due to the Earth rotating the other way.
1440	Nicholas of Cusa proposes an infinite, centre-less universe, a moving Earth and non-uniform, non-circular motions for the planets in his *De Docta Ignorantia* (The Ignorance of the Learned).
1543	*De revolutionibus orbium coelestium* (On the Revolutions of the Heavenly Spheres) published (though written from 1514 onwards) by Nicolaus Copernicus in which the Sun-centred universe is realistically proposed for the first time.
~1570–1601	Tycho (Tyge) Brahe makes large numbers of highly accurate observations of the positions of planets and stars. He used naked eye instruments but by taking great care in their construction and in his observing methods (e.g. conceived the idea of determining and correcting for errors in the instruments) was able to achieve positional accuracies of around 3 arc-minutes – far, far better than any previous work. His data were used by Johannes Kepler to deduce his three laws of planetary motion (1609).
1572	Tycho (Tyge) Brahe cannot measure any parallax for the new star (supernova) observed in Cassiopeia, thus showing that is distance must be at least 1000 times that of the Moon and demonstrating that changes could occur within the 'immutable spheres' of the planets and stars.
1577	Tycho (Tyge) Brahe cannot measure any parallax for the comet observed this year, thus showing that it is not a phenomenon occurring within the Earth's atmosphere.

TABLE B.2   The Era of the Telescope (~1609 – 4 October 1957)

Year	Discovery, event, fact, idea, possibility or wild speculation
c. 1609	The invention of the telescope – probably by Hans Lipperhey (or Lippershey) – although several other opticians may have made the same discovery at much the same time or even centuries earlier, including Jacob Metius and Zacharias Jansen (~1609), Leonard Digges (~1540–1559) and Bishop Robert Grosteste (~1220 to ~1235).
1609–1621	Johannes Kepler establishes a firm scientific footing for Copernicus' Sun-centred model of the universe by discovering the three laws of planetary motion. The three laws are the following:

1. The orbit of a planet is an ellipse with the Sun occupying the position of one of the (two) foci of the ellipse.
2. The planet's speed around its orbit varies in such a way that the line joining the planet and the Sun sweeps out equal areas in equal times (i.e. the planet moves faster when closer to the Sun than when it is further away).
3. The radius of a planet's orbit cubed is proportional to its orbital period squared (strictly the relationship is with half the length of the longest axis of the ellipse – known as the semi-major axis – but the orbits for most solar-system planets are near enough circular for this approximation to be useful).

Year	Discovery, event, fact, idea, possibility or wild speculation
1610	Galileo Galilei publishes his *Siderius Nuncius* (Starry Messenger) outlining his early telescopic observations of the heavens. The discoveries announced in this work, such as Jupiter's four satellites (given their current names of Io, Europa, Ganymede and Callisto by Simon Marius after the lovers of Zeus – Jupiter is the Roman name for Zeus), together with later observations, such as of the phases of Venus, convince him of the correctness of Copernicus' Sun-centred model of the universe. Not everyone though, especially the ecclesiastical authorities, concurs, leading to his subsequent trial for heresy.
1610	Galileo Galilei observes Saturn and interprets his observations as showing three planets in a line with the central object about three times the size of the ones on each side. The true structure (Saturn's rings) was determined in 1655 by Christiaan Huygens.
c. 1611	Johannes Kepler uses the idea of space travel to the Moon in his book *Somnium* (Dreams – not published until 1635). The space traveller, Duracotus, is able to observe the Earth moving through space, thus confirming (in fiction at least) Copernicus' heliocentric solar system model.
c. 1630	Godefroy Wendelin determines the distance of the Sun from the Earth to be about 95 million km – about 60% of the true value.
1631	Pierre Gassendi makes the first unequivocal observation of a planetary transit across a star's disk. Gassendi observes Mercury as it crosses the face of the Sun following predictions of the event by Kepler.
1639	Jeremiah Horrocks makes the first observations of transit of Venus by using a small telescope to project the solar image onto a white sheet of paper.
1644	In his *Principia Philosophiae* (Principles of Philosophy), René Descartes proposes his vortex theory for the motion of the planets.
1655	Christiaan Huygens discovers Saturn's natural satellite, Titan. He used the then common practice of establishing his precedence in the discovery, but not announcing the discovery itself, by sending an anagram of the announcement to various friends. The discovery was thus not published until the following year when Huygens translated the anagram's meaning as '*Saturno luna sua circunducitur diebus sexdecim horis quatuor*' (A moon revolves around Saturn in 16 days and 4 hours). He simply named the satellite '*Luna Saturni*' (Saturn's moon). The name Titan was proposed in 1847 by Sir John Herschel.

(Continued)

TABLE B.2 (*Continued*)   The Era of the Telescope (~1609 – 4 October 1957)

Year	Discovery, event, fact, idea, possibility or wild speculation
1655	Christiaan Huygens observes that the odd structure of Saturn (first noted by Galileo in 1610) is a thin ring centred upon a spherical planet. He again uses an anagram to establish his precedence in the discovery. He provided the translation of the anagram in his book, *Systema Saturnium* (The Saturn System, published 1659) as "*Annulo cingitur, tenui, plano, nusquam cohaerente, ad eclipticam inclinato*" (It is encircled by a ring, thin and flat, nowhere touching, inclined to the ecliptic).
1655	Giovanni Cassini is the first to see Jupiter's Great Red Spot (Robert Hooke described a spot on Jupiter a year earlier, but this was probably not the Great Red Spot).
1671–1673	Giovanni Cassini (working from Paris) and Jean Richer (working at Cayenne) measure the parallax of Mars. They found a value of 25″, suggesting that Mars' distance from the Earth was then about 55,000,000 km. Using Kepler's laws enables the Earth–Sun distance (the Astronomical Unit, AU) also to be estimated and Cassini and Richer found a value of ~140,000,000 km for this (actual value 149,600,000 km).
	John Flamsteed obtains similar values at the same time by taking advantage of the Earth's rotation which, over a period of 6 hours or so, changed the position of the his observatory in Derby by some 6000 km.
1687	Isaac Newton's *Philosophiae Naturalis Principia Mathematica* (Mathematical Principles of Natural Philosophy – better known simply as the *Principia*) is published, setting out Newton's law of gravity and laws of motion and deriving Kepler's three laws of planetary motion from the former. The orbital motions of all planets (solar system or exoplanets) are governed by these laws.
1705	Edmond Halley predicts (correctly) that the comet of 1682 will reappear in 1758.
1716	Edmond Halley suggests that the distance of the Earth from the Sun (the Astronomical Unit) may be found accurately by making observations of Venus' (or Mercury's) parallax during its transit of the Sun (having tried to use the method by observing the 1676 transit of Mercury from St. Helena, but without success).
1728	James Bradley discovers the aberration of starlight – which is the first experimental proof that the Earth is orbiting the Sun.
1734	Emanuel Swedenborg proposes in his *Opera Philosophica et Mineralia* that the solar system was formed from a disk of material that had condensed out of the primordial matter – essentially the modern model for the formation of planetary systems (and many other astronomical objects). The model was developed further in 1755 by Immanuel Kant. In 1796, apparently independently, Pierre Laplace proposes the idea again as his nebular hypothesis.
1761	The transit of Venus is observed from many sites in order to try to determine the value of the Earth–Sun distance via Halley's 1716 method. The accuracy of the method was limited by the difficulty of accurate timings due to the blurring of the edge of the planet by Venus' atmosphere.
1766	Johann Titius notices a simple relationship between the relative distances of the known planets from the Sun. Now known as the Titius–Bode rule (Johann Bode publicised the relationship), it gives (roughly) the semi-major axes of the planets' orbits, in AU, via the formula $$a = 0.4 + 0.3 \times 2^m$$ where a is the semi-major axis of a planet's orbit and m takes the values successively of $-\infty$, 0, 1, 2 for Mercury, Venus, Earth and Mars and 4 and 5 for Jupiter and Saturn. The value $m = 3$ did not give the semi-major axis of any then known planet; however, the discovery of Uranus (1781 – $m = 6$ – *q.v.*) and of Ceres in 1801 which neatly filled the $m = 3$ gap seemed to give the relationship a high significance. Subsequent discoveries (Neptune, Pluto, Eris, etc.) have not fitted the rule and it is now generally regarded as an interesting, but not very significant, coincidence.

(*Continued*)

TABLE B.2 (*Continued*)   The Era of the Telescope (~1609 – 4 October 1957)

Year	Discovery, event, fact, idea, possibility or wild speculation
1769	The transit of Venus is observed again from many sites in order to try to determine the value of the Earth–Sun distance via Halley's 1716 method.
1781	First 'new' planet discovered (i.e. the discovery of a planet not known since pre-historical times). Uranus was discovered through visual telescopic observations by William Herschel.
1787	Sir William Herschel discovers Uranus' satellites Titania and Oberon. They were only given these names, though, in 1852 (by Sir John Herschel).
1798	Sir William Herschel publishes an account of observations of distortions to the shape of Uranus which he interprets as rings around the planet. The earliest such observation occurred in 1787 when he noted "When I see it most distinctly it appears to have double, opposite points . . . perhaps a double ring . . .". The discovery of the rings, however, is more generally accepted as being by James Elliot in 1977.
1800	Discovery of infrared radiation by Sir William Herschel.
1801	Discovery of ultraviolet radiation by Johann Ritter.
1801	Discovery of the first asteroid (now classed as a dwarf planet), Ceres, by Giuseppe Piazzi. Named after the Roman goddess of agriculture, it fits nicely into the $m = 3$ gap in the Titius–Bode rule (1766).
1802	Heinrich Olbers discovers Pallas (named for the Greek goddess Pallas Athena). For many years classed as the second largest asteroid, with the re-classification of Ceres as a dwarf planet, it is now the largest asteroid (though slightly less massive than Vesta).
1802	William Wollaston observes seven narrow dark regions within the solar spectrum. Wollaston regarded these as the natural boundaries between the colours, but we now know them to be absorption lines produced by some of the chemical elements.
1807	Heinrich Olbers discovers Vesta (named for the Roman goddess of the home). For many years classed as the third largest asteroid, with the re-classification of Ceres as a dwarf planet, it is now the second largest (after Pallas) and the most massive asteroid.
1813	William Moore in his *A Treatise on the Motion Rockets* publishes the first derivation of the rocket equation: $$\Delta v = v_e \ln\left[\frac{m_i}{m_f}\right]$$ where Δv is the change in velocity for a rocket with an initial mass, m_i, final mass, m_f and exhaust velocity, v_e.
1838	Stellar parallax (the change in position in the sky of a nearby star relative to the positions of very distant stars as the Earth moves around its orbit) is observed for the first time. Freidrich Bessel measures the parallax of 61 Cyg as 0.3136″ – giving its distance as 10.6 light years (just 6% smaller than the modern value). This is the second experimental demonstration (after the aberration of starlight – 1728) of the Earth's orbital motion.
1842	Christian Doppler publishes his *Über das farbige Licht der Doppelsterne und einiger Gestirne des Himmels* (On the coloured light of double stars and of some of the other stars of heaven) in which he proposes that the different colours sometimes observed for stars in a double or binary system arise from the different velocities of the two stars along the line of sight. The suggestion for the colours of stars was incorrect but the change in wavelength of light towards longer wavelengths when the light emitter moves away from us and to shorter wavelengths when the light emitter moves towards us is real. The effect, now known as the Doppler shift, enables astronomical objects' velocities along the line of sight to be measured.

(*Continued*)

TABLE B.2 (*Continued*)   The Era of the Telescope (~1609 – 4 October 1957)

Year	Discovery, event, fact, idea, possibility or wild speculation
1846	Discovery of Neptune by Johann Gottfried Galle working at the Berlin observatory and using predictions of its position by Urbain Le Verrier. Le Verrier (and also John Couch Adams) based their predictions upon the observed deviations of Neptune from its predicted orbit. Pre-discovery observations were (possibly) recorded by Galileo (1613), Jérôme Lalande (1795) and Sir John Herschel (1830).
1846	Neptune's satellite Triton discovered by William Lassell. It was named by Camille Flammarion in 1880 for the son of the Greek god Poseidon (known to the Romans as Neptune).
1847	Sir James Herschel proposes the names Mimas, Enceladus, Tethys, Dione, Rhea, Titan and Iapetus (based upon the Greek mythological Titans – Saturn is the Roman name for the Titan, Cronus) for the then known satellites of Saturn.
1849	Edouard Roche determines the limiting distance from a primary object of a secondary orbiting object (satellite), that is held together purely by gravity, before the latter will be disrupted by the tidal effects of the primary.
1851	Léon Foucault's pendulum at the Paris observatory is the first experimental proof of the Earth's rotation. The pendulum is free to oscillate in any direction and acts like a gyroscope. It continues to swing in the same direction in space (i.e. relative to the distant galaxies) while the Earth rotates beneath it. The pendulum constructed by Foucault at the Panthéon in Paris a few weeks later demonstrated the effect to anyone who cared to watch it for a few minutes. This pendulum, which was 67 m in length, had an apparent rotation of 11° per hour (not 15° because of Paris' latitude of 49° – a Foucault pendulum apparently rotates at 15° at the poles and does not rotate with respect to the Earth at the equator).
1852	Sir James Herschel proposes the names Ariel, Umbriel, Titania and Oberon for the then known satellites of Uranus (after magical spirits/fairies in Sir William Shakespeare's *A Midsummer Night's Dream* and Alexander Pope's *The Rape of the Lock* – Uranus was the Greek god of the sky).
1859	James Clerk Maxwell awarded the Adams prize for demonstrating that if Saturn's rings were solid or liquid, then they would be torn apart by tides and so they must consist of a myriad of small solid satellites in co-planar orbits.
1859	Gustav Kirchhoff and Robert Bunsen showed that some elements, when heated strongly in a gas flame, could emit bright spectrum lines at the same wavelengths as some of the dark spectrum lines found by William Wollaston in 1802 in the Sun's spectrum. By this means, in the next few years Kirchhoff showed that 16 elements found on the Earth were also a part of the Sun's composition.
1865	Jules Verne's *De la Terre à la Lune* (From the Earth to the Moon) presages realistic exploration of planets and other solar system objects with some surprisingly good estimates of what will be needed, including the correct calculation of the escape velocity (11 km/s) required to leave the Earth.
1866	Giovanni Schiaparelli suggests that meteor showers result from the trail of debris left behind by comets as Earth passes through the comet's orbit.
1868	Discovery of helium by Sir Norman Lockyer and Jules Janssen from observations of a solar chromospheric yellow emission line whose wavelength could not be matched with the lines arising from any element then known on the Earth. This was the first time that a chemical element had been discovered extraterrestrially (it was eventually found on the Earth in 1895 by Per Cleve and Nils Langlet).
1870	William Parsons detects infrared radiation from the Moon.
1874	The transit of Venus is observed from many sites in order to try to determine the value of the Earth–Sun distance via Halley's 1716 method.

(*Continued*)

TABLE B.2 (*Continued*) The Era of the Telescope (~1609 – 4 October 1957)

Year	Discovery, event, fact, idea, possibility or wild speculation
1877	Mars' satellites Phobos and Deimos discovered by Asaph Hall. They are named, at the suggestion of Henry Madan, for the sons of Ares (known to the Romans as Mars).
1877	Giovanni Schiaparelli observes dark lines on Mars which he calls channels. Unfortunately, the mistranslation of the Italian for channels, '*canali*', into the English *canals* is taken to imply a completely unintended artificial origin for the features. Schiaparelli's canali are now known to be optical illusions; the affair, though, sparked wholesale but generally ill-informed speculation about Martian and other forms of alien life which continues to the present day.
1882	The transit of Venus is observed from many sites in order to try to determine the value of the Earth–Sun distance via Halley's 1716 method. The observations from this transit and from that of 1874 give a value of the AU of 149,590,000 km – very close to the modern determinations.
1895	Konstantin Tsiolkovsky proposes the idea of a 'space elevator', a tower at least 35,800 km high on the equator, which could be used to launch spacecraft into orbit, because, at its top, the Earth's rotation gives it a velocity equal to the orbital velocity at that distance (see also geostationary orbit – 1945)
1903	Konstantin Tsiolkovsky published *Исследование мировых пространств реактивными приборами* (The Exploration of Cosmic Space by means of Reaction Devices). In this and in subsequent publications he showed that rockets would work in space (i.e. that they did not need the atmosphere to push against) and discussed multi-stage rockets, liquid rocket propellants, space suits and colonisation of the planets.
1905	Forest Ray Moulton and Thomas Chrowder Chamberlin propose that the planets formed from solar material ejected from the Sun by tidal forces from another star combined with solar prominence activity. Although later discarded, the theory also included the idea of the condensation of planetesimals and their coalescence into larger objects which is still a part of the current theory of planet formation.
1906	The first Trojan asteroid, Achilles, associated with Jupiter, is discovered by Max Wolf.
1907	Josep Comas i Solà suggests that Titan may have an atmosphere because he thought he could see limb darkening on the satellite. The existence of the atmosphere was confirmed in 1944 by Gerard Kuiper.
1919	James Jeans proposes in his *Problems of Cosmogony and Stellar Dynamics* that the planets were formed from a strand of material tidally wrenched from the proto-Sun by a close encounter with another star. Since close passages between two stars are extremely rare this would make the existence of exoplanetary systems extremely rare as well. Subsequent work, however, has shown that Jeans' proposal is most unlikely to lead to the formation of planets and so contraction from a nebulosity is the currently favoured process.
1926	Robert Goddard builds and successfully launches the first liquid-fuelled rocket.
1930	Discovery of Pluto by Clyde Tombaugh. For a long time classed as the ninth planet of the solar system, Pluto is now designated as a dwarf planet and is probably just one of the larger members of the Kuiper belt objects (KBOs).
1930	Edison Pettit and Seth Nicholson measure the lunar surface temperature by using a thermocouple to detect its infrared emission. They find the sub-solar point of the full Moon to be at about 407 K (134 °C).
1932	Ernst Öpik suggests that a cloud of icy objects extending out to 50,000 AU, surrounds the inner solar system and is the source of the long period comets. Jan Oort independently makes the same suggestion in 1950. The region is now known as the Öpik–Oort cloud or rather more commonly (and unfairly) as the Oort cloud.
1933	Karl Jansky makes the first observation of radio emission from an extraterrestrial object (the centre of the Milky Way Galaxy).

(Continued)

TABLE B.2 (*Continued*)   The Era of the Telescope (~1609 – 4 October 1957)

Year	Discovery, event, fact, idea, possibility or wild speculation
1935	Captains Albert Stevens and Orvil Anderson become the first people in history to see the curvature of the Earth's horizon, when the Explorer II balloon lifts them to an altitude of 22,000 m.
1942	A V2 A4 rocket reaches an altitude of 85–90 km (53–56 miles) thus becoming the first man-made object to reach outer space (based upon the United States' definition of an 'Astronaut' as someone who has 'travelled to an altitude of 50 miles or more' – however, the Kármán line at 100 km (62 miles) is a more scientifically based definition since it is the point in the Earth's atmosphere above which aerodynamic vehicles would have to travel faster than the orbital velocity in order to fly).
1942	James Hey makes the first detection of radio emissions from the Sun.
1943	Kenneth Edgeworth suggests the presence of a concentration of icy objects (analogous to the asteroid belt) which is the source of the short period comets in the region of space beyond Neptune. The suggestion is repeated by Gerard Kuiper in 1951, Alastair Cameron in 1962, Fred Whipple in 1964 and Julio Fernández in 1980. The region is now known as the Kuiper belt and the first object belonging to it – (15760) 1992 QB1 – was discovered in 1992 by David Jewitt and Jane Luu. Pluto is now also regarded as being a member of the belt and well over a thousand KBOs have now been found.
1944	Carl Friedrich von Weizsäcker revises Kant's nebular hypothesis of the formation of planetary systems to more or less its present form. The modern nebular hypothesis, while not without its problems still, is now generally accepted as the probable way in which planets are created.
1944	A V-2 rocket reaches an altitude of 189 km (117 miles). This is above the Kármán line which is the modern definition of the beginning of outer space.
1944	Gerard Kuiper proves Titan to have an atmosphere by detecting the presence of methane. The existence of the atmosphere had previously been suggested in 1907 by Josep Comas i Solà because of Titan's apparent limb darkening.
1945	Sir Arthur C. Clarke points out that a spacecraft at an altitude of 35,790 km would orbit the Earth in 23 hours 56 minutes. This is the Earth's rotational period (not 24 hours – because of the Earth's motion around its orbit). The spacecraft, if in an equatorial orbit, would therefore remain fixed above the same point on the Earth. Such geostationary spacecraft are now much used for meteorological and communication purposes.
1946	The first detection of the Moon by radar.
1949	Herbert Friedman detects solar x-ray emissions using a Geiger counter carried on a V2 rocket.
1955	Kenneth Franklin and Bernard Burke detect decametric radio emission from Jupiter. The observations imply that Jupiter has a general magnetic field with a strength of about 1 mT (cf. 30–60 μT for the Earth's magnetic field).

TABLE B.3   The Era of Space Exploration (4 October 1957 – to date)

Year	Discovery, event, fact, idea, possibility or wild speculation
1957	4 October – *Sputnik 1* launched – the first successful Earth orbiter.
1958	*Explorer 1* launched leading to the discovery of the Van Allen radiation belts,
1959	*Luna-2* becomes the first spacecraft to land on another celestial object when it impacts the lunar surface on September 14.
1959	*Luna-3* obtains the first images of the far side of the Moon. The Moon's gravitational field was used to send the spacecraft back to Earth – the first of many gravitational assist manoeuvres used by spacecraft to save on fuel requirements.
1961	The *Explorer-11* spacecraft detects gamma rays from astronomical sources.
1961	The first confirmed detection of Venus by radar. Walter Victor and Robert Stevens of JPL used two 26 m radio dishes at Goldstone operating at 2.4 GHz and determined a value of 1.49599×10^8 km for the astronomical unit – 130,000 km larger than the previously accepted value of 1.49467×10^8 km.
1962	The first successful visit to another planet occurred when *Mariner 2* was launched on a fly–by mission to Venus. The fly–by occurred on 14 December 1962, some 35,000 km from Venus and six experiments made various measurements of the planet.
1964	*Ranger 7* (launched in July) impacts the Moon's surface, sending back images as it approached that have a resolution (at best) that is less than half a metre.
1964	*Mariner 4* launched in November. It completed a successful fly-by of Mars in July 1965. The spacecraft sent back 21 images which show numerous impact craters as well as other data.
1965	Radar observations by Robert Dyce and Gordon Pettengill using the Arecibo radio telescope show that Mercury's rotation period is 58.65 days – two-thirds of its orbital period of 88 days. Previously, the planet's rotation period had been thought to equal its orbital period.
1965	*Pioneer 6*, the first of four spacecraft studying the solar wind and magnetic fields and cosmic rays, launched (December).
1966	First successful soft landing on the Moon by the *Luna-9* spacecraft with 27 images of the lunar surface being transmitted back to Earth.
1966	The first successful lunar orbiter. *Luna-10*, makes measurements of gamma rays, magnetic fields and micro-meteorites, operating for nearly 2 months.
1967	*Venera 4* successfully aerobrakes into Venus' atmosphere and during the subsequent parachute descent through the atmosphere measures its composition, temperature and pressure. It may also be the first spacecraft to have landed on another planet although *Venera 3* (whose communications systems failed) may have crashed into Venus in 1966.
1969	*Apollo 11* makes the first manned landing on the Moon and returns with over 20 kg of lunar soil samples to the Earth.
1970	*Venera 7* makes a partially successful soft landing on Venus and measures the surface temperature as 475°C.
1970	*Luna-16* obtains soil samples from the Moon robotically and returns them to the Earth.
1970	*Luna-17* soft lands on the Moon and successfully deploys the first robotic rover, Lunokhod 1, which travels for 10 months over the lunar surface, covering 10.5 km and obtaining images and samples.
1971	*Mariner 9* becomes the first Mars orbiter and obtains over 7000 images of the Martian surface.
1971	The Soviet Union's *Mars 2* orbiter and lander launched towards Mars. Although the lander crashes, it becomes the first man-made object to reach the surface of Mars.
1971	The Soviet Union's *Mars 3* orbiter and lander launched towards Mars. Although the lander reaches the surface successfully, it only broadcasts back to Earth for 15 seconds – possibly due to the effects of an intense dust storm on Mars at the time. Nonetheless it was the first successful soft landing on Mars.

(Continued)

TABLE B.3 (*Continued*)　The Era of Space Exploration (4 October 1957 – to date)

Year	Discovery, event, fact, idea, possibility or wild speculation
1972	*Pioneer 10* launched to fly-by Jupiter in 1973. The spacecraft obtained some 500 images and its other instruments measured magnetic fields, charged particles, temperatures and micro-meteorites.
1972	Launch of the first of many *Landsat* Earth resources spacecraft.
1973	*Pioneer 11* launched. It flies by Jupiter in 1974 and Saturn in 1979 sending back images and measuring magnetic fields, charged particles, temperatures and micro-meteorites.
1973	*Luna-21* soft lands on the Moon and successfully deploys the second robotic rover, Lunokhod 2, which travels for 4 months over the lunar surface, covering 39 km and obtaining images and samples.
1973	*Mariner 10* launched. It flies by Venus (once – 1974) and Mercury (three times – 1974, 1974 and 1975). Some 2800 images were obtained and measurements made of magnetic fields and charged particles.
1975	The first images obtained of Venus' surface by *Venera 9* – a spacecraft which comprised both the first successful orbiter of Venus and its lander.
1975	*Viking 1* launched to go into orbit around Mars in 1976 and to make the second soft landing on Mars' surface. The orbiter continued operating until 1980 and the lander until 1982. As well as images from orbit and the surface, soil analyses were undertaken, including searching for organic compounds.
1977	James Elliot and his team discover the rings of Uranus. The rings, however, may have been seen by Sir William Herschel as early as 1787 (account published 1798 – *q.v.*).
1977	*Voyager 2* launched in August to fly by Jupiter (1979), Saturn (1981), Uranus (1986) and Neptune (1989). The spacecraft continues to operate at the time of writing making measurements of plasma density and temperature and entering the heliosheath in 2007. *Voyager 1* was launched in September to fly by Jupiter (1979) and Saturn (1980).
1978	James Christy discovers Pluto's satellite Charon. The name originated from that of Christy's wife, Charlene. It is also, however, the name of the ferryman of the dead across the rivers Styx and Acheron in the Greek underworld and who is associated with the god, Hades (known to the Romans as Pluto).
1978	The *ICE* (International Cometary Explorer) spacecraft launched to study the solar wind, the Earth's magnetosphere, cosmic rays and solar flares. In 1985, it passed though the tail of comet Giacobinni–Zinner about 8000 km away from the nucleus.
1978	Launch of the *Nimbus 7* Earth-observation spacecraft which *inter alia* monitors the atmospheric concentrations of ozone. In 1984, it confirms the existence of the Antarctic ozone hole.
1984	*Vega 1* launched to fly by Venus (1985) and comet Halley (1986). The Venus part of the mission also included a lander and a balloon which floated at a height of about 50 km making measurements of the winds and temperature and pressure and which travelled a third of the way around Venus before communications from it were lost.
1985	The *Giotto* spacecraft launched to fly by (at a distance of 600 km from the nucleus) comet Halley in 1986.
1986	The *Giotto, Vega 1, Vega 2* and *Suisei* space craft make relatively close approaches (600–150,000 km) to comet Halley obtaining images and measuring gas and particle compositions, dust densities and temperatures.
1989	The *Magellan* spacecraft launched to orbit Venus from 1990 to 1994. Its synthetic aperture radar maps 98% of the planet to a resolution (at best) of ~100 m.

(*Continued*)

TABLE B.3 (*Continued*)   The Era of Space Exploration (4 October 1957 – to date)

Year	Discovery, event, fact, idea, possibility or wild speculation
1989	*Galileo* spacecraft launched to orbit Jupiter in 1995 and to release an atmospheric probe. The probe collected data on Jovian weather and atmospheric composition for nearly an hour. The orbiter continued to obtain images and measure magnetic fields, particles and dust properties until 2003. On its journey to Jupiter *Galileo* made the first fly-bys of asteroids and obtained close-up images – Gaspra (closest approach 1600 km, 1991) and Ida (closest approach 2400 km, 1993). Ida was found to have a small (~1 km) moon, now named Dactyl.
1990	Solar x-rays scattered from the Moon's surface detected by *ROSAT*.
1991	Using the 70 m Goldstone radio telescope, Martin Slade finds material with a high radar reflectivity within permanently shadowed craters at Mercury's north polar region. The material is suggested to be water ice since the temperature in regions of permanent shadow on Mercury is 100 K or less. The *MESSENGER* spacecraft confirmed the presence of the water in 2014.
1991	The EGRET (Energetic Gamma Ray Experiment Telescope) instrument on board the *Compton Gamma Ray Observatory* detects γ-rays from the Moon that have originated during cosmic-ray interactions with its surface.
1992	David Jewitt and Jane Luu discover the first object in a planetary-type orbit further out from the Sun than Pluto. Given the designation (15760) 1992 QB1, it is about 160 km across and takes 290 years to complete a single circuit of its 44 astronomical unit orbit. It is now classed, along with Pluto, Charon and over thousand similar entities detected since 1992, as a Trans-Neptunian Object (TNO) or KBO.
1996	Launch of *NEAR Shoemaker* to orbit the Asteroid Eros in 2000 and to touch down on its surface in 2001. This was the first soft landing on an asteroid. As well as imaging, the spacecraft measured compositions, mineralogy and magnetic fields. On its way to Eros, the spacecraft also flew by the asteroid Mathilde in 1997.
1996	Launch of *Mars Pathfinder* which landed on Mars in 1997 and deployed a small vehicle (rover) called Sojourner that carried an alpha proton x-ray spectrometer to analyse the soil composition. The rover travelled about 100 m over the Martian surface in total.
1996	David McKay et al. announce the discovery of biosignatures, including possible bacterial remains, within meteorite ALH 84001. The meteorite is thought to have come from Mars, being blasted off the planet during a large meteorite impact. Subsequent work has failed to confirm the claim, with all of the features explainable by inorganic processes.
1997	Launch of the *Cassini–Huygens* spacecraft to orbit Saturn (*Cassini* – 2004–2017) and to land on Titan (*Huygens* – 2005). *Cassini* carried imagers, synthetic aperture radar, spectrometers and dust analysers and magnetometers. *Huygens* made an aero-brake and parachute entry into Titan's atmosphere and measured its structure, winds and composition and the physical properties of Titan's surface.
1999	Launch of *Stardust* to collect samples from comet Wild 2 and return them to Earth. The samples were returned successfully in 2006. *Stardust* went on to fly by comet Tempel 1 at a distance of 180 km in 2011.
2001	Launch of *Genesis*, a sample-return mission to study the solar wind. The sample container crash-landed upon its return in 2004, but most of the mission's objectives were still able to be achieved.
2003	Launch of *Mars Exploration Rover A* which landed on Mars in 2004 and deployed a small vehicle (rover) called Spirit that carried cameras and six instruments for analysing the Martian soil. Spirit travelled a total distance of 7.7 km before becoming trapped in sand. The mission ended in 2010.
2003	Launch of *Mars Exploration Rover B* which landed on Mars in 2004 and deployed a small vehicle (rover) called Opportunity that carried cameras and six instruments for analysing the Martian soil. Opportunity is still operating at the time of writing and has travelled a total distance of 40.3 km so far.

(*Continued*)

TABLE B.3 (*Continued*)   The Era of Space Exploration (4 October 1957 – to date)

Year	Discovery, event, fact, idea, possibility or wild speculation
2003	Discovery of Eris (announced in 2005) by Mike Brown et al. The object is the largest dwarf planet known within the solar system and a member of the Kuiper belt. Its name comes from the Greek goddess of strife.
2003	Launch of *Hayabusa*, the first sample-return mission to an asteroid (Itokawa). The spacecraft successfully collected samples in 2005 and returned them to Earth in 2010. A separate small lander that was designed move around by hopping over the surface failed to reach the asteroid.
2003	Discovery of Sedna – possibly a dwarf planet and possibly also the first known member of the inner Oort cloud. The discovery was made by Mike Brown et al. and named for the Inuit goddess of the sea. Its orbital period is ~11,400 years and at aphelion it is ~936 AU from the Sun. It is just under 1000 km in diameter.
2004	Launch of the *Rosetta* spacecraft to orbit comet 67P/Churyumov–Gerasimenko (67P) in 2014. Instruments to measure compositions, temperatures, particles and dust as well as imaging cameras were carried. A lander, called *Philae*, failed to anchor to the surface and bounced to a site where its solar panels were mostly in shadow; however, some measurements were still able to be made. The spacecraft was crashed into the comet in September 2016 to end the mission.
2004	Launch of the *MESSENGER* spacecraft to orbit Mercury in 2011 (after several fly-bys of Venus and Mercury). The objectives of the mission were to study the geology of the planet, along with its composition and internal structure. In 2014, it confirmed the existence of water ice within the regions in permanent shadow near Mercury's north pole (1991). The spacecraft was crashed into Mercury in 2015 to end the mission.
2004	Mike Brown and his team discover Haumea (although the first public announcement was made by José Moreno in 2005); it is named for the Hawaiian goddess of fertility and childbirth.
2005	Haumea's satellite, Hi'iaka discovered by Mike Brown and his team (named for the daughter of Haumea).
2005	Mike Brown and his team discover Makemake (named for the creator of humanity in the pantheon of the Rapa Nui peoples).
2005	Eris' satellite, Dysnomia, discovered by Mike Brown and his team (named for the daughter of Eris).
2005	Launch of the *Deep Impact* spacecraft to fly by comet Tempel 16 months later and comet Hartley 2 in 2010. During the fly-by of Tempel 1, the mother ship launched a smaller, self-guided, 'missile' to impact the comet at 10.2 km/s in order to observe the effects of the impact (a crater) and to analyse the ejecta to determine the comet's composition.
2006	The International Astronomical Union (IAU) revises the definitions of planets, asteroids and satellites, controversially 'demoting' Pluto to a dwarf planet alongside Ceres, Eris, Makemake, Haumea and (possibly) Sedna.
2006	Launch of the *New Horizons* spacecraft to fly by Pluto in 2015 and the KBO MU69 in 2019.
2007	Launch of the *Dawn* spacecraft to orbit and survey Vesta and Ceres. The spacecraft successfully observed Vesta between July 2011 and September 2012 and entered an orbit around Ceres in March 2015.
2008	Launch of *Chandrayaan-1* spacecraft, a lunar orbiter and impactor that made the first definitive detection of the presence of water molecules on the Moon.
2009	Jamie Elsila et al. announce that glycine, the simplest amino acid and one of the building blocks of proteins and DNA, has been discovered in samples recovered from comet Wild 2 by the *Stardust* spacecraft. The discovery lends support to the possibility of life originating in space.
2010	*Chang'e-2* orbiter reaches the Moon and later (2012) successfully flies by asteroid Toutatis.
2011	Launch of the two *GRAIL* (Gravity Recovery and Interior Laboratory) spacecraft to orbit the Moon and to map its gravitational field in detail.

(Continued)

TABLE B.3 (*Continued*)   The Era of Space Exploration (4 October 1957 – to date)

Year	Discovery, event, fact, idea, possibility or wild speculation
2011	Launch of the *Juno* spacecraft – a Jupiter orbiter that went into orbit around the planet in July 2016 – mission continuing at the time of writing.
2011	Launch of the *Mars Science Laboratory* – a Mars lander and rover (Curiosity) – mission continuing at the time of writing; Curiosity has travelled a distance of 8.6 km so far.
2012	The LOFAR (low-frequency array for radio astronomy) radio array commences operations later obtaining resolved images of Jupiter's radiation belts at150 MHz.
2012	Launch of the two *van Allen probes* to study the Earth's radiation belts – mission continuing at the time of writing.
2013	*Chang'e-3* lander and rover successfully reaches the Moon.
2013	Launch of the *Mars Orbiter Mission* spacecraft – mission continuing at the time of writing.
2013	Launch of the Mars Orbiter *MAVEN* spacecraft – mission continuing at the time of writing.
2013	Launch of the *LADEE* (Lunar Atmosphere and Dust Explorer).
2014	Launch of *Hayabusa* 2, a sample-return mission to the asteroid Ryugu. Return to Earth is expected in 2020.
2016	Launch of *ExoMars Trace Gas* orbiter and *EDM Schiaparelli* lander to Mars. Schiaparelli crashed onto the Martian surface and was destroyed.
2016	Launch of *OSIRIS-REX*, an asteroid sample-return mission to asteroid 101955 Bennu, by NASA.
2017	Probable launch of *ADM-Aeolus* by ESA to study Earth's winds.
2017	Planned launch of *Chang'e-5* – a lunar sample-return mission by China.
2017	Possible lunar landers and rovers to be launched by a private consortia (Astrobotic Technology, Barcelona Moon Team, SpaceIL and Team Indus) competing for the Google Lunar X Prize.
2017	Possible lunar lander launch by Moon Express.
2018	Planned launch of the *James Webb Space Telescope*.
2018	Planned launch of *BepiColombo* – a pair of orbiters to study Mercury.
2018	Planned launch of *Chandrayaan-2* – a lunar orbiter, lander and rover.
2018	Planned launch of *Chang'e 4* lunar lander and rover. To be the first lander on the Moon's far side.
2018	Planned launch of *InSight* – a Mars lander.
2018	Planned launch of *SELENE-2* – a lunar orbiter, lander and rover by Japan.
2018	Planned launch of the *Lunar Polar Hydrogen Mapper* by NASA.
2018	Possible launch of the first node of the International Lunar Network – a linked network of lunar landers.
2018	Possible launch of *Red Dragon*, a Mars lander, by the private company, Space-X.
2018	Possible launch of several lunar missions by NASA.
2018–2020	Possible launch of a Venus orbiter and atmosphere probe (balloons) by India.
2019	*New Horizons* due to fly by the KBO 2014 Mu69 in January.
2019–2022	Possible launch of *Prospector-1* – an asteroid lander by Deep Space Industries.
2020	Planned launch of *ExoMars 2020* – a Mars rover by ESA.
2020	Planned launch of *AIDA* to land on the Asteroid Didymos and observe the impact of *DART* (planned to launch in 2021) with the small (150 m) satellite of the asteroid by NASA.
2020	Possible launch of *Chang'e-6* – a lunar sample-return mission by China.
2020	Possible manned lunar landing by India.
2020	Possible manned lunar landing by Japan.
2020	Possible Mars rover mission by China.
2020	Possible Mars orbiter, *Hope*, by the UAE.
2020	Possible Mars orbiter, *Mangalyaan 2*, by India.

(*Continued*)

TABLE B.3 (*Continued*) The Era of Space Exploration (4 October 1957 – to date)

Year	Discovery, event, fact, idea, possibility or wild speculation
2020/2021	Possible launch of *Luna 28* and *Luna 29*, a Lunar rover and a (separate) mission to return samples by RKA. (Federal'noye kosmicheskoye agentstvo Rossii [Russian Federal Space Agency])
2021	Planned launch of *Psyche*, an asteroid orbiter, to asteroid 16 Psyche by NASA.
2021	Possible launch of *DAVINCI*, a Venus atmosphere probe, by NASA.
2021	Possible launch of *NEOCam*, (Near Earth Asteroid Cam) an NEO (near earth object) probe, by NASA.
2021	Possible launch of *VERITAS*, a Venus orbiter, by NASA.
2022	Possible launch of *Mars Sample Return*, a Mars orbiter, lander and sample-return mission by ESA.
2022	Possible lander and sample-return mission to Phobos or Deimos by Japan.
2022	Possible launch of *Mars 2022*, a Mars orbiting communications and mapping satellite to link with Martian landers and rovers.
2022	Possible Launch of *FLEX* (Fluorescence Observer) to map Earth's vegetation by ESA.
2023	Planned launch of *Lucy*, a Jovian Trojan asteroid probe, by NASA.
2023	Possible launch of a manned version of the *Orion* module into a lunar orbit.
2024	Possible lander and sample-return mission to Mars by NASA.
2024	Possible launch of *Venus in situ*, a Venus atmosphere probe and/or lander, by NASA.
2024	Possible launch of *Phobos-Grunt-2*, a Phobos lander and sample-return mission by RKA.
2024	Possible launch of a crowd-funded lunar probe – *Lunar Mission One*.
2025	Possible launch of *Luna-Glob* – a lunar lander and orbiter pair of spacecraft.
2025	Possible launch of *Luna-Resurs* – a lunar lander.
2025	Possible launch of *Venera-D*, a Venus orbiter by RKA.
2025–2030	Possible missions to Neptune or Uranus being considered by several agencies.
2025–2030	Possible launch of one or more lunar missions by South Korea.
2026	Possible launch of *Mars-Grunt*, a Mars lander and sample-return mission by RKA.
2026	Possible launch of a lunar mission by North Korea.
2026?	Possible launch of the *Turbulence Heating Observer* by ESA.
2029–2030	Possible Saturn/Titan mission – orbiter, lander, balloon??
2030–2040	Possible launch of the *Venus Long Life Surface Package* by ESA.

Details of the Spacecraft, Rockets, Observatories and Other Missions Mentioned in This Book

Mission/Instrument*	Agency	Launch date	Current status†	Target‡	Mission type	Instruments§	Page(s)
ACE (Advanced Composition Explorer)	NASA	1997 August	Still active	Inter-Planetary Medium (IPM)	Lagrange L1 point	SEPICA	148
ADM-Aeolus	ESA	2017?	Under construction	Earth winds	Earth orbiter	ALADIN	84
Aerobee	USAF	1962 June	Ended 1962 June	Moon	Sub-orbital	X-ray detector	96, 150
Aeronomy of Ice in the Mesosphere	NASA	2007 April	Still active	Earth – noctilucent clouds	Earth orbiter	Dust detector	265
AGILE (Astro-Rivelatore Gamma a Immagini Leggero)	ASI	2007 April	Still active	Earth (TGFs) Universe	Earth orbiter	GRID MCAL SuperAGILE	98, 104 115, 131 132
AIDA (Asteroid Impact and Deflection Assessment) comprising AIM (Asteroid Impact Mission – orbiter and lander) and DART (Double Asteroid Re-direction Test – impactor).	ESA/NASA	2020/2021	Planned	Asteroid Didymos	Orbiter lander and impactor	Bi-static radar MASCOT-2	205, 231, 245
AIM – see AIDA and Aeronomy of Ice in the Mesosphere							
Akatsuki (Japanese for 'Dawn') – also known as Venus Climate Orbiter	JAXA	2010 May	Still active	Venus	Venus orbiter	NIR camera	23
ALMA	ESO	2011	Still active	Universe	Terrestrial instrument	Microwave observations of comets	70

(Continued)

Mission/Instrument*	Agency	Launch date	Current status†	Target‡	Mission type	Instruments§	Page(s)
Apollo 11	NASA	1969 July	Ended 1969 July	Moon	Manned lunar landing	Radar Seismometer	79, 238, 247
Apollo 12	NASA	1969 November	Ended 1969 November	-	Manned lunar landing	-	233, 238
Apollo 13	NASA	1970 April	Ended 1970 April	-	Manned lunar landing attempt	-	238
Apollo 14	NASA	1971 January	Ended 1971 February	Moon	Manned lunar landing	Seismometer	238, 244, 247
Apollo 15	NASA	1971 July	Ended 1971 August	Moon	Manned lunar landing	X-ray fluorescence Seismometer	108, 204, 237, 247
Apollo 16	NASA	1972 April	Ended April 1972	Earth	Manned lunar landing	FUCVS X-ray fluorescence Seismometer SSB	15, 16, 41, 59, 108, 163, 204, 237, 247
Apollo 17	NASA	1972 December	Ended 1972 December	Earth	Manned lunar landing	Ground penetrating radar Seismometer SSB	163, 204, 237, 238, 245, 247
Arecibo	SRI/USRA/UMET	1963	Still active	Planets, asteroids, universe	Terrestrial instrument	Radar, 305 m fixed parabolic dish	63, 77, 79, 88
ARGOS (Advanced Research and Global Observation Satellite)	AFRL/NRL/STP	1999 February	2003 July	Earth	Earth orbiter	Dust detector	264
ARM	NASA	2025?	Planned	Asteroid	Asteroid lander + placing a boulder into lunar orbit	-	241

(Continued)

Mission/Instrument*	Agency	Launch date	Current status†	Target‡	Mission type	Instruments§	Page(s)
ARTEMIS P1 (=THEMIS B)	NASA	2007 February	Still active	Interplanetary medium	Lunar orbit	Magnetometer Top hat analyser	175, 242, 250, 256
ARTEMIS P2 (=THEMIS C)	NASA	2007 February	Still active	Interplanetary medium	Lunar orbit	Magnetometer Top hat analyser	175, 242, 250, 256
Astrosat	ISRO	2015 September	Still active	Universe	Earth orbiter	CZTI LAXPC SXT	102, 107, 115, 131, 156
ATHENA (Advanced Telescope for High Energy Astrophysics)	ESA	2028?	Planned	Universe	Lagrange L2 point	HXI WFT Micro calorimeter	116, 125
BARREL	NASA	2013 January	Last campaign 2016 August	X-rays	Multiple high altitude balloons	NaI scintillators	104
Beagle-2 lander – see Mars Express Orbiter							
Bepi-Colombo – named after Giuseppe (Bepi) Colombo – comprising the Mercury Planetary Orbiter (MPO) and the Mercury Magnetospheric Orbiter (MMO)	ESA/ JAXA	2018 October?	Under construction	Mercury	Two Mercury orbiters	MIXS-C MIXS-T MORE MSASI	43, 54, 58, 90, 123
Black Brandt IX	NASA	2012 December	Ended 2012 December	Demonstration	Sub-orbital	Lobster-eye collimator	123
CALIPSO	NASA/ CNES**	2006 April	Still active	Earth clouds	Earth orbiter	Lidar	87

(Continued)

Mission/Instrument*	Agency	Launch date	Current status†	Target‡	Mission type	Instruments§	Page(s)
Cassini and Huygens – named after Giovanni Cassini	NASA/ESA/ASI††	1997 October	Ended 2017 September (de-orbited into Saturn) Huygens landed on Titan in 2005 January	Saturn	Saturn orbiter and lander	Atmospheric pressure Atmospheric temperature CIRS Density Dust detector GCMS INMS NAC Radar Radio occultation Refractive index RPWS UVIS VIMS WAC	13, 21, 46, 53, 55, 57, 67, 84, 86, 89, 91, 177, 190, 221, 222, 229, 231, 238, 251, 252, 254, 255, 258, 263, 265,, 296
Chandra X-Ray Observatory (= AXAF, Advanced X-ray Astrophysics Facility) – named after Subrahmanyan Chandrasekhar	NASA	1999 July	Still active	Comet LINEAR Pluto Universe	Earth orbiter	ACIS HRC	106, 107, 132, 205
Chandrayaan-1 – Chandrayaan is Sanskrit for 'Moon vehicle'.	ISRO	2008 October	Ended 2009 August	Moon	Lunar orbiter	C1XS MIP	113, 189, 205
Chandrayaan-2 – Chandrayaan is Sanskrit for 'Moon vehicle'.	ISRO	2018	Planned	Moon	Lunar orbiter, lander and rover	CLASS	113, 118, 304
Change 3 – Change is the Chinese Moon goddess + the Yutu (Jade Rabbit) rover	CNSA	2013 December	Still active. Yutu ceased functioning in Mar 2015	Moon	Lunar lander and rover	LUT VNIS APXS	12, 210, 215, 234, 242, 244

(Continued)

Mission/Instrument*	Agency	Launch date	Current status†	Target‡	Mission type	Instruments§	Page(s)
Chang'è 4	CNSA	2018?	Under construction	Moon, Jupiter and Saturn?	Lunar far side soft lander plus relay communications satellite at Earth–Moon L2 point	Low-frequency radio receiver	71, 204
Chang'è 5	CNSA	2017?	–	Moon	Lunar lander, sample and return	–	238
Clementine	BMDO‡‡/ NASA	1994 January	1994 June (rocket malfunction)	Moon and Asteroid Geographos	Lunar orbiter, asteroid flyby	Lidar LWIC UV and visible camera NIR camera	11, 22, 86
Cluster II – FM5	ESA	2000 August	Still active	Interplanetary medium	Earth orbiter	WDB WHISPER EDI Magnetometer	70, 87, 257, 263
Cluster II – FM6	ESA	2000 July	Still active	Interplanetary medium	Earth orbiter	WDB WHISPER EDI Magnetometer	70, 87, 257, 263
Cluster II – FM7	ESA	2000 July	Still active	Interplanetary medium	Earth orbiter	WDB WHISPER EDI Magnetometer	70, 87, 257, 263
Cluster II – FM8	ESA	2000 August	Still active	Interplanetary medium	Earth orbiter	WDB WHISPER EDI Magnetometer	70, 87, 257, 263

(Continued)

Mission/Instrument*	Agency	Launch date	Current status†	Target‡	Mission type	Instruments§	Page(s)
Compton Gamma Ray Observatory – named after Arthur Compton	NASA	1991 April	Ended 2000 June	Earth and Moon, universe	Earth orbiter	BATSE EGRET	96, 98, 104, 105, 116, 127
Constellation X	NASA	–	Mission cancelled 2008	X-ray observatory	–	–	116, 132
CONTOUR (COMet Nucleus TOUR)	NASA	2002 July	Ended 2002 August	Comet	Solar orbit intended	NGIMS	190
Cryosat-2	ESA	2010 April	Still active	Earth ice caps	Earth orbiter	SIRAL	86
Curiosity – See Mars Science Laboratory							
DART – see AIDA							
Dawn	NASA	2007 September	Still active	Ceres/Vesta	Vesta and Ceres orbiter	FC GRaND Gravity measurements	20, 46, 176, 197, 198, 225, 271
Deep Impact	NASA	2005 January	Ended 2005 July (fly-by 2013 September)	Comet Tempel-1	Flyby and impactor	–	107, 121, 205, 206
Deep Space 1	NASA	1998 October	Ended 2001 December	Asteroid 9969 Braille, Comet 19P/Borelly	Flyby	Ion propulsion test	169, 176
DSCOVR	NASA	2015 February	Still active	Earth	Lagrange L1 point	EPIC	11
Einstein (= High Energy Astrophysics Observatory-HEAO-2, HEAO-B) – named after Albert Einstein	NASA	1978 November	Ended 1981 April (transmitter failure)	Jupiter, universe	Earth orbiter	HRI IPC MPC	96, 100, 106, 120, 132, 134

(Continued)

Mission/Instrument*	Agency	Launch date	Current status†	Target‡	Mission type	Instruments§	Page(s)
eLISA (evolved Laser Interferometer Space Antenna)	ESA	2034?	–	Gravitational wave observatory	Heliocentric	–	268
Equator-S	DARA	1997 December	Ended 1998 May	Magnetosphere	Earth orbiter	EDI	263
ERS-1	ESA	1991 July	Ended 2000 March	Earth surface	Earth polar orbiter	ATSR	56
ERTS-1 – see Landsat-1							
Europa Multiple Flyby (= Europa Clipper)	NASA	2023?	Planned	Europa (Jupiter)	Jupiter orbiter + (perhaps) Europa lander	E-THEMIS ICEMAG MASPEX PIMS REASON SUDA UVS	46, 47, 56, 57, 88, 193, 214, 245, 258, 265
EUVE (Extreme Ultra-Violet Explorer)	NASA	1992 June	Ended 2002 January	Comet Shoemaker-Levy/Jupiter collision, universe	Earth orbiter	EUV/x-ray telescopes	106
Exo Mars 2020 Lander	ESA/RKA	2020 July	Planned	Mars	Mars lander and rover	GCMS LDMS PanCam WAC	54, 204, 208, 212, 213, 216, 219, 222, 223, 225, 236, 251, 254
ExoMars Rover – see ExoMars 2020 Lander							

(Continued)

Mission/Instrument*	Agency	Launch date	Current status†	Target‡	Mission type	Instruments§	Page(s)
ExoMars Trace Gas Orbiter and Schiaparelli – named after Giovanni Schiaparelli	ESA/RKA	2016 February	Still active Lander crashed 2016 October End 2021 April planned	Mars	Mars orbiter and lander	CaSSIS ACS – MIR ACS – NIR ACS – TIRVIM MicroARES Atmospheric pressure Atmospheric temperature DECA	19, 36, 46, 93, 198, 230, 244, 259, 251, 252, 254
EXOSAT (European X-ray Observatory Satellite)	ESA	1983 May	Ended 1986 May	Universe (lunar occultations)	Earth orbiter	EUV/x-ray telescopes	132
Explorer 1	NASA	1958 February	Ended 1958 May (re-entered 1970 March)	van Allen belts	Earth orbiter	Geiger counter	139, 141
Explorer 2	NASA	1958 March	Launch failure	van Allen belts	Earth orbiter	Geiger counter	141
Explorer 3	NASA	1958 March	Ended 1958 June–May (re-entered 1958 June)	van Allen belts	Earth orbiter	Geiger counter	139, 141
Explorer 4	NASA	1958 October	Ended 1958 October May (re-entered 1959 October)	van Allen belts	Earth orbiter	Geiger counter	141
Explorer 11	NASA	1961 April	Ended 1961 November	Universe	Earth orbiter	γ ray detector	96
Explorer 38 (= RAE-A)	NASA	1968 July	Ended 1968 September	Solar system radio sources, universe	Earth orbiter	Super heterodyne radio receivers	71
Explorer 61 (= MagSat)	NASA	1979 October	1980 June (re-entry)	Earth	Earth orbiter	Magnetometer	258

(Continued)

Mission/Instrument*	Agency	Launch date	Current status†	Target‡	Mission type	Instruments§	Page(s)
Fermi Gamma Ray Space telescope (= GLAST – Gamma ray Large Area Space Telescope) – named after Enrico Fermi	NASA	2008 June	Still active	Earth (TGFs), lunar γ ray emissions, universe	Earth orbiter	GBM LAT	96, 98, 105, 151
FIREBIRD IA	Montana State and New Hampshire universities	2013 December	Ended 2014 June	Space weather	Earth orbiter	SSD electron detector	162
FIREBIRD IB	Montana State and New Hampshire universities	2013 December	Ended 2014 June	Space weather	Earth orbiter	SSD electron detector	162
FIREBIRD IIA	Montana State and New Hampshire universities	2015 January	Ended 2015 July	Space weather	Earth orbiter	SSD electron detector	162
FIREBIRD IIB	Montana State and New Hampshire universities	2015 January	Ended 2015 July	Space weather	Earth orbiter	SSD electron detector	162

(Continued)

Mission/Instrument*	Agency	Launch date	Current status†	Target‡	Mission type	Instruments§	Page(s)
Fobos-Grunt (or Phobos-Ground)	Roscosmos	2011 November	Failed to leave Earth orbit – ended 2012 January	Phobos	Phobos sample-return mission	MicrOmega	212, 239
Galileo – named after Galileo Galilei	NASA	1989 October	Ended 2003 September (atmosphere probe 1995 December)	Jupiter	Jupiter orbiter and atmospheric entry probe	Atmospheric pressure, Atmospheric temperature, NIMS, SSI	14, 15, 48, 189, 250–252, 265, 269
Gemini 6A	NASA	1965 December	Ended 1965 December	Earth	Manned Earth orbiter	Radar	79
Gemini 7	NASA	1965 December	Ended 1965 December	Earth	Manned Earth orbiter	Target for Gemini 6A	79
Gemini Observatory	AURA§§	2000	Still active	Universe	Terrestrial instrument Volcanoes on Io	2 × 8.2 m optical telescopes	13
Genesis	NASA	2001 August	2004 September (Earth return)	Solar wind/Interplanetary medium	Heliocentric orbit – Sample and return	Ion collectors	241, 242, 269
Giotto – named for Giotto di Bondone	ESA	1985 July	Ended 1992 July	Comets Halley and Grigg-Skjellerup	Fly-by	HMC, Doppler radio	11, 53, 55, 89
Global Precipitation Measurement Core Observatory	JAXA/NASA	2014 February	Still active	Earth precipitation	Earth orbiter	Dual band radar, Microwave radiometer	66, 87
GOCE (Gravity field and steady-state Ocean Circulation Explorer)	ESA	2009 March	Ended 2013 November (re-entry)	Earth	Earth orbiter – drag free	Gravity field	270

(Continued)

Mission/Instrument*	Agency	Launch date	Current status†	Target‡	Mission type	Instruments§	Page(s)
Goldstone	JPL	1950s onwards	Still active	Venus, universe	Terrestrial instrument	Radar – 2 × 26 m parabolic dishes, 70 m parabolic dish	75, 77, 78
GRACE-1 (Gravity Recovery And Climate Experiment = 'Tom')	NASA/DLR	2002 March	Still active	Earth	Earth orbiter	Gravity measurements	270, 271
GRACE-2 (Gravity Recovery And Climate Experiment = 'Jerry')	NASA/DLR	2002 March	Still active	Earth	Earth orbiter	Gravity measurements	270, 271
GRACE-FO (Gravity Recovery And Climate Experiment – Follow On)	NASA/DLR	2017?	–	Earth	Earth orbiter	Gravity measurements	270
GRAIL A (= Ebb)	NASA	2011 September	Ended 2012 December (lunar impact)	Moon	Lunar orbiter	Gravity measurements	270, 271
GRAIL B (= Flow)	NASA	2011 September	Ended 2012 December (lunar impact)	Moon	Lunar orbiter	Gravity measurements	270, 271
Gravity Probe-B	NASA	2004 April	Ended 2010 December	General relativity test	Earth orbiter		268
Green Bank	NRAO	1960s onwards	Still active	Ceres, Titan, comets, universe	Terrestrial instrument	Radio, 100 m parabolic dish, 3 × 26 m parabolic dish interferometer	63, 69, 77, 88
Hayabusa-1 (= Muses-C) – Hayabusa is Japanese for 'Peregrine Falcon'	JAXA	2003 May	Ended (re-entered) 2010 June	Asteroid Itokawa	Asteroid sample return	LIDAR	10, 55, 84, 239, 240

(Continued)

Mission/Instrument*	Agency	Launch date	Current status†	Target‡	Mission type	Instruments§	Page(s)
Hayabusa-2 – Hayabusa is Japanese for 'Peregrine Falcon' – MASCOT lander	JAXA	2014 December	Still active	Asteroid Ryugu	Asteroid sample return and lander	LIDAR Magnetometer MARA MASCOT MicrOmega	10, 23, 55, 84, 205, 209, 212, 227, 231, 239, 240–242, 245, 248, 258
HEAO-1 (High Energy Astrophysical Observatory)	NASA	1977 August	Ended 1977 March	Earth, universe	Earth orbiter	CXE (= A-2)	97
Hinode (= Solar-B) – Hinode is Japanese for 'Sunrise'	JAXA	2006 September	Still active	Venus 2012 transit and Sun	Earth orbiter	EIS SOT SXI	17, 118
Hitomi (= Astro-H) – Hitomi is Japanese for 'Pupil' (of the eye)	JAXA	2016 February	2016 March (broke up)	Universe	Earth orbiter	SXI	107
Hubble Space Telescope	NASA/ESA	1990 April	Still active	Solar system, universe	Earth orbiter	Deep Impact OPAL STIS WFPC	13, 21, 46, 48, 205
Huygens – see Cassini							
IBEX (Interstellar Boundary Explorer)	NASA	2008 October	Still active	Helio-pause	Earth orbiter	Star sensor	57
ICE	NASA	1978 August	Ended 1997 May	Interplanetary medium, comets	L1 Earth–Sun Lagrangian point – later heliocentric	Magnetometer	258
ICON	NASA	2017	Under construction	Earth's ionosphere	Earth orbiter	Ion drift meter	184
InSight	NASA	2018 May	Constructed	Mars	Mars lander and penetrator	HP³	228, 233, 238, 244, 248, 251
INTEGRAL (International Gamma Ray Astrophysics Laboratory)	ESA	2002 October	Still active	Earth, universe	Earth orbiter	IBIS JEM-X SPI	98, 100, 118, 131

(Continued)

Mission/Instrument*	Agency	Launch date	Current status†	Target‡	Mission type	Instruments§	Page(s)
International Space Station (ISS)	CSA, ESA, JAXA, NASA, Roscosmos	1998 November	Still active	High energy charged particles	Earth orbiter	AMS-02 – from May 2011 JEM-EUSO – Planned 2018	93
IRAM (Institut de Radioastronomie Millimétrique)	CNRS/MPG/IGN	1984	Still active	Comet, universe	Terrestrial instrument	Radio, 30 m parabolic dish EMIR WILMA	
Jason 2	NASA	2008 June	Still active	Earth oceans	Earth orbiter	Microwave altimeter.	86
Jason 3	NASA	2016 January	Still active	Earth oceans	Earth orbiter	Microwave altimeter.	86
JUICE (Jupiter Icy Moon Explorer)	ESA	2023?	Planned	Jupiter	Jupiter orbiter	Langmuir probe Magnetometers UVS	46, 57, 258, 263
Juno	NASA	2011 August	Still active	Jupiter	Jupiter orbiter	JEDI JIRAM JunoCam Magnetometer MHz Radio MW Radiometer	22
JWST	NASA/ESA/CSA	2018 October?	Under construction	Universe	Lagrange L2 point	NIRSpec	48, 71, 72, 103, 193, 257, 259, 260
Karl G, Jansky Very Large Array	NRAO	1975 September	Still active	Planets, asteroids	Terrestrial instrument	Radar & Radio interferometer – twenty-seven 25 m dishes	70
Keck Observatory	CARA***	1993	Still active	Universe	Terrestrial instrument Volcanoes on Io	2 × 10 m optical telescopes	13

(Continued)

Mission/Instrument*	Agency	Launch date	Current status†	Target‡	Mission type	Instruments§	Page(s)
Kepler	NASA	2009 March	Still active	Exoplanets, comet	CCD wide field imager		15
LADEE (Lunar Atmosphere and Dust Explorer)	NASA	2013 September	Ended 2014 April (lunar impact)	Moon	Lunar orbiter	Dust detector Mass spectrometer	190, 206, 265
LAGEOS-1 (Laser Geometric Environmental Observation Survey)	NASA	1976 May	Still active	Earth	Earth orbiter	Laser retro reflector	243, 268
LAGEOS-2 (Laser Geometric Environmental Observation Survey)	NASA	1992 October	Still active	Earth	Earth orbiter	Laser retro reflector	244, 268
Landsat-1 (=ERTS-1)	NASA	1972 July	Ended 1978 January	Earth	Earth orbiter		55
Landsat-7	NASA	1999 April	Still active	Earth	Earth orbiter	ETM+	19, 55, 242
LCROSS (Lunar Crater Observations and Sensing Satellite)	NASA	2009 June	2009 October (lunar impact)	Moon	Lunar double impactor	Cameras and spectrometers	205
LISA Pathfinder (Laser Interferometer Space Antenna)	ESA	2015 December	Still active	Gravitational wave observatory	Lagrange L2 point		268
Low Frequency Array for Radio Astronomy (LOFAR)	ASTRON	2012	Still active	Jupiter, universe	Terrestrial instrument	Radio, 10–240 MHz. Multiple arrays	68
Lucy	NASA	2021?	Planned	Trojan asteroids	Heliocentric orbit with asteroid fly-bys	L'LORRI	12
Luna-3	USSR	1959 October	1960 April (re-entered atmosphere)	Moon (far side)	Lunar fly-by, Earth orbiter	Photographic camera	59

(Continued)

Mission/Instrument*	Agency	Launch date	Current status†	Target‡	Mission type	Instruments§	Page(s)
Luna-9	USSR	1966 January	1966 February	Moon	Lunar lander	TV Camera Geiger counter	141, 203, 204, 232
Luna-10	USSR	1966 March	1966 May	Moon	Lunar orbiter	Orbit perturbations	269
Luna-13	USSR	1966 December	1996 December	Moon	Lunar soft lander	Densitometer	232
Luna-16	USSR	1970 September	Ended 1970 September (sample return)	Moon	Lunar lander – sample and return	Surface sampler	235, 236, 238
Luna-19	USSR	1971 September	Ended 1972 October	Moon	Lunar orbiter	Linear camera	24, 57
Luna-20	USSR	1972 February	Ended 1972 February (sample return)	Moon	Lunar lander – sample and return	Surface sampler	113, 235, 238
Luna-24	USSR	1976 August	Ended 1976 August (sample return)	Moon	Lunar lander – sample and return	Surface sampler	113, 235, 238, 242
Luna-25 (= Luna-Glob lander)	RKA	2025?	–	Moon – Polar region	Lunar lander	–	226, 248
Luna-26 (= Luna-Glob Orbiter)	RKA	2025?	–	Moon	Lunar orbiter	–	226, 248
Luna-27 (= Luna Resurs lander)	RKA	2020?	–	Moon	Moon South Pole lander	Drill (Prospect)	236
Luna-28 (= Luna-Grunt Rover)	RKA	2020?	–	Moon	Lunar rover	–	239
Luna-29 (= Luna-Grunt sample-return vehicle)	RKA	2021?	–	Moon	Lunar lander – sample and return	Return of samples from Luna-28 to Earth	239

(Continued)

Mission/Instrument*	Agency	Launch date	Current status†	Target‡	Mission type	Instruments§	Page(s)
Lunar Orbiter 1	NASA	1966 August	Ended 1966 October (lunar impact)	Moon	Lunar orbiter and impactor	Orbit perturbations	269
Lunar Orbiter 2	NASA	1966 November	Ended 1967 October (lunar impact)	Moon	Lunar orbiter and impactor	Orbit perturbations	269
Lunar Orbiter 3	NASA	1967 February	Ended 1967 October (lunar impact)	Moon	Lunar orbiter and impactor	Orbit perturbations	269
Lunar Orbiter 4	NASA	1967 May	Ended 1967 October (lunar impact)	Moon	Lunar orbiter and impactor	Orbit perturbations	269
Lunar Orbiter 5	NASA	1967 August	Ended 1968 January (lunar impact)	Moon	Lunar orbiter and impactor	Orbit perturbations	269
Lunar Polar Hydrogen Mapper	NASA	2018	Planned	Moon	Lunar orbiter	Neutron detector	199
Lunar Prospector	NASA	1998 January	Ended 1999 July (lunar impact)	Moon	Lunar orbiter	GRS NS	104, 197
Lunar Reconnaissance Orbiter	NASA	2009 September	Still active	Moon	Lunar orbiter	Diviner (a microwave radiometer) LAMP	43, 71, 198, 206, 227
Lunokhod 1 (launch vehicle Luna-17)	USSR	1970 November	Ended 1971 September	Moon	Lunar rover	Panoramic camera	26, 203, 204, 215, 237
Lunokhod 2 (launch vehicle Luna-21)	USSR	1973 January	Ended 1973 May	Moon	Lunar rover	Panoramic camera	25, 203, 204, 237
Magellan	NASA	1989 May	Ended 1994 October (entered Venus' atmosphere)	Venus	Venus orbiter	SAR	84, 85, 86

(Continued)

Mission/Instrument*	Agency	Launch date	Current status†	Target‡	Mission type	Instruments§	Page(s)
Magnetospheric Multiscale Mission A (MMS)	NASA	2015 March	Still active	Earth (Magnetosphere)	Earth orbiter	DIS HPCA EDI	153, 154, 174, 257, 258, 258, 262
Magnetospheric Multiscale Mission B (MMS)	NASA	2015 March	Still active	Earth (Magnetosphere)	Earth orbiter	DIS HPCA EDI	153, 154, 174, 256, 257, 258, 262
Magnetospheric Multiscale Mission C (MMS)	NASA	2015 March	Still active	Earth (Magnetosphere)	Earth orbiter	DIS HPCA EDI	153, 154, 174, 256, 257, 258, 262
Magnetospheric Multiscale Mission D (MMS)	NASA	2015 March	Still active	Earth (Magnetosphere)	Earth orbiter	DIS HPCA EDI	153, 154, 174, 256, 257, 258, 262
MagSat – see Explorer 61							
Mariner 2	NASA	1962 August	Ended 1063 January	Venus	Venus fly-by	Electrostatic analyser	173
Mariner 4	NASA	1964 November	Ended 1967 December	Mars	Mars fly-by	Radio occultation	89
Mariner 5	NASA	1967 June	Ended 1967 December	Venus	Venus fly-by	Radio occultation	89, 259
Mariner 6	NASA	1969 February	Ended 1975 March	Mars	Mars fly-by	Radio occultation	89
Mariner 7	NASA	1969 March	Ended1975 March	Mars	Mars fly-by	Radio occultation	89
Mariner 9	NASA	1971 May	Ended 1972 October	Mars	Mars orbiter	Atmospheric temperature IRIS Radio occultation	89, 92, 253

(Continued)

Mission/Instrument*	Agency	Launch date	Current status†	Target‡	Mission type	Instruments§	Page(s)
Mariner 10	NASA	1973 November	Ended 1975 March	Mercury and Venus	Mercury and Venus fly-by	TPE Radio occultation	13, 89
Mars 2	Soviet Union	1971 May	Ended 1972 August (orbiter) 1971 November (lander and rover – crashed)	Mars	Mars orbiter, lander and tethered rover	PrOP-M	203
Mars 3	Soviet Union	1971 May	Ended 1972 August (orbiter) 1971 December (lander and rover – transmission ceased 14 seconds after landing)	Mars	Mars orbiter, lander and tethered rover	Densitometer PrOP-M	203
Mars 96 (= Mars 8)	RKA	1996 November	Launch failed	Mars	Mars orbiter, lander and penetrator	HRSC	23, 248
Mars 2020	NASA	2020?	Planned	Mars	Mars lander/rover	RIMFAX TIRS	88, 204, 205, 211, 212, 214, 215, 223, 228, 234, 239, 244, 250, 251, 254, 255
Mars Exploration Rover A – Spirit	NASA	2003 June	Ended 2010 March	Mars	Mars rover	MI	204, 207, 210, 217, 218, 231, 235

(Continued)

Mission/Instrument*	Agency	Launch date	Current status†	Target‡	Mission type	Instruments§	Page(s)
Mars Exploration Rover B – Opportunity	NASA	2003 July	Still active	Mars	Mars rover	MI MiniTES	204, 207, 210, 211, 217, 218, 228, 231, 234, 236, 242
Mars Express Orbiter and Beagle-2	ESA	2003 June	Still active Lander crashed 2003 December	Mars	Mars orbiter	ASPERA Atmospheric pressure Atmospheric temperature HRSC MARSIS Microscope OMEGA-SWIR OMEGA-VNIR PLUTO SPICAM-IR SPICAM-UV SRC VMC	15, 24–44, 54, 55, 58, 88, 92, 175, 207, 215, 218, 219, 228, 235, 238, 245, 249, 250, 252
Mars Global Surveyor	NASA	1996 November	Ended 2006 November	Mars	Mars orbiter	TES	210, 253, 271
Mars Odyssey Orbiter (named for the film "2001 – A Space Odyssey").	NASA	2001 April	Still active	Mars	Mars orbiter	Atmospheric temperature GRS HEND MARIE THEMIS	17, 53, 56, 114, 167, 197, 198, 206, 210, 225, 232, 242, 253, 271

(Continued)

Mission/Instrument*	Agency	Launch date	Current status†	Target‡	Mission type	Instruments§	Page(s)
Mars Orbiter Mission (MOM) = Mangalyaan (from the Hindi 'mangala' (Mars) and 'yana' (vehicle)).	ISRO	2013 September	Still active	Mars	Mars orbiter	MCC	23
Mars Pathfinder – Sojourner	NASA	1996 December	Ended 1997 December	Mars	Mars lander and rover	APXS Atmospheric pressure Atmospheric temperature Magnets	204, 217, 231, 251, 254, 269
Mars Polar Lander	NASA/JPL	1999 January	Crashed 1999 December	Mars polar region	Mars lander	Atmospheric pressure Mars Microphone TEGA	234, 249, 251, 255
Mars Reconnaissance Orbiter (MRO)	NASA	2005 August	Still active	Mars	Mars orbiter	Atmospheric pressure SHARAD	88, 242, 245, 251, 253, 271
Mars Science Laboratory (MSL) – Curiosity rover	NASA	2011 November	Still active	Mars	Mars rover	Atmospheric pressure Atmospheric temperature DAN GCMS GHTS GTS MARDI MAHLI REMS RMI	22, 135, 223, 224, 234, 236, 249, 252, 254
Mars-Grunt	RKA	2026?	–	Mars	Mars sample and return	–	239
MASCOT – see Hayabusa-2							

(Continued)

Mission/Instrument*	Agency	Launch date	Current status†	Target‡	Mission type	Instruments§	Page(s)
MASCOT-2 – see AIDA							
MAVEN (Mars Atmosphere and Volatile Evolution)	NASA	2013 November	Still active	Mars	Mars orbiter	IUVS NGIMS	10, 26, 35, 36, 41, 43, 54, 58, 189, 190, 261, 262, 269
Mercury Magnetospheric Orbiter – see Bepi Colombo							
Mercury Planetary Orbiter – see Bepi Colombo							
MESSENGER (Mercury Surface, Space Environment, Geochemistry and Ranging)	NASA	2004 August	Ended 2015 April (Mercury impact)	Mercury	Mercury flyby and orbiter	EPS GRNS Laser altimeter NAC Ritchey-Chrétien telescope ToF mass spectrometer WAC XRS	13, 21, 86, 102,114, 115, 128, 153, 192, 193, 197, 198, 206, 225
Mir Space Station	Soviet Union	1986 February	Ended 2001 March – re-entry	Manned experimentation	Earth orbiter	ODC	92, 241
Nançay telescope	Observatoire de Paris	1965	Still active	Comet, universe	Terrestrial instrument	Radio, horizontal fixed parabolic reflector	93
NEAR-Shoemaker (Near Earth Asteroid Rendezvous) – named after Eugene Shoemaker	NASA	1996 February	Ended 2001 February	Asteroid Eros	Eros orbiter	MSI XGRS Radar	21, 84, 86, 101, 102, 104, 106, 128
NEOWISE – see WISE							

(Continued)

Mission/Instrument*	Agency	Launch date	Current status†	Target‡	Mission type	Instruments§	Page(s)
New Horizons	NASA	2006 January	Still active	Pluto	Pluto and 2014 MU69 fly-by. On a path into interstellar space	Alice LORRI PEPSSI Ralph Ralph-LEISA SWAP	11, 18, 43, 47, 175, 193, 259, 265
Nimbus 3	NASA	1969 April	Ended 1972 January	Earth	Earth orbiter	IRIS	92
Nimbus 4	NASA	1970 April	Ended 1980 September	Earth	Earth orbiter	IRIS	92
Nozomi (- Planet-B) – Nozomi is Japanese for 'Hope'	JAXA	1998 July	2003 December	Mars	Mars orbit intended	XUV	106
NuSTAR (= SMEX-11)	NASA JPL	2012 June	Still active	γ rays	Earth orbiter	Proportional counter	103
Ooty telescope	NCRA	1970	Still active	Comet, universe	Terrestrial Instrument	Radio, half-wave dipole array	90
Orbiting Carbon Observatory-2 (OCO-2)	NASA	2014 July	Still active	Earth atmosphere	Earth orbiter	Visible and NIR spectrometers	43, 55
Orion programme (several spacecraft planned)	NASA/ESA	2014 December (first model)	Under development	Mars and asteroids	Multi-purpose manned space flight		241
OSIRIS-REx	NASA	Sept 2016	Still active	Asteroid Bennu	Orbiter, sample and return	REXIS PolyCam	97, 208, 240
Parkes	CSIRO	1961 October	Still active	Uranus, Neptune, universe	Terrestrial instrument	Radio, 64 m parabolic dish	69, 90
Philae – see Rosetta							
Phobos-1	IKI	1988 July	Ended 1989 March (shut down by mistake)	Phobos	Phobos orbiter and lander	Seismometer	248

(Continued)

Mission/Instrument*	Agency	Launch date	Current status[†]	Target[‡]	Mission type	Instruments[§]	Page(s)
Phobos-2	IKI	1988 July	Ended 1988 September (contact lost)	Phobos	Phobos orbiter and lander	Seismometer	248
Phobos-Grunt – see Fobos-Grunt							
Phobos-Grunt (second attempt)	RKA	2024?	–	Phobos	Phobos lander and sample return	–	294
Phoenix	NASA	2007 August	Ended 2008 November	Mars	Mars lander	Atmospheric pressure Atmospheric temperature Optical microscope TECP TEGA Wet chemistry laboratory	188, 207, 208, 226, 228, 230, 234, 235, 250–255,
Pioneer 1	NASA	1958 October	Ended 1958 October (failed to orbit and re-entered)	Moon	Lunar orbiter	Ion chamber Magne tometer	147, 256
Pioneer 2	NASA	1958 November	Ended 1958 November (failed to orbit and re-entered)	Moon	Lunar orbiter	Ion chamber Magnetometer	147
Pioneer 5	NASA	1960 March	Ended 1960 April	Interplanetary medium	Heliocentric orbit	Ion chamber Magnetometer	147, 256

(Continued)

Mission/Instrument*	Agency	Launch date	Current status†	Target‡	Mission type	Instruments§	Page(s)
Pioneer 10	NASA	1972 March	Last contact 2003 January	Jupiter	Jupiter fly-by – On a path into interstellar space	Gravity measurements Magnetometer	259, 271
Pioneer 11	NASA	1973 April	Last contact 1995 September	Jupiter, Saturn	Jupiter and Saturn flybys – On a path into interstellar space	Gravity measurements Magnetometer	259, 271
Pioneer Venus Multi Probe	NASA	1978 May	Ended 1978 December	Venus	Venus atmospheric probe	Accelerometer Atmospheric pressure Atmospheric temperature Mass spectrometer	186–188, 220, 249–252, 254, 268
Pioneer Venus Orbiter (= Pioneer Venus 1, Pioneer 12)	NASA	1978 May	Ended 1992 October (entered Venus' atmosphere)	Venus	Venus orbiter	Mass spectrometer OEFD OETP Radar	84, 189, 260, 262
Polar	NASA	1996 February	Ended 2008 April	Earth	Earth orbiter	PIXIE	98, 101, 127
Proba-V	ESA	2013 May	Still active	Earth vegetation	Earth orbiter	Three visible and IR telescopes	19
Prospector 1	Deep Space Industries	2019–2022	Planned	Asteroid	Asteroid orbiter and lander	Water detection	293
Psyche Mission	NASA	2021	Planned	Asteroid 16 Psyche	Asteroid orbiter	GRNS	114, 198
Ranger 3	NASA	1962 January	Ended 1962 January (missed the Moon)	Moon	Lunar rough lander	Seismometer	247

(Continued)

Mission/Instrument*	Agency	Launch date	Current status†	Target‡	Mission type	Instruments§	Page(s)
Ranger 4	NASA	1962 April	1962 April (crashed on far side)	Moon	Lunar rough lander	Seismometer	247
Ranger 5	NASA	1962 October	1962 October (missed the Moon)	Moon	Lunar rough lander	Seismometer	247
Ranger 7	NASA	1964 July	Ended 1964 July	Moon	Lunar impactor	NAC WAC	20
Resurs-DK No. 1	RRC-EOM	2006 June	Ended 2016 February	High energy charged particles	Earth orbiter	PAMELA	162
RHESSI (Reuven Ramaty High Energy Solar Spectroscopic Imager.)	NASA	2002 February	Still active	Earth, universe	Earth orbiter	γ ray detector	98
ROSAT (Röntgensatellit)	DLR/NASA	1990 June	Ended 2012 October	Moon/comets	Earth orbiter	WFC XRT	96, 101, 105, 120, 121
Rosetta and Philae – named for the Rosetta stone and the Philae obelisk	ESA	2004 March	Philae ended 2014 November Rosetta deorbited 2016 September	Comet 67/P	Comet orbiter and lander	Alice CIVA M/I CIVA M/V CONSERT COPS COSIMA COSAC Doppler radio GIADA LAP MIRO MUPUS OSIRIS Ptolemy ROMAP ROSINA	17, 44, 46, 47, 70, 89, 188, 191, 193, 206, 208, 211, 212, 217, 221, 223, 228, 231, 236, 238, 242, 245, 252, 258, 262, 265

(Continued)

Mission/Instrument*	Agency	Launch date	Current status†	Target‡	Mission type	Instruments§	Page(s)
Schiaparelli – see ExoMars Trace Gas Orbiter							
Seasat	NASA	1978 June	Ended 1978 October	Earth	Earth orbiter	SAR	79, 83
Selene (= Kaguya)	JAXA	2007 September	Ended 2009 June (lunar impact)	Moon	Lunar orbiter – mothership + two sub-satellites	Gravity measurements LRS	245, 270
Selene-2	JAXA	2018?		Moon	Lunar orbiter, lander and rover	–	293
Sentinel 1A	ESA	2014 April	Still active	Earth resources	Earth orbiter	SAR	83
Sentinel 1B	ESA	2016 April	Still active	Earth resources	Earth orbiter	SAR	83
Sentinel-3	EUMETSAT†††	2016 February	Still active	Earth surface temperatures	Earth orbiter	SLSTR	56
Skylab-4	NASA	1973 November	Ended 1974 February	Comet Kohoutek	Manned Earth orbiter	UV Imager	16
SMAP	NASA	2015 January	Still active	Earth soil moisture	Earth orbiter	1.4 GHz radiometer	70
SMART-1 (Small Missions for Advance Research in Technology-1)	ESA	2003 September	Ended 2006 September	Moon	Lunar orbiter	AMIE D-CIXS NIR imager	108, 128, 242, 262
SOFIA	NASA/DLR	2010 May	Still active	Mars Pluto Universe	Aircraft_borne telescope	Pluto occultation Mars atmosphere	11
Sojourner – see Mars Pathfinder							
Solar Dynamics Observatory	NASA	2010 February	Still active	Solar observatory	Earth orbiter	Venus x-ray emission	98
Space Shuttle – Endeavour	NASA	1992 May	2011 May (last of 25 missions)	Many purposes	Earth orbiter	Ground penetrating radar	245

(Continued)

Mission/Instrument*	Agency	Launch date	Current status†	Target‡	Mission type	Instruments§	Page(s)
Spektr-RG – named from the Russian words for Spectrum, Röntgen and Gamma	Russia/Europe	2018 April		Interplanetary magnetic field, universe	Lagrange L-2 point	eROSITA ART-XC	120, 126
Spektr-RG- named from the Russian words for Spectrum, Röntgen and Gamma	Russia/Europe	–	Mission cancelled 2002	–	Earth orbiter	JET-X	120
Spirit – see Mars Exploration Rover B							
Spitzer	NASA	2003 August	Still active	Comet Tempel-1 (Deep Impact)	Solar orbit	IR camera IR spectrograph	205
Sputnik 1	USSR	1957 October	Ended 1958 January	High energy charged particles	Earth orbiter	Radio transmitter	203
Sputnik 2	USSR	1957 November	Ended 1958 April	High energy charged particles	Earth orbiter	Geiger counter	139, 141
Square Kilometre Array (SKA)	Australia, Canada, China, Germany, India, Italy, New Zealand, South Africa, Sweden, the Netherlands and the UK	2018–2030	Under construction	Solar system, universe	Terrestrial instrument	Radio, 50 MHz to 14 GHz. Multiple arrays	68
Stardust	NASA	1999 February	Ended 2011 March (2006 January – sample return to Earth)	Comets Wild and Tempel 1	Interplanetary dust sample and return	Aerogel dust collector	241, 264, 265

(Continued)

Mission/Instrument*	Agency	Launch date	Current status[†]	Target[‡]	Mission type	Instruments[§]	Page(s)
Stratoscope 2	NASA/ ONL/NRL	1963 March (first flight)	1971 September (last flight)	Planetary atmospheres	High altitude balloon	Infrared telescope	17
SuperTIGER	Universities of Washington & Minnesota, Caltech, JPL	2012 December	Ended 2013 January	Cosmic ray interactions in Earth's atmosphere	High altitude balloon	Čerenkov detectors	167, 168
Surveyor 3	NASA	1967 April	Ended 1967 May	Moon	Lunar lander	Sampler arm	233
Surveyor 5	NASA	1967 September	Ended 1967 December	Moon	Lunar lander	APS	216
Surveyors 1–7	NASA	1966 May (Surveyor 1)	1968 January (Surveyor 7)	Moon	Lunar lander	Radar Magnets	79, 216, 230, 231, 232, 237, 247
Susaku (= Astro-E2) – Susaku is the Japanese for 'Vermillion (or red) Bird'	JAXA/ NASA	July 2005	2015 September (XRS failed 2005 August)	Earth (TRFs)	Earth orbiter	XRS	116
Swarm A	ESA	2013 November	Still active	Earth magnetic field	Earth orbiter	Accelerometer Helium vector magnetometer Langmuir probe Retro-reflector	244, 259, 263, 269
Swarm B	ESA	2013 November	Still active	Earth magnetic field	Earth orbiter	Accelerometer Helium vector magnetometer Langmuir probe Retro-reflector	244, 259, 263, 269

(Continued)

Mission/Instrument*	Agency	Launch date	Current status†	Target‡	Mission type	Instruments§	Page(s)
Swarm C	ESA	2013 November	Still active	Earth magnetic field	Earth orbiter	Accelerometer Helium vector magnetometer Langmuir probe Retro-reflector	244, 259, 263, 269
Swift	NASA	2004 November	Still active	Comets, universe	Earth orbiter	BAT UVOT XRT	107, 115, 120
THEMIS A (see also ARTEMIS P1 and P2)	NASA	2007 February	Still active	Interplanetary medium	Earth orbiter	Magnetometer Top hat analyser	17, 18, 56, 175, 242, 250, 257
THEMIS D (see also ARTEMIS P1 and P2)	NASA	2007 February	Still active	Interplanetary medium	Earth orbiter	Magnetometer Top hat analyser	17, 18, 56, 175, 242, 250, 257
THEMIS E (see also ARTEMIS P1 and P2)	NASA	2007 February	Still active	Interplanetary medium	Earth orbiter	Magnetometer Top hat analyser	17, 18, 56, 175, 242, 250, 257
THOR	ESA	2026?	Proposed	Interplanetary medium	Earth orbiter	Top hat analyser Mass spectrometer	175, 194
Ulysses	NASA/ESA	1990 October	2009 June	Solar studies, Jupiter	Heliocentric orbit (Jupiter fly-by)	Dust detector	258, 265
V2 Rocket		1946 October	1946 October	Earth	Sub-orbital	35 mm cine camera	20, 96
V2 Rocket		1949 January	1949 January	Sun	Sub-orbital	X-ray detector	96
van Allen Probes (= Radiation Belt Storm Probes – RBSP) – named for James van Allen	NASA	2012 August	Still active	Earth	Two Earth orbiters	ECT-HOPE ECT-MagEIS ECT-REPT Magnetometers RBSPICE RPS	139, 154, 156, 162, 167, 182, 193, 258

(Continued)

Mission/Instrument*	Agency	Launch date	Current status[†]	Target[‡]	Mission type	Instruments[§]	Page(s)
Vega 1	USSR	1984 December	Ended 1987 January	Venus and comet P/Halley	Venus, Halley fly-by, Venus lander and balloon	Radio occultation	89, 215, 235, 249, 250, 252, 264
Vega 2	USSR	1984 December	Ended 1987 March	Venus and comet P/Halley	Venus, Halley fly-by, Venus lander and balloon	Radio occultation	89, 215, 235, 249, 250, 252, 264
Venera 3	USSR	1965 November	Ended 1966 March (crushed)	Venus	Venus probe		251
Venera 4	USSR	1967 June	Ended 1967 October (crushed)	Venus	Venus probe	Pressure sensor Temperature sensor	251
Venera 5	USSR	1969 January	Ended 1969 May (crushed)	Venus	Venus probe	Pressure sensor Temperature sensor	251, 252
Venera 6	USSR	1969 January	Ended 1969 May (crushed)	Venus	Venus probe	Pressure sensor Temperature sensor	251
Venera 7	USSR	1970 August	Ended 1970 December	Venus	Venus lander	Pressure sensor Temperature sensor	251
Venera 8	USSR	1972 March	Ended 1972 July	Venus	Venus lander	Pressure sensor Temperature sensor	251

(Continued)

Mission/Instrument*	Agency	Launch date	Current status†	Target‡	Mission type	Instruments§	Page(s)
Venera 9	USSR	1975 June	Ended 1975 December (orbiter), 1975 October (lander)	Venus	Venus orbiter and lander	Bi-static radar Cycloramic optical camera Pressure sensor Temperature sensor	25, 57, 83, 248, 250, 254
Venera 10	USSR	1975 June	Ended 1975 December (orbiter), 1975 October (lander)	Venus	Venus orbiter and lander	Bi-static radar Cycloramic optical camera Pressure sensor Temperature sensor	25, 83, 248, 250, 254
Venera 11	USSR	1978 September	Ended 1980 January (fly-by), 1978 December (lander)	Venus	Venus fly-by and lander	Pressure sensor	248, 250, 255
Venera 12	USSR	1978 September	Ended 1980 March (fly-by), < 1978 December (lander)	Venus	Venus fly-by and lander	Pressure sensor Temperature sensor	248, 250, 255
Venera 13	USSR	1981 October	Ended 1982 March (lander)	Venus	Venus fly-by and lander	Gas Chromatograph Pressure sensor Temperature sensor XRF	215, 220, 228, 235, 247, 248, 250, 254, 255
Venera 14	USSR	1981 November	Ended 1982 March (lander)	Venus	Venus fly-by and lander	Gas Chromatograph Pressure sensor Temperature sensor XRF	215, 220, 228, 235, 247, 248, 250, 254, 255

(Continued)

Mission/Instrument*	Agency	Launch date	Current status†	Target‡	Mission type	Instruments§	Page(s)
Venera 15	USSR	1983 June	Ended 1984 July	Venus	Venus orbiter	Radar	84, 86, 251
Venera 16	USSR	1983 June	Ended 1984 July	Venus	Venus orbiter	Radar	84, 86, 251
Venera-D	RKA	2025?	Planned	Venus	Venus orbiter	SAR	84
Venus Climate Orbiter – see Akatsuki							
Venus Express Orbiter	ESA	2005 November	Ended 2015 January	Venus	Venus orbiter	PFS	92, 93, 175, 260, 269
Viking 1	NASA	1975 August	Ended 1982 November	Mars	Mars orbiter and lander	Atmospheric pressure Atmospheric temperature Dust detector DUCMA GCMS GEX Radar Thermocouple temperature sensors	84, 215, 220, 221, 223, 226, 228, 234, 247, 251, 252, 269
Viking 2	NASA	1975 September	Ended 1980 April	Mars	Mars orbiter and lander	Dust detector DUCMA GCMS GEX Radar Thermocouple temperature sensors	84, 215, 220, 221, 223, 226, 228, 234, 247, 251, 252, 269

(Continued)

Mission/Instrument*	Agency	Launch date	Current status†	Target‡	Mission type	Instruments§	Page(s)
Voyager 1	NASA	1977 September	Still active	Jupiter, Saturn, Helio-pause	Jupiter and Saturn fly-bys. On a path into interstellar space	IRIS NAC PWS Radio occultation Super heterodyne radio receiver	13, 14, 21, 59, 71, 89, 91, 258–261, 264
Voyager 2	NASA	1977 August	Still active	Jupiter, Saturn, Uranus, Neptune, Helio-pause	Jupiter, Saturn, Uranus and Neptune fly bys. On a path into interstellar space	IRIS NAC Radio occultation PWS Super heterodyne radio receiver	13, 14, 21, 59, 71, 89, 91, 92, 258–261, 264
Wind	NASA	1984 November	Still active	IPM	Lagrange L1 point	ESA SWICS	100, 191
WISE (=NEOWISE)	NASA	2009 December	Ended 2011 February (WISE – mission complete) Still active (NEOWISE)	Universe (WISE) NEOs (NEOWISE)	Earth orbiter	IR imager	11, 55
XMM-Newton (= High Throughput X-ray Spectroscopy mission – HTXS)	ESA	1999 December	Still active	Comets	Earth orbiter	EPIC RGS X-ray telescope	132, 133
Yutu – see Change-3							
Zond-5	Soviet Union	1968 September	Re-entry 1968 September	Moon	Moon fly-by/ Earth return	Film camera	20

(Continued)

Mission/Instrument*	Agency	Launch date	Current status[†]	Target[‡]	Mission type	Instruments[§]	Page(s)
Zond-6	Soviet Union	1968 November	Re-entry 1968 November	Moon	Moon fly-by/ Earth return	Film camera	20
Zond-7	Soviet Union	1969 August	Re-entry 1969 August	Moon	Moon fly-by/ Earth return	Film camera	20
Zond-8	Soviet Union	1970 October	Re-entry 1970 October	Moon	Moon fly-by/ Earth return	Film camera	20

*Only missions mentioned in this book are included. Missions that have never observed objects within the solar system are thus not listed.

[†]As of the start of 2017.

[‡]Specific solar system observations are mentioned. The entry 'universe' indicates that the mission also has/had targets outside the solar system.

[§]Only instruments discussed in this book are included.

**Centre National d'Études Spatiales.

[††]Agenzia Spaziale Italiana.

[‡‡]Ballistic Missile Defence Organisation.

[§§]Association of Universities for Research in Astronomy.

***California Association for Research in Astronomy.

[†††]European Organisation for the Exploitation of Meteorological Satellites.

Data and Units

Symbols and Constants

Symbol	Meaning/Definition/Name	Definition/SI/Other value*	Working definition/SI working value/Other working value (as generally used for calculations in this book)/Conversions to other units
A	ampere (common abbreviation – amp)	SI base unit of electric current. Defined as the constant electric current which, if maintained in two straight parallel conductors, of negligible circular cross section and placed 1 m apart in a vacuum, would produce a force equal to 2×10^{-7} N per metre of length.	A current of 6.2415×10^{18} electrons per second
A	Atomic mass	Units (usually) Da	
A_e	Effective area of a radar transmitting or receiving dish		
AU or au	Astronomical Unit	$1.499578707 \times 10^{11}$ m (exact) A non-SI measure of length widely used by astronomers. Originally defined as the mean distance between the centres of the Sun and Earth, it is now defined in terms of the above number of metres (but is still in practice the mean Sun–Earth distance).	1.4996×10^{11} m or 1.5×10^{11} m = 4.8481×10^{-6} pc
B	Magnetic flux density	Units: T or equivalently V s·m^{-2}.	
C	coulomb	SI derived unit of electric charge. 1 C = 1 A s. The charge on 6.24150934 (14) $\times$ 10^{18} electrons	The charge on 6.2415×10^{18} electrons
c	The speed of light in a vacuum	2.99792458×10^{8} m·s^{-1} (exact)	2.9979×10^{8} m·s^{-1} (often 3×10^{8} m·s^{-1} used)

(Continued)

Symbols and Constants (Continued)

Symbol	Meaning/Definition/Name	Definition/SI/Other value*	Working definition/SI working value/Other working value (as generally used for calculations in this book)/Conversions to other units
cm^{-1}	Frequency	A non-SI unit of the frequency of e-m radiation. $1\ cm^{-1} = 3.33564095 \times 10^{-11}\ Hz$ Defined as the number of wavelengths in 1 cm.	
Da or u or amu†	dalton or unified atomic mass unit	$1.660538921(73) \times 10^{-27}\ kg$ A non-SI unit widely used for masses in the region of that of a proton. Defined as one-twelfth of the mass of an unbound neutral carbon-12 atom in its nuclear and electronic ground state.	$1.6605 \times 10^{-27}\ kg$ $= 9.3149 \times 10^{8}\ eV$
E	Electric field strength	Units: $V{\cdot}m^{-1}$ or, equivalently, $N{\cdot}C^{-1}$.	
E	Energy	Units: J or eV	
e^{-} or e^{+}	Charge on the electron or positron, anti-proton or proton	$1.602176565\ (35) \times 10^{-19}\ C$	$1.6022 \times 10^{-19}\ C$
eV	electron-volt	$1.602176565\ (35) \times 10^{-19}\ J.$ A non-SI measure of energy widely used for energies of sub-atomic particles, x-rays and gamma rays. Also used for the masses of sub-atomic particles via $E = Mc^2$. Defined as the amount of energy gained (or lost) by an electron when moving between points with a potential difference of one volt.	$1.6022 \times 10^{-19}\ J$ $= 1.0735 \times 10^{-9}\ Da$ (mass equivalent)
F	farad	SI derived unit of electric capacitance $1\ F = 1\ C{\cdot}V^{-1}$ Defined as the capacitance of a capacitor that has a potential difference of 1 V when charged with 1 C.	
F(s)	Fourier transform		
f	Frequency (usually of e-m radiation)	SI unit: Hz (= 1 cycle per second) In some sources the unit cm^{-1} may be encountered. This equals the number of wavelengths in a centimetre. The conversion is: $1\ Hz = 2.99792458 \times 10^{10}\ cm^{-1}$ $1\ cm^{-1} = 3.33564095 \times 10^{-11}\ Hz$	

(Continued)

Symbols and Constants (Continued)

Symbol	Meaning/Definition/ Name	Definition/SI/Other value*	Working definition/SI working value/Other working value (as generally used for calculations in this book)/Conversions to other units
f(x)	Inverse Fourier transform		
G	Newton's gravitational constant	$6.67384\,(80) \times 10^{-11}$ m³·kg⁻¹·s⁻²	6.6738×10^{-11} m³·kg⁻¹·s⁻²
Gy	gray	SI derived unit of ionising radiation dose, 1 Gy = 1 J·kg⁻¹. Defined as the absorption of one joule of energy per kilogram of matter.	
h	Planck's constant	$6.626070040(81) \times 10^{-34}$ m² kg·s⁻¹. Relates the energy (E) of an electromagnetic photon to its frequency (ν) when acting as a wave – via the Planck equation: $E = h\,\nu$	6.62607×10^{-34} m² kg·s⁻¹
Hz	hertz	SI derived unit of frequency. $1\ \mathrm{Hz} = 1\ \mathrm{s}^{-1}$ Defined as the number of cycles per second	
I, I(λ,T), I(ν,T)	Intensity of e-m radiation. Sometimes specified at a particular wavelength or frequency or for a particular temperature, per unit area, unit solid angle, etc.	Units: W, W·m⁻¹, W·hz⁻¹, W·str⁻¹, W·m⁻², etc. The SI quantities of luminous intensity (unit: candela, cd), luminous flux (unit: lumen, lm) and illuminance (unit: lux, lx) may also be encountered.	
J	joule	SI derived unit of energy $1\ \mathrm{J} = 1\ \mathrm{N\,m}$ Defined as the energy expended when a force of one newton acts over a distance of one metre.	$= 6.2415 \times 10^{18}$ eV
Jy	jansky	10^{-26} W·m⁻²·Hz⁻¹ A non-SI unit of radiation intensity used within the microwave and radio regions.	10^{-26} W·m⁻²·Hz⁻¹
k	Boltzmann's constant	$1.38064852(79) \times 10^{-23}$ m² kg·s⁻¹·K⁻¹ It is the gas constant per particle.	1.3806×10^{-23} m² kg·s⁻¹·K⁻¹
K	kelvin	SI base unit of thermodynamic temperature. 1 K = 1/273.16 of the thermodynamic temperature of the triple point of water	

(Continued)

Symbols and Constants (Continued)

Symbol	Meaning/Definition/Name	Definition/SI/Other value*	Working definition/SI working value/Other working value (as generally used for calculations in this book)/Conversions to other units
kg	kilogram	SI base unit of mass. Defined as the mass of the International Prototype of the Kilogram (IPK) held at the Bureau International des Poids et Mesures (BIPM) at Sévres in France.	$= 6.0221 \times 10^{26}$ Da
m	metre or meter	SI base unit of length. Defined as the distance travelled by light in a vacuum during 3.335641×10^{-9} s.	
m	The order of a spectrum		
m_e or m_p	Rest mass of the electron or positron	$9.10938215\ (45) \times 10^{-31}$ kg	9.1094×10^{-31} kg $= 5.4859 \times 10^{-5}$ Da $= 5.1098 \times 10^{5}$ eV (often $= 500$ keV used)
m_n	Rest mass of the neutron or anti-neutron	$1.674927351\ (74) \times 10^{-27}$ kg	1.6749×10^{-27} kg $= 1.0097$ Da $= 9.3957 \times 10^{8}$ eV (often $= 1,000$ MeV used)
m_p	Rest mass of the proton or anti-proton	$1.672621777\ (74) \times 10^{-27}$ kg	1.6726×10^{-27} kg $= 1.0073$ Da $= 9.3823 \times 10^{8}$ eV (often $= 1,000$ MeV used)
N	newton	SI derived unit of force $1\ N = 1\ m\ kg\ s^{-2}$ Defined as the force required to accelerate a mass of 1 kg by 1 m s^{-2}.	
n_e	Electron number density	Units: m^{-3}	
p	Momentum	Units: $kg\ m \cdot s^{-1}$	
Pa	pascal	SI derived unit of gas pressure and stress. $1\ Pa = 1\ N \cdot m^{-2} = 1\ kg \cdot m^{-1} \cdot s^{-2}$	$1\ Pa = 1 \times 10^{-5}$ bar $= 1.45 \times 10^{-4}$ psi $= 7.501 \times 10^{-3}$ torr $= 9.869 \times 10^{-6}$ Atm
pc	parsec	3.0857×10^{16} m A non-SI measure of length widely used by astronomers. Defined as the distance at which one AU subtends an angle of one arc-second.	3.0857×10^{16} m $= 2.0626 \times 10^{5}$ AU
q	Electric charge on a charged particle. Often expressed as multiples of the electron charge (e)	C or $n \times e$	

(Continued)

Symbols and Constants (Continued)

Symbol	Meaning/Definition/Name	Definition/SI/Other value*	Working definition/SI working value/Other working value (as generally used for calculations in this book)/Conversions to other units
R_L	Larmor radius	The 'orbital' radius of a charged particle moving in a magnetic field.	
s	second (common abbreviation – sec)	SI base unit of time. Defined as the duration of 9.192631770×10^9 periods of the radiation from the transition between the two hyperfine levels of the ground state of the caesium-133 atom.	
T	Temperature	SI base unit of temperature Unit: kelvin (K) Defined as 1/273.16 of the thermodynamic temperature of the triple point water 0 K is often called the absolute zero temperature	
T	tesla	SI derived unit of magnetic flux density. $1\ T = 1\ kg \cdot s^{-2} \cdot A^{-1}$. Defined as the magnetic field strength when a particle carrying a charge of 1 C experiences a force of 1 N when passing perpendicularly through the magnetic field at a speed of $1\ m \cdot s^{-1}$.	
T_s	The noise temperature of a radar system	Unit: kelvin	
V	volt	SI derived unit of electric potential. $1\ V = 1\ kg\ m^2 \cdot s^{-3} \cdot A^{-1}$ Defined as the difference of potential between two points on a conductor when the power dissipated between the two points is 1 W and the current is 1 A.	
v	velocity	Unit: $m \cdot s^{-1}$	
α	The fine structure constant	A dimension-less constant $\alpha = 7.2973525698\ (24) \times 10^{-3}$	$7.2974 \times 10^{-3} = \sim 1/137$
α	Radar cross section		
α	Spectral index of synchrotron radiation		

(Continued)

Symbols and Constants (Continued)

Symbol	Meaning/Definition/Name	Definition/SI/Other value*	Working definition/SI working value/Other working value (as generally used for calculations in this book)/Conversions to other units
γ	The Lorentz factor	$\gamma = \left(1 - \dfrac{v^2}{c^2}\right)^{-\frac{1}{2}} = \dfrac{E}{(m\,c^2)}$ where v is the particle's velocity, c is velocity of light in a vacuum, E is particle's energy and m is the particle's rest mass.	
ε_o	Absolute permittivity	A measure of the resistance that a medium exerts to the formation of an electric field within it. The absolute permittivity of a vacuum (free space) has a value of $8.854187817620 \times 10^{-12}$ F·m⁻¹.	8.8542×10^{-12} F·m⁻¹.
ε_r	Relative permittivity	The permittivity of a medium relative to the absolute permittivity of a vacuum (free space). It is a dimensionless quantity.	
λ	Wavelength (usually of e-m radiation)	SI unit: m	
μ, $\mu(\lambda)$ $\mu(\nu)$	Refractive index (the symbol, 'n', is also in widespread use)	Dimensionless. Defined as the ratio of the speed of light in a vacuum to the speed of light within the substance concerned.	
ν	Frequency (usually of e-m radiation)	SI unit: Hz (= 1 cycle per second) In some sources the unit, cm⁻¹, may be encountered. This equals the number of wavelengths in a centimetre. The conversion is: 1 Hz = $2.99792458 \times 10^{10}$ cm⁻¹ 1 cm⁻¹ = $3.33564095 \times 10^{-11}$ Hz	
ν_L	Larmor frequency	The frequency of the 'orbital' motion of a charged particle moving in a magnetic field.	

*Where applicable/possible the numbers in parenthesis are the 1-σ uncertainty in the last two quoted digits of the value. Thus the value of e⁻ could also be written $1.602176565 \pm 0.000000035 \times 10^{-19}$ C. The SI unit definitions are the current ones. The Comité International des Poids et Mesures (CIPM) is considering redefining the definitions of the SI base units for possible implementation in 2018.

†Strictly the atomic mass unit – an obsolete unit based upon a 16th of the mass of the oxygen-16 atom but widely used now based upon the unified atomic mass unit's definition.

Abbreviations and Acronyms

ACE	Advanced Composition Explorer
ACIS	Advanced CCD Imaging Spectrometer
ACS	Atmospheric Chemistry Suite
ADM Aeolus	Advanced Dynamics Mission Aeolus
AFM	Atomic Force Microscope
AGILE	Astro-Rivelatore Gamma a Immagini Leggero
AIDA	Asteroid Impact and Deflection Assessment
AIM	Aeronomy of Ice in the Mesosphere
AIM	Asteroid Impact Mission
AKR	Auroral Kilometric Radiation
ALADIN	Atmospheric Laser Doppler Instrument
ALMA	Atacama Large Millimeter Array
ATSR	Along Track Scanning Radiometer
AMIE	Asteroid-Moon Micro-Imager Experiment
AMS-02	Alpha Magnetic Spectrometer No 2
AOTF	Acousto-Optical Tunable Filter
APD	Avalanche Photo-Diode
APS	Active Pixel Sensor
APS	Alpha Particle Spectrometer
APXS	Alpha Particle X-ray Spectrometer
ARGOS	Advanced Research and Global Observation Satellite
ARM	Asteroid Retrieval Mission
ARTEMIS	Acceleration, Reconnection, Turbulence and Electrodynamics of the Moon's Interaction with the Sun
ART-XC	Astronomical Röentgen Telescope – X-ray Concentrator
ASPERA	Analyser of Space Plasmas and Energetic Atoms
ASEM	Atmospheric Scanning Electron Microscope
ASI	Agenzia Spaziale Italiana (Italian Space Agency)
ASTRON	Netherlands Institute for Radio Astronomy
ATHENA	Advanced Telescope for High Energy Astrophysics
ATHLETE	All-Terrain Hex-Limbed Extra-Terrestrial Explorer
AURA	Association of Universities for Research in Astronomy

AXAF	Advanced X-ray Astrophysics Facility
BARREL	Balloon Array for Radiation belt Relativistic Electron Losses
BAT	Burst Alert Telescope
BATSE	Burst and Transient Source Experiment
BGO	Bismuth/germanium/oxygen
BIPM	Bureau International des Poids et Mesures
BMDO	Ballistic Missile Defence Organisation
C1XS	Chandrayaan-1 X-ray Spectrometer
CALIPSO	Cloud Aerosol Lidar and Infrared Pathfinder Satellite Observations
Caltech	California Institute of Technology
CARA	California Association for Research in Astronomy
CaSSIS	Colour and Stereo Surface Imaging System
CCD	Charge Coupled Device
CCM	Continuous Channel Multiplier
CEM	Channel Electron Multiplier
CIPM	Comité International des Poids et Mesures
CIRS	Composite Infrared Spectrometer
CIVA	Comet Infrared and Visible Analyser
CLASS	Chandrayaan Large Area Soft x-ray Spectrometer
CMOS	Complementary Metal-Oxide Semiconductor
CNES	Centre National d'Études Spatiales
CNRS	Centre National de la Recherche Scientifique
CNSA	Chinese National Space Administration
CONSERT	Comet Nucleus Sounding Experiment by Radiowave Transmission
CONTOUR	Comet Nucleus TOUR
COPS	Comet Pressure Sensor
COSAC	Cometary Sampling and Composition
COSIMA	Cometary Secondary Ion Mass Analyser
CSA	Canadian Space Agency
CSIRO	Commonwealth Scientific and Industrial Research Organisation
CZT	Cadmium–Zinc–Telluride
DAN	Dynamic Albedo of Neutrons
DART	Double Asteroid Re-direction Test
DAVINCI	Deep Atmosphere Venus Investigation of Noble gases, Chemistry, and Imaging
D-CIXS	Demonstration of a Compact Imaging X-ray Spectrometer
DECA	Descent Camera
DGNAA	Delayed Gamma Neutron Activation Analysis
DIS	Dual Ion Spectrometer
DLR	Deutsches Zentrum für Luft und Raumfahrt
DSCOVR	Deep Space Climate Observatory
DUCMA	Dust Counter and Mass Analyser
EAS	Extensive Air Shower

ECAL	Electromagnetic calorimeter
ECT	Energetic particle, Composition and Plasma suite
EFW	Electric Field and Waves suite
EGRET	Energetic Gamma Ray Experiment Telescope
EIS	Extreme Ultraviolet imaging Spectrometer
ELF	Extremely low frequency radiation
eLISA	Evolved Laser Interferometer Space Antenna
EM	Electron Multiplier
e-m radiation	Electro-magnetic radiation
EMFISIS	Electric and Magnetic Field Instrument Suite and Integrated Science
EMIR	Eight Mixer Receiver
EOR	Extended Optical Region
EPIC	European Photon-Imaging Camera
EPIC	Earth Polychromatic Imaging Camera
EPS	Energetic Particle Spectrometer
eROSITA	Extended Röentgen Survey with an Imaging Telescope Array
ERS-1	European Remote Sensing satellite-1
ERTS-1	Earth Resources Technology Satellite-1
ESA	Electrostatic Analyser
ESA	European Space Agency
ESEM	Environmental Scanning Electron Microscope
ESO	European Southern Observatory
ESPRESSO	Echelle Spectrograph for Rocky Exoplanet and Stable Spectroscopic Observations
ET	Extra-Terrestrial
E-THEMIS	Europa Thermal Imaging System
ETM+	Enhanced Thematic Mapper plus
EUV	Extreme Ultraviolet
EUVE	Extreme Ultraviolet Explorer
EXOSAT	European X-ray Observatory Satellite
FAST	Five hundred meter Aperture Spherical Telescope
FIR	Far infrared
FIREBIRD	Focused Investigations of Relativistic Burst Intensity, Range and Dynamics
FLEX	Fluorescence Observer
FTS	Fourier Transform Spectroscope
FUV	Far Ultraviolet
FUVCS	Far Ultraviolet Camera/Spectrograph
GBM	Gamma ray Burst Monitor
GCMS	Gas Chromatograph – Mass Spectrometer
GEX	Gas Exchange Experiment
GIADA	Grain Impact And Dust Accumulator

GOCE	Gravity field and steady-state Ocean Circulation Explorer
GRACE	Gravity Recovery And Climate Experiment
GRACE-FO	Gravity Recovery And Climate Experiment – Follow On
GRAIL	Gravity Recovery and Interior Laboratory
GRaND	Gamma Ray and Neutron Detector
GRID	Gamma Ray Imaging Detector
GRNS	Gamma Ray and Neutron Spectrometer
GRS	Gamma Ray Spectrometer
GTS	Ground Temperature Sensor
HCAL	Hadronic calorimeter
HEAO	High Energy Astrophysical Observatory
HEB	Hot Electron Bolometer
HEMT	High Electron Mobility Transistors
HEND	High Energy Neutron Detector
HMC	Halley Multicolor Camera
HOPE	Helium Oxygen Proton Electron detector
HPCA	Hot Plasma Composition Analyser
HRC	High Resolution Camera
HRD	High Rate Detector
HRI	High Resolution Imager
HRSC	High Resolution Stereo Camera
HST	Hubble Space Telescope
HTXS	High Throughput X-ray Spectroscopy
HXI	Hard X-Ray Imager
IBEX	Interstellar Boundary Explorer
IBIS	Imager on Board the Integral Satellite
ICE	International Cometary Explorer
ICEMAG	Interior Characterisation of Europa using Magnetometry
ICON	Ionospheric Connection Explorer
IGFoV	Instantaneous Geometrical Field of View
IGN	Instituto Geográfico Nacional
IKI	Institut Komicheskih Issledovanyi
INMS	Ion and Neutral Mass Spectrometer
INTEGRAL	International Gamma Ray Astrophysics Laboratory
IPC	Imaging Proportional Counter
IPK	International Prototype of the Kilogram
IPM	Inter-Planetary Medium
IRAM	Institut de Radioastronomie Millimétrique
IRIS	Infrared Radiometer Interferometer and Spectrometer
ISON	International Scientific Optical Network
ISRO	Indian Space Research Organisation
ISS	Imaging Science Sub-system
ISS	International Space Station

IUPAC	International Union of Pure and Applied Chemistry
IUVS	Imaging Ultraviolet Spectrometer
IXO	International X-ray Observatory
JAXA	Japan Aerospace Exploration Agency
JEDI	Jupiter Energetic particle Detector Instrument
JEM-EUSO	Japanese Experiment Module – Extreme Universe Space Observatory
JEM-X	Joint European X-ray Monitor
JET-X	Joint European Telescope for X-ray astronomy
JHUAPL	John Hopkins University Applied Physics Laboratory
JIRAM	Jovian Infrared Auroral Mapper
JPL	Jet Propulsion Laboratory
JUICE	Jupiter Icy Moon Explorer
JWST	James Webb Space Telescope
LADEE	Lunar Atmosphere and Dust Explorer
LAGEOS	Laser Geometric Environmental Observation Survey
LAMP	Lyman-α Mapping Project
LAP	Langmuir Probe
LAT	Large Area Telescope
LAXPC	Large Area X-ray Proportional Counter
LCROSS	Lunar Crater Observations and Sensing Satellite
LDMS	Laser Desorption Mass Spectrometer
LED	Light Emitting Diode
LEISA	Linear Etalon Imaging Spectral Array
LIBS	Laser Induced Breakdown Spectrometer
LIDAR	Light detection and ranging
LINEAR	Lincoln Near Earth Asteroid Research
LISA	Laser Interferometer Space Antenna
L'LORRI	Lucy Long-Range Reconnaissance Imager
LOFAR	Low Frequency Array for Radio Astronomy
LORRI	Long-Range Reconnaissance Imager
LRO	Lunar Reconnaissance Orbiter
LRS	Lunar Radar Sounder
LunaH-Map	Lunar Polar Hydrogen Mapper
LUT	Lunar-based Ultraviolet Telescope
MagEIS	Magnetic Electron Ion Spectrometer
MAHLI	Mars Hand Lens Imager
MAMA	Multi Anode Microchannel Array
MaMISS	Multispectral Imager for Subsurface Studies
MARA	MASCOT Radiometer
MARDI	Mars Descent Imager camera
MARIE	Mars Radiation Experiments
MARSIS	Mars Advanced Radar for Subsurface and Ionospheric Sounding

MASCOT	Mobile Asteroid Surface Scout
MASPEX	Mass Spectrometer for Planetary Exploration
MAVEN	Mars Atmosphere and Volatile Evolution
MCAL	Mini Calorimeter
MCC	Mars Colour Camera
MCP	Micro Channel Plate
MESSENGER	Mercury Surface, Space Environment, Geochemistry and Ranging
MI	Microscopic Imager
MicroARES	Micro Atmospheric Radiation and Electricity Sensor
MicroMegas	Micro-Mesh gaseous structure
MIDAS	Micro Imaging Dust Analysis System
MIMOS	Miniature Mössbauer Spectrometer
Mini-TES	Miniature Thermal Emission Spectrometer
MIP	Moon Impact Probe
MIR	Medium Infrared
MIRO	Microwave Instrument for the Rosetta Orbiter
MIXS	Mercury Imaging X-ray Spectrometer
MMO	Mercury Magnetospheric Orbiter
MMS	Magnetospheric Multiscale Spacecraft
MOM	Mars Orbiter Mission
MORE	Mercury Orbiter Radio Science Experiment
MPC	Monitor Proportional Counter
MPG	Max Planck Gesellschaft
MPGD	Micropattern Gaseous Detectors
MPO	Mercury Planetary Orbiter
MRO	Mars Reconnaissance Orbiter
MSASI	Mercury Sodium Atmosphere Spectral Imager
MSI	Multi-Spectral Imaging
MSL	Mars Science Laboratory
MUPUS	Multi-Purpose Sensors for surface and subsurface science
MURA	Modified Uniformly Redundant Array
MVP-SEM	Miniaturized Variable Pressure Scanning Electron Microscope
NAA	Neutron Activation Analysis
NAC	Narrow Angle Camera
NASA	National Aeronautics and Space Administration
NCRA	National Centre for Radio Astrophysics
NEAR	Near Earth Asteroid Rendezvous
NEOCam	Near Earth Asteroid Cam
NEOWISE	Near Earth Object Wide field Infrared Survey Explorer
NGIMS	Neutral Gas and Ion Mass Spectrometer
NIMS	Near Infrared Mapping Spectrometer
NIR	Near Infrared
NIRSpec	Near Infrared Spectrograph

NRAO	National Radio Astronomy Observatory
NRL	Naval Research Laboratory
OCO	Orbiting Carbon Observatory
ODC	Orbiting Debris Collector
OEFD	Orbiter Electric Field Detector
OETP	Orbiter Electron Temperature Probe
OMEGA	Observatoire pour la Minéralogie, l'Eau, les Glaces et l'Activité
ONL	Office of Naval Research
OPAL	Outer Planet Atmospheres Legacy
OSIRIS	Optical, Spectroscopic and Infrared Remote Imaging System
OSIRIS-REx	Origins, Spectral Interpretation, Resource Identification, Security, Regolith Explorer
PAMELA	Payload for Antimatter Exploration and Light-nuclei Astrophysics
PANIC	Pico Autonomous Near-Earth asteroid In situ Characteriser
PanSTARRS	Panoramic Survey Telescope and Rapid Response System
PEPSSI	Pluto Energetic Particle Spectrometer Science Investigation
PFS	Planetary Fourier Spectrometer
PIMS	Plasma Instrument for Magnetic Sounding
PIXIE	Polar Ionospheric X-ray Imaging Experiment
PIXL	Planetary Instrument for X-ray Lithochemistry
PMT	Photo-Multiplier Tube
PNAA	Prompt Neutron Activation Analysis
PRA	Planetary Radio Astronomy
PrOP-M	Pribori Otchenki Prokhodimosti-Mars
PVDF	Polyvinylidene Fluoride
PWS	Plasma Wave Subsystem
QBD	Quadrupole Beam Deflector
Radar	Radio detection and ranging
RAE-A	Radio Astronomy Explorer-A
RAR	Real Aperture Radar
RBSPICE	Radiation Belt Storm Probes Ion Composition Experiment
REASON	Radar for Europa Assessment and Sounding: Ocean to Near-surface
REMS	Rover Environmental Monitoring System
REPT	Relativistic Electron Proton Telescope
REXIS	Regolith X-ray Imaging Spectrometer
RGS	Reflection Grating Spectrometer
RHESSI	Reuven Ramaty High Energy Solar Spectroscopic Imager.
RICH	Ring Imaging Čerenkov detector
RIMFAX	Radar Imager for Mars Subsurface Exploration
RKA	Federal'noye kosmicheskoye agentstvo Rossii (Russian Federal Space Agency)
RLS	Raman Laser Spectrometer
RMI	Remote Micro-Imager

ROLIS	Rosetta Lander Imaging System
ROMAP	Rosetta lander Magnetometer and Plasma monitor
ROSAT	Derived from Röntgensatellit
ROSINA	Rosetta Spectrometer for Ion and Neutral Analysis
RPS	Relativistic Proton Spectrometer
RPWS	Radio and Plasma Wave Science
RRC-EOM	Russian Research Centre for Earth Operative Monitoring
SAR	Synthetic Aperture Radar
SCD	Swept Charge Device
SDC	Student Dust Collector
SEIS	Seismic Experiment for Interior Structure
SEM	Scanning Electron Microscope
SEPICA	Solar Energetic Particle Ionic Charge Analyser
SETI	Search for Extra-Terrestrial Intelligence
SHARAD	Shallow Radar
SHERLOC	Scanning Habitable Environments with Raman & Luminescence for Organics & Chemicals
SIRAL	SAR and Interferometric Radar Altimeter
SIS	Superconductor-Insulator-Superconductor
SKA	Square Kilometre Array
SLR	Single Lens Reflex
SMAP	Soil Moisture Active Passive
SMART-1	Small Missions for Advance Research in Technology-1
SMEX	Small Explorer
SOFIA	Stratospheric Observatory For Infrared Astronomy
SOT	Solar Optical Telescope
SPADUS	Space Dust
SPEDE	Spacecraft Potential, Electron and Dust Experiment
SPI	Spectrometer for Integral
SPICAM	Ultraviolet and Infrared Atmospheric Spectrometer
SRC	Super Resolution Camera
SRI	Stanford Research Institute
SSB	Silicon Surface Barrier
SSI	Solid State Imager
STIS	Space Telescope Imaging Spectrograph
SUDA	Surface Dust Analyzer
SuperTIGER	Super Trans Iron Galactic Element recorder
SWAP	Solar Wind Around Pluto
SWCX	Solar Wind Charge Exchange
SWICX	Solar Wind and Suprathermal Ion Composition Spectrometer
SWIR	Short Wavelength Infrared
SXI	Soft X-ray Imager
SXT	Soft X-ray imaging Telescope

TEGA	Thermal and Evolved Gas Analyser
TES	Thermal Emission Spectrometer
TDI	Time-delayed integration
TGF	Terrestrial Gamma Flashes
THEMIS	Thermal Emission Imaging System
THEMIS	Time History of Events and Macroscale Interactions during Substorms
TIRS	Thermal Infrared Sensor
TIRVIM	Thermal Infrared V-shape Interferometer Mounting
THOR	Turbulence Heating Observer
ToF	Time of Flight
TPC	Time Projection Chamber
TPE	Television Photography Experiment
TRD	Transition Radiation Detector
UMET	Metropolitan University
URA	Uniformly Redundant Array
USAF	US Air Force
USRA	Universities Space Research Association
UVIS	Ultraviolet Imaging Spectrograph
UVOT	Ultraviolet and Optical Telescope
VERITAS	Venus Emissivity, Radio Science, InSAR, Topography, and Spectroscopy
VCAM	Vehicle Cabin Atmosphere Monitor
VIMS	Visible and Infrared Mapping Spectrometer
VLA	Very Large Array
VL²SP	Venus Long Life Surface Package
VMC	Visual Monitoring Cameras
VME	Viking Meteorology Experiment
VNIR	Visible and Near Infrared
WAC	Wide Angle Camera
WFC	Wide Field Camera
WFI	Wide Field Imager
WFPC	Wide Field Planetary Camera
WHISPER	Waves of High Frequency and Sounder for Probing of Electron Density by Relaxation Experiment
WILMA	Wideband Line Multiple Autocorrelator
WISE	Wide field Infrared Survey Explorer
XEUS	X-ray Evolving Universe Spectroscopy
XGRS	X-ray/Gamma Ray Spectrometer
XMM	X-ray Multi-Mirror
XRD	X-Ray Diffraction
XRS	X-Ray Spectrometer
XRT	X-Ray Telescope

Index